AGRICULTURE ENGINEERING

AGRICULTURE ENGINEERING

Editor

Maqsood Kader

scitus
academics

Agriculture Engineering

Edited by **Maqsood Kader**

This Book is Printed on Acid-Free Paper.

ISBN: 9781681179063

Library of Congress Control Number: 2017948934

SCITUS Academics LLC,

616, Corporate Way, Suite 2, 4766,

Valley Cottage, NY 10989

www.scitusacademics.com

Notice

Printed in United States of America.

Preface

Agriculture Engineering is a technologically sophisticated field that offers bright career prospects and opportunity in any country around the world, and in various sectors. Agriculture is the world's largest and most important industry feeding and clothing an ever growing world population. However, agriculture is growing in many new ways as energy from biofuels become important to our economy and new biotechnologies create additional uses for agricultural waste and byproducts. Biofuel use and other on-farm sustainable operations along with better technologies bring new hope that engineering agriculture can provide for more people. Engineering is the input that permits this industry to produce our most basic needs without requiring the efforts of the whole population. Rural areas, especially in developing countries, are facing with numerous social, economical, cultural and environmental problems. In the most cases there is a growing concern because only very few farmers or inhabitants are able to follow contemporary, sustainable and environment-friendly methods in farming systems. The consequences of this are numerous, such as: poverty, significant migration to urban areas and unplanned suburban settlements. This has a significant impact on the national economy, demography, and environment.

This book entitled 'Agriculture Engineering' provides a scientific discussion of techniques and approaches leading to a higher degree of efficiency in agricultural production, the responsible use of natural resources, a cleaner environment, greater sustainability, value-added processing and other subjects of interest to agricultural engineers. It also includes economic considerations and policy issues of interest to agricultural engineers and other disciplines. It disseminates fundamental and applied knowledge in almost all area of agricultural sciences from agronomy to animal sciences, and particularly agricultural engineering.

Table of Contents

CHAPTER 1

Improving Water-Efficient Irrigation: Prospects and Difficulties of Innovative Practices

Les Levidow[a,*], Daniele Zaccaria[b], Rodrigo Maia[c], Eduardo Vivas[c], Mladen Todorovic[d], Alessandra Scardigno[d]

[a]*Development Policy and Practice, Open University, UK, Milton Keynes MK7 6AA, UK*

[b]*University of California, Davis, CA, United States*

[c]*Universidade do Porto, Praça de Gomes Teixeira, Portugal*

[d]*CIHEAM—Mediterranean Agronomic Institute of Bari (IAMB), Bari, Italy*

ABSTRACT

Innovative irrigation practices can enhance water efficiency, gaining an economic advantage while also reducing environmental burdens. In some cases the necessary knowledge has been provided by extension services, helping farmers to adapt and implement viable solutions, thus gaining more benefits from irrigation technology. Often investment in technological improvements has incurred higher water prices, however, without gaining the full potential benefits through water efficiency. Farmers generally lack adequate means and incentives to know crops' water use, actual irrigation applications, crops' yield response to different water management practices, and thus current on-farm water-efficiency levels.

Those general difficulties are illustrated by our two case studies investigating options, stimuli and difficulties to improve water-efficient practices. The two areas have strong stimuli for improvement but lack a knowledge-exchange system to help farmers and resource managers identify scope for improvements. Partly for this reason, farmers' responsibility for efficient water management has been displaced to hypothetical prospects, e.g. extra supplies from reuse of treated wastewater or a long-term low water pricing. In both cases a displaced responsibility complements the default assumption that farmers' irrigation practices already have adequate water-use efficiency. Under current circumstances, agricultural water management will maintain the unknown water-efficiency level and farmers will have weaker incentives to make efforts for more efficient practices. A continuous knowledge-exchange is necessary so that all relevant stakeholders can share greater responsibility across the entire water-supply chain. On this basis, more water-efficient management could combine wider environmental benefits with economic advantage for farmers.

Keywords: Innovative irrigation practices Water-efficiency Water users organizations Shared responsibility Knowledge-exchange system

1. INTRODUCTION

Irrigation systems have been under pressure to produce more with lower supplies of water. Various innovative practices can gain an economic advantage while also reducing environmental burdens such as water abstraction, energy use, pollutants, etc. (Faurès and Svendsen, 2007). Farmers can better use technological systems already installed, adopt extra technologies, enhance their skills in soil and water management, tailor cropping patterns to lower water demand and usage, reduce agrochemical inputs, etc. Water-efficient practices potentially enhance the economic viability and environmental sustainability of irrigated agriculture, without necessarily reducing water usage. To inform such practices, experts have developed various models of water efficiency, yet these are little used by farmers.

Through two case studies in the EU context, this paper will address the following questions:

- When an irrigation area invests in innovative technology, how can its operation help farmers to achieve the full potential benefits together, e.g. an economic advantage, greater water-use efficiency and lower resource burdens?
- Why are innovative technologies often applied in ways which miss the full potential benefits?

- What tensions arise among various objectives and potential benefits?
- How can these difficulties be addressed?

The paper first surveys analytical perspectives on irrigation efficiency – especially the means, incentives and limitations – as a basis to analyse two cases and draw general conclusions.

2. INNOVATIVE IRRIGATION PRACTICES: ANALYTICAL PERSPECTIVES

Innovative irrigation technology is generally promoted as raising water-use efficiency along with multiple benefits, but these remain elusive in practice, as outlined in the first sub-section below. The limitations have fundamental reasons, as outlined in Section 2.2. To address these issues, our case studies are introduced in Section 2.3.

2.1. Practical limitations of water-efficient irrigation technology

EU policy frameworks place great expectations upon technologies to improve water efficiency. The European Commission emphasises 'technological innovation in the field of water, given that water efficiency will be an increasingly important factor for competitiveness' (CEC, 2008). According to the European Parliament, solutions should be found in 'clean technologies that facilitate the efficient use of water' (EP, 2008).

Such technological expectations arise in expert reports on agricultural water use:

Water-efficient irrigation, irrigation on demand and irrigation using brackish water are technologies that will enable the better husbandry of more scarce freshwater resources. Technological developments in respect of irrigation will encompass sensors and communication, intelligent watering systems and high-efficiency delivery mechanisms for water and nutrients, as well as the means of incorporating all of these elements into irrigation 'packages' (EIO, 2011: 25).

Likewise water efficiency can be enhanced by better using current installations and/or by adopting new equipment (WssTP, 2012: 9).

The main European farmers' organisation has likewise advocated technological means to increase water efficiency. In particular this needs 'investments in more efficient irrigation systems, use of new technologies (e.g. soil moisture and canopy sensors) to better match irrigation with plant needs, and good agricultural practices', such as conservation tillage, management of soil fertility and water retention capacity, and scheduling of irrigation during night to reduce evaporation (COPA-COGECA, 2007: 4). The basis for improvement is described as follows:

... water efficiency measures that provide complementary benefits, such as reduced energy needs or other environmental benefits, will also deliver better results. In many Member States, efforts are being made to increase the water storage capacity of soil under agricultural land use. The modernisation of irrigation systems has steadily progressed and water productivity has also improved considerably (COPA-COGECA, 2013: 3)

As indicated above, greater water-use efficiency depends on better agricultural practices alongside extra technology. Yet companies generally promote irrigation technology as if it inherently brings all the benefits (interview, COPA-COGECA, 08.07.13). Improperly managed 'hi-tech' systems can be as wasteful and unproductive as poorly managed traditional systems (Perry et al., 2009). When incorrectly applied, irrigation technology 'can cause losses arising on investments made by farmers, thus decreasing the economic water productivity index and the overall sustainability' (Battilani, 2012).

Beyond a problem-diagnosis of inefficiency, moreover, intensive farming practices can degrade soil and water resources, especially through more input-intensive farming in crops such as maize, vegetables, orchard and vine cultivation:

Intensive arable production is partly responsible for poor soil structure, soil erosion, loss of soil OM [organic matter] and pollution from fertilisers and pesticides.... The expansion of maize cropping and the move to growing winter cereals in particular have contributed to soil erosion even further (Miller, 2007: 44–45).

Such harmful practices have been driven and supported by EU policies. In past decades CAP subsidies have tended to favour crops with high water demands, such as maize, thus increasing the risk of water shortages under climate-uncertain conditions (Garcia-Vila and Fereres, 2012). Either as price-support or area-based, CAP subsidies likewise have ensured the profitability of some water-intensive crops such as cotton which otherwise would be phased out under a market-orientated scenario; likewise water-price subsidies.

In some cases, water-price increases have induced farmers to adopt technology and appropriate practices for conserving water (Caswell and Zilberman, 1985). Yet water-pricing policies often have been ineffective means to reduce water demand (Molle and Berkoff, 2007; Molle, 2008).

Farmers experience rising water prices as an extra penalty. Rather than higher water prices, administrative water allocation or re-allocation lowering the supply often has led farmers to adopt water-efficiency practices (Molden et al., 2010). If agricultural water demand is inelastic, then policies which encourage changes in cropping patterns can be more effective than higher prices (Fraiture and Perry, 2007; Iglesias and Blanco, 2008; Kampas, 2012).

Inelastic water demand results from farmers' perspectives on water benefits. Water-use efficiency (WUE) and water productivity (WP) are often used interchangeably but have different meanings. WUE specifically means the ratio of biomass produced per unit of irrigation water used, i.e. the sum of transpiration by the crop and evaporation from the soil (Sinclair et al., 1984). By contrast, WP means the ratio of above-ground biomass per unit of water transpired by the crop (Steduto, 2007). Both terms have relevance to farmers' economic goals. WUE interests mainly the water districts or management agencies, while WP interests more farmers and research community. WP better speaks to perspectives linking water usage with production levels and economic benefit (interview, COPA-COGECA, 08.07.13).

Yet even WP remains distant from farmers' perspectives. They generally perceive 'irrigation efficiency' as maximising net revenue rather than saving water (Knox et al., 2012). Policies seek to lower water usage, and river basin managers try to allocate limited supplies, yet water-saving is not a priority for most farmers (Luquet et al., 2005). They manage labour and other inputs to get better economic gains (Molden et al., 2010). Towards that economic aim, most growers make irrigation decisions by relying on subjective judgements, based only on their practical experience and observation (Knox et al., 2012). Consequently, there have been limited benefits from irrigation technology, as well documented in the technical literature; the following examples compare various techniques.

For example, mobile-laboratory evaluations compared the distribution uniformity and irrigation efficiency of various irrigation systems in California. Although microirrigation systems are seen as 'efficient technologies', they were performing less well than traditional surface irrigation methods such as furrows and borders. To gain the extra benefits of such technology, most important is adequate system design, alongside proper installation, operation and maintenance, regardless of the irrigation method used (Hanson et al., 1995).

Howell (2003) and Irmak et al. (2011) reported the attainable application efficiencies for different irrigation methods, assuming irrigations are applied to meet the crops' water needs. Microirrigation has the potential to achieve the highest uniformity (90%) in water applied to each plant, yet poor uniformity and application efficiency can result from various causes, e.g., inadequate maintenance, low inlet pressure or pressure fluctuations, emitter clogging and inadequate system design (Hsiao et al., 2007). Consequently, microirrigation technology has on-farm efficiencies varying from 0.7 to 0.95 (Howell, 2003).

As another example, a Spanish study compared various irrigation methods via the annual relative irrigation supply index (ARIS), i.e. a ratio of water applied versus water required. It found a greater efficiency of solid-set and drip than surface irrigation. But average annual figures conceal great variations in water applied to a given crop and irrigation efficiency at farm level, partly for lack of adequate knowledge. A remedy would be 'actions to improve farmers' water management via a combination of irrigation advisory services and policy measures' (Salvador et al., 2011: 586).

2.2. Reasons for those limitations and ways to overcome them

Given the above water-efficiency limitations in applying irrigation technology, the literature has outlined some fundamental reasons. They include the following: irrigation equipment is promoted as if the technology per se brings various benefits, farmers seek to maximise net income rather than water productivity per se, innovative technologies can achieve the full potential benefits only through appropriate technical advice, and farmers lack a knowledge-system for anticipating effects of specific irrigation practices or for retrospectively evaluating their irrigation efficiency.

Although research has developed technical scheduling procedures to improve agricultural water management, these have been little adopted, for many reasons.

The one most frequently mentioned by growers is the lack of perceived [financial] benefits relative to their current practices, which they consider adequate. Ease of use and the expenses involved are also important grower considerations (FAO, 2012).

Technical advice on irrigation scheduling is little used at farm level; at most, it helps retrospectively to evaluate seasonal approaches (ibid.).

One obstacle is inadequate knowledge about proper irrigation levels and scheduling over a growing season. Farmers generally lack adequate assistance to develop and adopt better approaches for environmental sustainability, while also maintaining their financial and social objectives (Pereira et al., 2012: 39). For example, sub-surface moisture sensors can improve knowledge about a crop's need for water. But the technology has limitations, so farmers need technical advice to interpret the measurements; for example, 'soil humidity sensors are still neither easy to handle nor reliable' (WssTP, 2012: 33). Moreover, these sensors are not well adapted to all soil types; their installation and maintenance requires the employment of specialised technical staff. The same is true for the canopy sensors, whose proper application is limited to some crops and during specific growing stages, periods of day and climatic conditions.

Improvements in irrigation practices depend on quantitative knowledge of farmers' current practices in relation to actual and potential crop water use:

Any effort to improve water use efficiency needs to start with the assessment of the actual and attainable efficiencies for the given situation, as quantitatively as possible. This information is fundamental for making rational improvements aiming at raising the overall efficiency to the attainable level (Hsiao et al., 2007: 228, 218).

But such information is rarely available to farmers.

Such difficulties arise for water-management improvements through expert systems. Decision Support Systems (DSS) have aimed to improve crop water use efficiency at farm and water basin scale, but few are widely applied, given the necessary specialised skills. For a DSS to be successful, the key elements have been: giving farmers a simple, timely, user friendly, free-of-charge, informative system helpful to decide how much to irrigate in everyday practice; tailoring the tools for a large number of crops; calculating the irrigation profitability; and assessing the economic benefit, especially its relevance to the next irrigation. Such benefits have been demonstrated by the Irrinet project in Italy's Emilia Romagna (Battilani, 2012). Thus more reliable information systems and expert capacity are necessary to guide farmers in using water more efficiently (Battilani, 2013). This exemplifies the broader need for farmer training and education in order to improve modern irrigation management.

As a way forward in the UK, expert support has been recently linking farmers' responsibility, economic benefits and practical knowledge. A 'pathway to efficiency' improves the irrigation network, alongside better practices of soil and water management, e.g. by monitoring whether the right amounts are used at the right place and time. 'Using financial criteria for water efficiency rather than an engineering one appears a sensible approach when assessing irrigation performance at the farm level, since any managerial (e.g. scheduling) and operational (e.g. equipment) inefficiencies associated with irrigation are implicitly included in the assessment' (Knox et al., 2012: 3). In particular, 'On-farm water auditing and benchmarking have the potential to provide useful information to farmer decision making, with respect to identifying operational and management changes to improve irrigation system performance and water productivity, and evaluating potential investments in new technology (and advanced practices) or infrastructures' (ibid: 7).

Such approaches have addressed various obstacles to water-efficiency measures. To exploit the full technological potential requires a broader dissemination of their benefits, specific training of farmers, and coupling properly-designed technological solutions with more precise operational practices to benefit farm economic performance (e.g. Tollefson and Wahab, 1994). In particular, advisory-extension services have enhanced irrigation practices which better fulfil potential benefits of irrigation technology (Hergert et al., 1994; Benham et al., 2000; Ahearn et al., 2003; Genius et al., 2014; Parker et al., 2000; Gold et al., 2013).

Beyond the farm level, greater resource efficiency also depends on shared responsibility among stakeholders, according to the World Business Council for Sustainable Development:

Business undoubtedly has many opportunities to increase its eco-efficient performance and thereby to help de-couple use of nature from overall economic growth… Establishing framework conditions which foster innovation and transparency and which allow sharing responsibility among stakeholders will amplify eco-efficiency for the entire economy and deliver progress toward sustainability (WBCSD, 2000: 6–7).

Analogous issues arise for service-oriented irrigation schemes, designed so that farmers can flexibly obtain water at their convenience, e.g. through on-demand delivery schedules. Here responsibility has institutional complexities. For example, a water users' organisation (WUO) bears largely fixed costs, as well as somewhat variable energy costs from drainage, excess water application, reuse, disposal, etc. If a WUO or water district relies on gravity-fed water conveyance and delivery systems, then its costs do not vary according to water-volume delivery. In such contexts, if farmers decrease water use, then the WUO must increase water prices to recover its fixed costs. Facing higher water prices, farmers may increase groundwater pumping, thus abstracting more water from aquifers, while distancing their individual practices from any group responsibility. Paradoxically, fostering greater water-use efficiency can generate a financial, environmental and institutional problem.

Given those difficulties for water-efficient techniques, their effective adoption depends on several enabling conditions, especially a policy and institutional context aligning incentives of producers, resource managers and society. Significant synergistic effects can emerge when water-efficiency practices are combined with other agronomic practices such as maintaining soil health and fertility, controlling weeds and avoiding diseases (Molden et al., 2010).

2.3. Methods and sources: EcoWater project

The above issues and earlier questions have been explored through two case studies of service-oriented irrigation schemes within a larger EU-funded research project, EcoWater (see Acknowledgements). It develops a methodology for assessing eco-efficiency at the meso level. The latter is defined as interactions among heterogeneous actors, e.g. between water-service users and providers. As generally understood, eco-efficiency means a ratio between economic advantage and resource burdens, as a basis to evaluate past or potential changes in a system.

The project uses eco-efficiency indicators to evaluate potential innovative practices including technology adoption. The project aims to: assess various options for innovative practices within a specific system; analyse the factors influencing decisions to adopt such practices; and improve understanding of the socio-technical dynamics that influence such decisions.

In the project's two agricultural case studies, farmers and/or their organisations have already invested in water-efficient technology, going beyond state subsidy alone. The irrigation distribution systems were designed for on-demand water delivery. SCADA technology at hydrants allow farmers to abstract water on demand any time and charges them according to a volumetric tiered water pricing. Each case-study area has strong stimuli for farmers to improve water efficiency, yet the full potential benefits of the technology investment were not being realised, for reasons analysed in the next two sections.

3. SINISTRA OFANTO CASE

Dating from the 1980s, the Sinistra Ofanto irrigation scheme is among the largest multi-cropped irrigated areas in Italy. It is located in south-eastern Foggia province within the Apulia region. Irrigation is crucial for the region's agricultural production and income, but it also generates resource burdens. Nearly 18.5% of Apulia's agricultural area is under irrigation; consequently, irrigated crops have contributed 69% of the total value of regional agricultural production, recently quantified as 3.8bn Euros (Fabiani, 2010). The entire study area is characterised by a high number of small land-holdings with intensive, market-oriented practices. The main crops are vineyards, olives, vegetables and fruit orchards (in descending order). The pedo-climatic conditions are favourable for intensive cropping, but profitable farming is strongly dependent on irrigation, due to the scant rainfall and its uneven distribution across the year.

The Sinistra Ofanto system commands an area of 40,500 ha stretching along the left side of the Ofanto River, of which 38,815 ha are irrigable lands and 28,165 ha are serviced with irrigation distribution. Designed and constructed for pressurised on-demand delivery schedule, the irrigation system is managed by a large water users' organisation (WUO), the *Consorzio per la Bonifica della Capitanata* (CBC, 1984; Altieri, 1995). The system diverts water from the Ofanto River and supplies it to growers both by gravity and lifting/pumping, ensuring a pressure head of at least 2 bar at each hydrant to enable farmers using micro-irrigation methods.

The system is already equipped with modern technologies to deliver and use water efficiently. From the diversion structure on the Ofanto River, water is conveyed to the Capacciotti reservoir through concrete-lined canals and pipe conduits, along which the flow regulation devices are downstream-controlled, thus manually or automatically adjusted through calibrated control devices enabling Supervisory Control and Data Acquisition (SCADA). The Capacciotti reservoir supplies seven concrete-lined storage and compensation reservoirs equipped with downstream-control flow regulation devices that adjust inflows and outflows to feed the district's piped distribution networks based on the downstream water demand.

PVC buried pipes comprise the open-branched distribution networks. Each sector's inlet has a control unit, equipped with flow and pressure metering-control devices. Water is supplied to farms on demand by means of multi-users' electronically-fed hydrants that control and regulate the deliveries, as well as the discharges demanded and thus flowing in the pipe distribution network. These technologies installed along the main infrastructure help keeping conveyance and distribution losses within 5–10% of the total water abstracted from the Ofanto River, as reported by the WUO's engineering staff.

Although the main water supply is surface water, during recurrent water shortages farmers pump groundwater from medium-depth (100–150 m) aquifers, especially since the late 1990s (Portoghese et al., 2013). Furthermore, studies found qualitative degradation of groundwater resources, most likely resulting from seawater intrusion into the coastal aquifer and to deep percolation of pollutants, such as fertilisers and pesticides, from intensive farming activities. Given the urgent need to assess these processes and to avoid their adverse environmental impacts, what are the prospects for water-efficiency improvements of irrigated agriculture in the Sinistra Ofanto area?

3.1. Irrigation patterns and resource burdens

The water users organisation (WUO), Consorzio per la Bonifica della Capitanata (henceforth the CBC), is the main irrigation management agency. It is responsible for all the sequential steps along the agriculture water supply chain, i.e. abstraction, conveyance, storage, distribution and final water delivery to farm gates. Established in 1933 by a national law of public interest, the CBC is by statute a non-profit organisation; it bears all the costs for performing its functions, and these costs are recovered through the water tariffs paid by farmers.

The CBC enforces the principle of solidarity among the different service areas. Even though the costs for supplying irrigation water differ significantly among areas supplied by gravity and by pumping, the tariff structure does not make such a distinction. Rather, as a tool to manage water use, water fees vary according to demand: volumetric tiered water tariffs progressively increase with the seasonal cumulative volumes withdrawn by each farmer. This structure is enforced through individual water metering at the delivery points; all farm hydrants are equipped with an electro-mechanical delivery device allowing the supply of water only to authorised users and storing information of each irrigation event. Besides simplifying the network operations, this technology proved to be very useful for accurate monitoring and control of water distribution, and for achieving better understanding of the irrigation management practices followed by farmers, especially through the possibility to retrieve and analyse historical data series (Zaccaria et al., 2013).

As an irrigation service provider, the CBC is composed of irrigation service users, i.e. farmers. In performing its daily activities, the CBC attempts to reconcile objectives which may be in conflict. Its technical and administrative choices aim to achieve high water-distribution efficiency in order to maximise the economic benefit to farmers. It aims to improve water distribution and use—at the farm level, through an effective operation of the delivery network, and at field and crop level through the technical support to growers aiming at improved water management skills (ibid). Technical support to farmers was effective in the 1980s–1990s but has declined in the last decade, due to WUO budget constraints and lower revenues from Italy's farm activities.

Irrigation water demand is driven mainly by farmers' perceptions, by the climatic conditions, and by the economic value of crop yields and production factors. Even beyond periods of water shortage, in some areas farmers pump groundwater in order to avoid the following problems: (i) the restricted-flow demand delivery-schedule that prevents the quick completion of irrigation cycles in medium-large farms, (ii) the restricted-frequency demand-delivery often imposed by the CBC during water shortage periods, (iii) the need to arrange water withdrawals with neighbour farmers supplied by the same hydrants, or (iv) the tiered water fees enforced by the CBC. Also, many farmers still perceive groundwater pumping as somewhat cheaper than water supplied by the CBC, even though the contrary was shown by economic analyses (e.g. Portoghese et al., 2013).

As a more fundamental problem, both the farmers' perception and the CBC's analyses ignore the ecological costs of groundwater degradation and remediation. The CBC accepts no responsibility for water-management practices beyond the farm gate. From the growers' standpoint, groundwater pumping aims to increase and/or stabilise the economic benefits of farming activities. Often farmers combine surface water and groundwater for various reasons such as to maximise crop yields and farm net benefit, or to minimise the seasonal water fees payable to the CBC, or to prevent yield reduction arising from high salinity in the groundwater during peak-demand periods. However, this conjunctive use of surface and groundwater is based solely on farmers' economic and technical considerations, regardless of environmental burdens such as aquifer depletion and degradation. Furthermore, fields close to the river banks are often irrigated by growers with water pumped out the river. In all these situations, return flows may result from run-off through the drainage networks, as well as from percolation through the soil profile, finally reaching the downstream reaches of the river, wetlands or the aquifer.

Farm activities generate various pressures on land and water resources, including quantitative depletion and qualitative degradation, especially biodiversity loss in farmland and in the natural environment. This harm has several sources: (i) intensive farming and tillage practices, (ii) fertilisers and pesticides application on cultivated fields, (iii) water abstraction from the Ofanto River, (iv) return flows of degraded water to downstream wetlands and aquifers, (v) over-drafting of groundwater, (vi) salinity build-up in cultivated soils, (vii) energy consumption for water pumping, and (viii) increased CO2 emissions from the energy usage related to pumping, transport, machinery, etc.

Relative to those ecological problems, much greater impetus for innovative practices comes from recurrent scarcity of water supply and the prospect of even greater future scarcity and uncertainty. Those problems in turn result from high water-demanding crops and from irrigation scheduling practices. Such decisions are often based solely on farmers' perceptions; their systems and practices are not monitored to assess the actual performance and efficiency achievements. No systematic technical support is available to growers for their daily or seasonal irrigation planning and scheduling.

Moreover there is detailed evidence of water-use inefficiency at farm level. According to a study of farmers' irrigation practices in a nearby irrigated area with similar features, there were often mis-matches between crops' water demand and irrigation applications on several occasions during the season. Although the overall seasonal applied irrigation depths may match a crop's water demand, farmers often under-irrigate during the early crop stages and over-irrigate during later stages; many choose inadequate timings and application depths. Such inadequate applications may be combined with uneven in-field water distribution, often due to the average low uniformity of irrigation systems—especially when not properly designed, evaluated and maintained; consequently, the farm may have up to 20% lower crop yields and income, along with inefficiencies between 20 and 40% due to excessive water applications (Zaccaria et al., 2010). As the main reason for the mismatch, irrigation scheduling practices are based only on farmers' perceptions and experiences; missing is status monitoring of soil or plant water, use of ET-based irrigation scheduling, or any other quantitative techniques (ibid). This study confirms a general problem of water-inefficient practices, as also found in the wider technical literature (e.g. Hanson et al., 1995, 1996; Burt, 2004; Salvador et al., 2011; *see Section* 1*).*

3.2. Innovative practices for stakeholders' consideration

Resource-efficiency could be enhanced by properly utilising several innovative technologies and practices. As listed in Table 1, several feasible options are already installed and implemented in the Sinistra Ofanto area, i.e. along the water conveyance and distribution system or on some progressive farms, but require either some refinements or significant operational improvements to gain their full economic and environmental benefits.

Investment costs for the installed technologies were paid by the WUO either from budget surpluses (in years when water demand and delivery were high), or by funds obtained from different sources, and by some individual farmers at the farm level. Their maintenance and operational costs are paid by growers within the regular water fees. The improved irrigation methods, systems and practices are implemented only to a small extent by some farmers.

The first two technologies were recently installed along the distribution networks to ease operation and flow control, and thus increase its efficiency, but they still need further testing and tune-ups to fully exploit their potential. For example, multi-user hydrants could be hooked up with

Table 1. Technologies, impacts and necessary improvements.

Technology	Level of application	State of implementation	Action	Impacts	Needs and Improvments
Multi-user electronic hydrants	Distribution network/hydrants	Installed and operated	Supply water to authorised users Control, record and store irrigation events Bill farmers according to used volumes Implement tiered water tariff Programmable to supply water during specific time slots	Simplify water distribution and water billing Reduce WUAoperational cost Responsible water use by farmers Reduce water demand	Data transmission Low-pressure alert High-salinity alert Integration with IAS
Variable-speed pumps	Pumping plant	Recently installed & operated	Adjust their operation dynamically Supply the discharges and pressure heads required by the network	Energy saving Reduce WUO operational costs	Data recording and transmission Accurate system analysis Tuning operational regulation Improve energy-efficiency
Shift to improved irrigation methods (sprinkle to mini-sprinkle and trickle)	Farm irrigation systems/cropped plots	Recently implemented by some farmers	Reduce soil wetting Reduce soil evaporation losses Reduce energy requirements	Reduce water demand Energy saving Reduce irrigation cost to farmers Reduce overall water abstraction	Better visualisation of benefits Technical support and extension Irrigation advisory service Improve application efficiency
Sub-surface drip irrigation (SDI)	Farm irrigation systems/cropped plots	Very limited	Reduce/minimise soil wetting Reduce application losses Increased WP	Reduce water demand Energy saving Reduce irrigation cost to farmers Reduce overall water abstraction	Precision operation Better visualisation of benefits Improve irrigation scheduling Improve application efficiency Technical support and extension Irrigation advisory service
Regulated deficit irrigation (RDI)	Cropped plots	Very limited	Increased WUE Increased WP	Reduce water demand Energy saving Reduce irrigation cost to farmers Reduce overall water abstraction	Soil and plant water status monitoring Precision irrigation Better visualisation of benefits Technical support and extension Irrigation advisory service

soil moisture sensors or plant–water status sensors to schedule irrigation applications on the basis of needed or user-defined thresholds and frequency. Also, they could be set to cease water withdrawals and irrigation when the pressure-head values at delivery are insufficient to achieve the target level of efficiency and distribution uniformity. Moreover, relevant datasets from the field (on irrigation timings, flow rates, pressure heads, volume and duration) could be automatically transmitted to a central WUO workstation, thus easing the data-gathering necessary for billing farmers, for evaluating the irrigation delivery service provided and growers' irrigation scheduling practices. These further adjustments of the available technologies could also enable the performance evaluation and benchmarking of irrigation at on-farm and project level.

As indicated by those examples, upgraded technologies alone do not ensure the anticipated resource- efficiency improvements. Compared to less technology-intensive systems, the past investments incur extra costs but bring uncertain gains. The full potential benefits depend on adequate management practices, in turn dependent on better expertise, skill-gap assessment, auditing monitoring and evaluation, control systems and extension programmes, including mission-oriented research and outreach.

The above technology options (in Table 1) were presented at the EcoWater project workshop, held in the case study area and attended by diverse stakeholders involved in the Sinistra Ofanto water-system value chain. These included: the River Basin Authority, the WUO, the Apulia Regional Administration, the regional Department of Water Resources, the Ente Irrigazione, ARPA Puglia, CNR-IRSA (Water Research Institute), University of Bari, etc. Stakeholders, decision-makers and experts were asked to share their visions about water management in the area. The discussion identified several issues, emphasising the need for measures to protect groundwater.

As many comments there highlighted, in the last three decades contradictory interests have generated conflicts over water allocation and use. Surface-water availability for agricultural use has been reduced by several factors: expansion of the command areas, incentivised partly by irrigation systems and by regional policies; greater water demands of the municipal, industrial and tourist sectors; and adverse environmental-climatic changes. Participants mentioned several solutions, in particular: monitoring and control of water use should be strengthened throughout the Apulia region, especially in the study area, in order to avoid aquifer over-exploitation and uncontrolled withdrawals from water courses; deficit irrigation management, already practiced in the study area by some innovative growers, offers a feasible option to enhance water-use efficiency and productivity; and, finally, farmers should share their personal experiences and knowledge in order to help others to use irrigation water in a sustainable and efficient way (reported in Ecowater, 2013).

In addition, CBC (local WUO) representatives emphasised the importance of using non-conventional water sources as an additional supply, especially the re-use of treated wastewater (TWW) as a valid alternative to aquifer exploitation for irrigation purposes. TWW re-use has several barriers: stringent quality criteria set by regulations for water reuse in agriculture; the high cost of treatment and water conveyance from the treatment plants to agricultural areas; and inadequate research dissemination, even disinformation; and reluctance by farmers. Consequently, TWW re-use should be fostered by several means – lowering the water-quality restrictions, installing the latest technologies available on the market for wastewater treatment to reduce costs of treatment plants, and broadly disseminating research findings to growers, especially through extension activities and farmers' advisers – according to the WUO representatives (reported in Ecowater, 2013). No stakeholder raised concerns about aquifer characterisation, safe-yield assessment, groundwater recharge or agrochemicals' fate in soil and aquifers, or farm-level performance assessment of agricultural water management practices.

As emerged during the stakeholder discussion, some future visions may be based on doubtful assumptions. In particular the CBC staff and farmers' representatives assumed that Sinistra Ofanto farmers already achieve high irrigation and water-use efficiency, simply on grounds that they use microirrigation methods. From this assumption, there would be little scope or incentive to growers for further improvement in farm-level water management.

From CBC technical reports over at least a decade, no recent information is available on whether the farm irrigation systems are properly designed, installed, operated and maintained—nor on whether irrigations are adequately scheduled and conducted by growers. Likewise the WUO has not recently fostered irrigation systems evaluations regarding the actual application efficiency and distribution uniformity achieved by growers, or WUE/WP for the various crops and irrigation methods, on the basis of quantitative measurements. Agricultural extension activities have been significantly limited in the last decade by budget constraints. So the CBC lacks an empirical basis for its efficiency assumptions about current performance of on-farm water management. In all those ways, the CBC's practical responsibility ends at the farm gate.

Moreover, responsibility for water-supply problems is displaced onto public-sector agencies, especially for additional water supplies via wastewater reuse. This option is simpler and perhaps more profitable for farmers and the CBC, especially if public bodies carry the main financial burden. Water reuse involves complex issues needing careful investigation and management, in particular: different water qualities may be required, contingent on the final water uses. Water that needs to be buffered or stored for seasonal reuse (e.g. groundwater banking or surface storage) may have other quality requirements than water used for a direct agricultural reuse. And there are potential risks to health and environment (WssTP, 2013: 19–20).

4. MONTE NOVO CASE

In Portugal's Alentejo the Monte Novo irrigation perimeter provides abundant water for a rising number of farmers, served by the larger Alqueva reservoir and irrigation project. Dating from 2009, the scheme set initially low water prices to incentivise irrigation by farmers—but without their involvement in discussing implications for crop choices, irrigation practices, farming profitability and expected income. The highly-discounted water price has incentivised crops with high water demands as well as agrochemical inputs, thus creating pollution problems. As mandated by law, the water price will rise significantly towards full-cost recovery by 2017, thus potentially stimulating practices which need less water. What are the prospects for water efficiency improvements?

4.1. Irrigation patterns and future price rise

Since the 1960s Portugal has had plans to irrigate relatively arid areas of the Alentejo, especially to increase employment and livelihoods. Spain's 1993 National Hydrologic Plan was proposing to abstract more of the shared water from the adjoining region, partly on grounds that Portugal was not using the water, but Portugal objected. To avoid such trans-border conflicts, the 1998 Albufeira Convention aimed to protect freshwater and groundwater, as well as the sustainable use of shared water resources, in the context of the negotiation process for the Water Framework Directive (Maia, 2000; Pulwarty and Maia, 2013). Under the Convention, for example, Spain guarantees annual stream flows to Portugal in normal years (Costa, 2003).

The Albufeira Convention facilitated Portugal's Alqueva Multipurpose Project—the Empreendimento de Fins Múltiplos de Alqueva (EFMA). The EU Structural Funds were expected to pay more than half the investment cost (Vergés, 2001). The 1993 EFMA plan combined a reservoir, hydroelectric dam and irrigation networks, thereby helping to justify Portugal's claim on the water. The project aimed to promote the development of a poor, deprived region through irrigated agriculture, electricity production and tourism. Together these were meant to ensure the sustainability of the project in its early phase (Santos et al., 2011).

From the start the EFMA was criticised as a hydrodinosaur—for accepting engineers' assumptions about future water needs, requiring enormous electricity to transport water and so raising the full cost (Costa, 2003). Before EFMA's operation, maize production had greatly risen alongside greater dependence on CAP subsidies for economic viability. The Agriculture Minister hoped that the Alqueva would irrigate industrial crops such as tomato and sugar beet (Vergés, 2001: 7). By using cheap water from the Alqueva dam, some Alentejo farmers were expecting to capture greater payments from the CAP 1st pillar. So they opposed mid-term reforms decoupling such payments from production levels; so did the Portuguese government (Costa, 2003: 26).

According to its promoters, the Alqueva system will foster significant regional development in Alentejo, both in social and economic terms, due to the strategic water reserve and the associated hydraulic infrastructures. The main objective is to guarantee an adequate volume and quality of water to users, especially irrigation water. Further objectives include the energy hydropower production, urban water supply, ecological conservation and recreational uses (EDIA, 2011).

With its geographical topography and high elevation, the Alqueva scheme was an expensive investment (Costa, 2003), especially relative to the small number of farmers who initially joined the scheme. This arrangement resulted in a high unit cost for the water supply. This has high political-economic stakes, given that full-cost recovery of irrigation systems is required by the Water Framework Directive (EC, 2000).

Irrigation delivery service is provided by the public company EDIA (Empresa para o Desenvolvimento das Infraestruturas de Alqueva), the agency responsible for the Alqueva project development and exploitation. Water is abstracted from the Alqueva reservoir and transported through a network of canals and ducts, from primary network to secondary network through hydrants to farmers. Within the larger Alqueva project, the Monte Novo irrigation perimeter is located in Alentejo district, near Évora municipality. The perimeter provides water for irrigation to an area of at least 7800 ha, while the Alqueva Project will have a total 115,000 ha expected capacity by 2015.

In the Guadiana region including Monte Novo, the tertiary sector has 76% of gross value added (GVA) and 67% of employment; the most important sources are tourist accommodation, catering and public services. The agricultural sector has 10% of the region's GVA and about 15% of the jobs; average farm size is large (55 ha). These characteristics, associated with the high proportion of farmers in a company structure, indicate the region's high potential to develop agricultural activities, especially with the full operation of the Alqueva project. But its agricultural activities have a very low competitiveness and a low productivity per unit area; incomes are greatly supported by public subsidies, which comprise about 65% of the total gross margin (ARH-Alentejo, 2011).

In the study area, the major irrigated crops are olives, maize, arable crops (mostly pasture) and horticultures (mostly tomatoes). Farm size ranges from less than 50 ha to more than 500 ha. The largest areas belong to important multi-crop farms, such as the ones owned by the Fundação Eugénio de Almeida (FEA), or the olive farms owned by Olivais do Sul (ODS); together they comprise more than 30% of the irrigated area in Monte Novo irrigation perimeter.

Fertilisers and pesticides are applied to all crops, especially tomatoes and maize. Greater irrigation stimulates run-off, leach-outs of fertilisers and pesticides and soil erosion. Together these practices reduce soil organic matter, soil fertility and its capacity to retain water, in turn increasing potential irrigation inefficiencies and need for better water management for the same cultivation level as before. Indeed, land degradation and nutrient depletion significantly constrain opportunities to increase water efficiency (Molden et al., 2010; IST, 2013).

In 2009 the WUO Associação de Beneficiários de Monte Novo (henceforth ABMN) was established, representing all farmers connected to the Alqueva water distribution system from EDIA. According to the regulatory Decree Law 84/82, the Association has formal recognition by the Ministry of Agriculture, Trade and Fisheries. The ABMN has the role to promote the administration of constructing the hydro-agricultural development. Nonetheless EDIA has carried out the management and operation of the irrigation perimeter, as well as the investments in constructing the irrigation network.

Water prices are differentiated according to the pressure heads provided at farm-gate delivery. Prices are lower in low-pressure blocks, where water is conveyed and distributed by gravity, generally to larger farmers; irrigation inside farms is within farmers' management. Prices are higher in high-pressure blocks, which are supplied by pumping stations; these reduce the need for farmers to bear the costs of farm-level booster pumps. However, in the high-pressure blocks some large farms need their own investment in irrigation equipment and distribution infrastructures. In some cases farmers already invested in important irrigation infrastructures prior to the construction of the Monte Novo irrigation perimeter. In both cases a higher water price is imposed on important farmers who will not necessarily benefit from higher pressure heads at hydrants (Rita and Capelo, 2005). Moreover, the difference in water prices between low-pressure and high-pressure blocks is insufficient for covering the necessary investments for constructing the main irrigation infrastructures within low-pressure farms' areas, as well as the corresponding exploitation costs, mainly energy and maintenance (Froes and Rodrigues, 2005).

Given those high costs, full-cost recovery would impede a full transition from rain-fed to irrigated agriculture. To attract farmers into the scheme, a 2010 law set the initial water price at only 30% of full cost. This initial price shifts more of the total cost onto the supplier, while incentivising maize cultivation, which demands relatively large amount of water but gains a higher market price. According to the ABMN (Monte Novo's WUO), the discount price gives farmers 'a certain hope', but many have difficulties to obtain credit, which is necessary for farm-level investment to electrify the cultivated zones (Silva, 2010).

As originally set by law, the low water price will be increased by 10% per year towards full-cost recovery by 2017 (Nuncio and Arranja, 2011). This legislative requirement has become contentious. At its full-cost price, water would comprise 13% of the total cost of maize production. Through their WUOs, farmers have been lobbying the authorities to delay the price rise. According to the Federação das Associações de Agricultores do Baixo Alentejo (regional federation of farmers' associations), the higher price will not allow farmers to be economically competitive, especially for a large proportion of maize cultivation, estimated as more than 36,000 tonnes aimed at export markets. According to the former Agriculture Minister, however, the higher price will be competitive with nearby regions, as an adequate basis for more farmers to join the scheme (Maneta, 2010; Rosado, 2012). The future increase could stimulate water-conserving measures and alternative cropping patterns which would be more profitable and/or need less water, while also reducing other resource burdens. But continuous support measures would be necessary for growers to evaluate and implement such options.

4.2. Innovative practices for stakeholders' consideration

The on-demand water-delivery system was meant to incentivise water-efficient practices by farmers and thus to reconcile tensions between various objectives of the Alqueva project. Such a strategy put great expectations upon a technological system. At the irrigation-perimeter level the eventual implementation of innovative practices is meant to be supported by the ABMN (WUO), with investment costs if necessary to be covered from budget surpluses. On the other hand, the farm-level water-use improvements are dependent on farmers' own investment or eventually on government subsidies fostering efficiency increases. Some technologies have been applied at farm level, e.g. sub-surface irrigation at FEA and ODS farms, but only to a small extent. Buried drip-irrigation systems need even more careful design, monitoring and operation than surface-drip or other micro-irrigation systems. Beyond the technological investment, greater application would depend on applied research, demonstration activities, outreach, farmers' training and skills development.

The study team assessed various options for lowering input demands (of energy, water, agrochemicals) or for increasing farmers' income per unit water used (Ecowater, 2012), in two main categories as follows. At the irrigation perimeter level, i.e. the distribution network:

- tiered volumetric water tariffs according to actual water use by growers;
- tiered water tariffs according to timing of withdrawals and energy costs, e.g. lower rates at night-time; and
- pressure head delivery.

And at the farm level:

- drip irrigation, reducing water evaporation (especially relevant to maize);
- sub-surface drip irrigation, minimising soil evaporation and facilitating mechanical weed-control or conservation tillage or minimum-tillage methods (especially relevant to vineyards);

- super-high density olive orchards;
- variable-irrigation practices, e.g. through regulated deficit irrigation; and
- alternative crops demanding less water.

The study team presented the above options at an April 2012 workshop. It was attended by representatives and experts from relevant local bodies. They included: EDIA, farmers' representatives (FEA and ODS), the ABMN (WUO), ARH-Alentejo, and the Centro Operativo de Tecnologia do Regadio (COTR), an advisory body for scientific research regarding agricultural development.

Stakeholders' comments converged around the following points:

- All the proposed technologies could add value to the Alqueva scheme.
- Farmers are interested in any technological configuration that might increase their profit margins, which are currently low.
- Given the high investment costs of the irrigation scheme, a successful operation is important in order to lower the unit cost of water through access to more growers (Ecowater, 2013).

Discussion focused on the knowledge lacking for farmers to minimise irrigation intensity and to conserve soil resources. In particular participants made these comments:

- Irrigation intensity must remain within the carrying capacity of the soil (infiltration rate and water-holding capacity), especially in order to prevent surface run-off, leach-outs and erosion—significant environmental impacts that must be taken into consideration.
- Without adequate knowledge for such judgements, farmers may intensify resource usage, thus increasing costs, leach-outs and soil erosion.
- Root-zone soil moisture conditions are not measured to identify in-field variability and vulnerable areas, so most farmers base decisions on their past experience and daily observations of farm conditions.
- Although a network of meteorological stations already exist in the area, relevant information is not available for irrigation planning and scheduling purposes; farmers' access is still under development (Ecowater, 2013).

Regarding the above point about the soil's water-capacity, a COTR expert emphasised soil organic carbon (SOC) as an environmental indicator and objective. Greater irrigation flows undermine the soil's capacity to retain water and nutrients, while also increasing GHG emissions. SOC conservation is more difficult for annual arable crops than for permanent pasture. To avoid further SOC loss and to enhance its role, farmers could incorporate crop residues into the soil and/or choose crops which need less irrigation and chemical inputs. Before any technological change, decision-makers should evaluate the implications for SOC, he argued. This issue was not taken up by other stakeholders at the workshop.

The workshop discussion also considered whether the cultivation of organic crops could be an alternative option, along with bio-labelling to gain a higher market price. According to the ODS representative, farmers would use organic cultivation methods if they could be convinced that their profit would increase. But owners of small farms would not be easily convinced (Ecowater, 2013). Indeed, small-scale growers lack an advisory service and systematic support for linking organic methods with higher-value markets. This institutional gap illustrates wider difficulties for farmers adjusting to new challenges and gaining the full potential benefits of the EFMA.

5. DISCUSSION AND CONCLUSIONS

Innovative irrigation practices can enhance water efficiency, gaining an economic advantage for farmers while also reducing environmental burdens. Water-efficient methods and better irrigation scheduling could also integrate water and nutrient management, thus minimising agrochemical runoff and leaching problems. To help fulfil this potential, experts have developed various models of water efficiency and environmental benefits. Yet these models are little used for irrigation scheduling; at most, they help retrospectively to evaluate seasonal approaches (FAO, 2012).

In that generally adverse context, our two case studies of irrigation schemes investigated options, stimuli and difficulties to improve water-efficient practices. In both cases, individual farmers and/or their organisations already made significant investments in irrigation technologies, but their implementation has not been systematically evaluated for effectiveness through better knowledge—regarding crops' water use, soil-moisture conditions, irrigation scheduling techniques, crops' yield response to different irrigation management strategies, etc. Farmers there have no formal responsibility to demonstrate efficient water use (unlike in the UK; Knox et al., 2012). Consequently, farmers pay higher water prices yet do not obtain the full potential benefits through water-efficient practices. Reasons why have been investigated by exploring perspectives of water users organisations (WUOs) and relevant agencies, especially through multi-stakeholder workshops.

The two cases have strong stimuli for more water-efficient practices. In the Sinistra Ofanto case, a main stimulus is the prospect of greater water scarcity and erratic supply, leading some farmers to rely on groundwater, which may progressively become more scarce, over-drafted, more expensive due to higher pumping costs, degraded and/or saline. In the Monte Novo case, a water-abundant system, a main stimulus is the statutory rise in water price towards full-cost recovery; this would weaken the commercial viability of some crops with high water demand, such as current maize cultivation.

Anticipating such future difficulties, stakeholders have discussed options for better practices. In each area, stakeholder representatives attended our project workshop on how farmers' economic advantage could be combined with environmental benefits. Despite the stimuli for more water-efficient practices, however, actors' responsibility has been somewhat displaced onto hypothetical prospects. In particular, for the two areas:

At the Sinistra Ofanto workshop, stakeholder representatives emphasised problems of water scarcity and aquifer salinity. Their comments suggested various possible measures such as region-wide control and regulation, farm-level practices and farmers' knowledge-exchange systems. At the same time, however, the WUO displaced responsibility onto the prospect of additional supply from wastewater reuse; this would depend on significant institutional, policy and technical changes, with an uncertain timescale and cost.

At the Monte Novo workshop, stakeholder representatives expressed interest in farm-level technological improvements which could improve water efficiency. Yet WUO representatives also emphasised the need to maintain a low unit cost of water. Moreover, the WUO publicly advocates a longer-term low water pricing, which would depend on a legislative change contradicting the country's EU obligations and the original mandate for the irrigation scheme.

In both cases responsibility is displaced, thus complementing WUOs' and farmers' assumption that their agricultural water management practices already have adequate efficiency. This attitude arises by default—from seeing no systematic means or incentive for more water-efficient practices, beyond the modern technologies already adopted. Experts generally have sought to overcome such obstacle through better models of evapotranspiration and efficient water use (e.g. Pereira et al., 2012). Yet such models cannot reconcile stakeholders' different understandings of irrigation efficiency: farmers seek to maximise their economic benefits, while water experts or agencies seek to conserve water (Knox et al., 2012).

Therefore a solution would need a continuous knowledge-exchange with three components:

i. expert scientific knowledge of crops' water needs, their yield-response to water and the actual on-farm versus attainable efficiency (Hsiao et al., 2007);

ii. links between farmers' perspectives, innovative practices and their income benefits (Knox et al., 2012); and

iii. means to lower resource burdens from inputs and pollutants (Miller, 2007).

Relative to such a solution, our two case-study areas have a nominal extension service which lacks the necessary resources, expertise and operational capability in agricultural water management; these services have no external auditing. A knowledge-exchange system does not exist in those areas, so farmers may miss the full potential benefits of the technological investment already adopted. Under current circumstances, irrigation practices will maintain the unknown efficiency level; along with their WUOs, farmers will have weak capacities to make extra efforts for more water-efficient practices, as well as weak incentives to pay for any further investments.

An adequate knowledge-exchange system would depend on greater institutional responsibility for water-related policies and strategies, for example, through an extension service or a functional equivalent through a WUO. This in turn would provide an enabling condition and incentive for all relevant stakeholders (especially WUOs) to share greater responsibility for agricultural water management across the entire water-supply chain, including farm-level practices as well as drainage and leach-out management. On this basis, more water-efficient practices could combine wider environmental benefits with economic advantage for farmers.

ACKNOWLEDGEMENTS

This work was supported by the European Commission through the project, 'EcoWater: Meso-level eco-efficiency indicators to assess technologies & their uptake in water use sectors', a collaborative research project of the 7th Framework Programme, grant agreement no. 282882, coordinated by the National Technical University of Athens (NTUA) during 2011–2014, http://environ.chemeng.ntua.gr/ecowater/. We would like to thank two anonymous referees of this journal for helpful comments.

REFERENCES

1. Ahearn, M., Yee, J., Bottom, J., 2003. Regional trends in extension system resources.In: Agricultural Information Bulletin No. 781. U.S. Department of Agriculture,Economic Research Service, Washington, DC.
2. Altieri, S., 1995. The Sinistra Ofanto irrigation scheme: management and maintenance problems. In: Pereira, L.S. (Ed.), Bonifica, vols. 1–2. Bastogi Editrice Italiana,Foggia, pp. 40–47.
3. ARH-Alentejo, 2011. Planos de Gestão das Bacias Hidrográficas–RegiãoHidrográfica 7–Relatório técnico para efeitos de Participaç̧ ão Pública,Nemus–Ecossistema–Agro.ges (River basin management plan—Guadiana Tech-nical report for public participation). Administraç̧ ão da Região Hidrográfica(ARH) do Alentejo.
4. Battilani, A., 2012. Sustainable Knowledge-based Irrigation Management: TheIRRINET package, http://ec.europa.eu/environment/archives/greenweek 2012/adriano-battilani.html, http://ec.europa.eu/environment/archives/greenweek2012/sites/default/files/3-9 battilani.pdf.
5. Battilani, A., 2013. Approaches to Improve the Productivity of Water Use in Irrigation[with diagrams], http://www.wsstp.eu/files/WSSTPX0001/II 2 %20Adriano%20Battilani.pdf.
6. Benham, B.L., Linder, G.M., Hinrichs, M.D., 2000. Improving irrigation managementin Nebraska's central Platte valley. J. Nat. Res. Life Sci. Educ. 29, 111–115.
7. Burt, C., 2004. Rapid field evaluation of drip and microspray distribution uniformity.Irrig. Drain. Syst. 18, 275–297.
8. Caswell, M., Zilberman, D., 1985. The choices of irrigation technologies in California.Am. J. Agric. Econ. 67 (2), 224–234.
9. CBC, 1984. Consorzio di Bonifica per la Capitanata, Cinquant'anni di bonifica nelTavoliere. Bastogi Editrice, Foggia.
10. CEC, December, 2008. Follow up Communication on Water Scarcity and Droughtsin the European Union COM (2007) 414 final, [SEC(2008) 3069]. Commission ofthe European Communities, Brussels.COPA-COGECA, 2007.
11. COPA-COGECA Position on Water Scarcity and Drought.COPA (Committee of Professional Agricultural Organisations), General Con-federation of Agricultural Cooperatives (COGECA), Brussels, http://www.copa-cogeca.be/Menu.aspx.
12. COPA-COGECA, 2013. Water and Agriculture under a Changing Climate. COPA-COGECA.
13. Costa, L., 2003. A Portuguese Hydrodinosaur: The Hidden Costs of the Alqueva Dam, http://www.hydrologie.org/hydrodinosaures/download/portugal ang.pdf.
14. EC, 2000. Directive 2000/60/EC of the European Parliament and of the Council of 23October 2000 Establishing a Framework for Community Action in the Field ofWater Policy. EC.
15. Ecowater, 2012. Deliverable 1.2: Value Chain Mapping of the AgriculturalWater Systems. Ecowater, EcoWater project, http://environ.chemeng.ntua.gr /ecoWater/Default.aspx?t=238.
16. Ecowater, 2013. Deliverable 6.1: Synthesis report from the 1st Round ofCase Study Events. Ecowater, http://environ.chemeng.ntua.gr/ecoWater/Default.aspx?t=238.
17. EDIA, 2011. Portal de mapas (Map Portal). EDIA, http:\\www.edia.pt (last visit inDecember 2011).
18. EIO, 2011. Water Innovation: How Eco-innovation can Contribute to the Sustainability of Europe's Water Resources. Eco Innovation Observatory.
19. EP, 2008. Addressing the challenge of water scarcity and droughts in the EuropeanUnion. European Parliament resolution of 9 October 2008 on addressing thechallenge of water scarcity and droughts in the European Union (2008/2074(INI).

20. Fabiani, S. (ed.), 2010. Aspetti economici dell'agricoltura irrigua in Puglia.Istituto Nazionale di Economia Agraria (INEA), http://dspace.inea.it/bitstream /inea/520/1/SE5-1164.pdf
21. FAO, 2012. FAO Irrigation and Drainage Paper 66: Crop Yield Response to Water, http://www.fao.org/nr/water/infores.html.
22. Faurès, J., Svendsen, M., Turral, H., 2007. Reinventing irrigation. In: Molden, D. (Ed.),Water for Food, Water for Life: A Comprehensive Assessment of Water Manage-ment in Agriculture. Earthscan and International Water Management Institute,London, Colombo (Chapter 9).
23. Fraiture, C., Perry, C., 2007. Why is agricultural water demand unresponsive at lowprice ranges? In: Molle, F., Berkof, J. (Eds.), Irrigation Water Pricing: The Gapbetween Theory and Practice. CABI, Wallingford, pp. 94–107.
24. Froes, J., Rodrigues, P.N., 2005. Distribuic, ão de água em baixa pressão no Alqueva.In: Paper in I Congresso da Rega e Drenagem. Beja, Portugal, Available in www.cotr.pt/informacao/web2/Papers/48.pdf.
25. Garcia-Vila, M., Fereres, E., 2012. Combining the simulation crop model AquaCropwith an economic model for the optimization of irrigation management at farmlevel. Eur. J. Agron. 36, 21–31.
26. Genius, M., Koundouri, P., Nauges, C., Tzouvelekas, V., 2014. Information transmis-sion in irrigation technology adoption and diffusion: social learning, extensionservices, and spatial effects. Am. J. Agric. Econ. 96 (1), 328–334.
27. Gold, A.J., Parker, D., Waskom, R.M., Dobrowolski, J., O'Neill, M., Groffman, P.M.,Addy, K., 2013. Advancing water resource management in agricultural, rural,and urbanizing watersheds: why land-grant universities matter. J. Soil WaterConserv. 68 (4), 337–348.
28. Hanson, B., Bowers, W., Davidoff, B., 1995. Field performance of microirrigation sys-tems. In: Microirrigation for a Changing World, Proceedings of Fifth InternationalMicroirrigation Congress, April 2–6, 1995. American Society of Agricultural Engi-neers, Orlando, FL, pp. 769–774.
29. Hergert, G.W., Penas, E.J., Ferguson, R.B., Cahoon, J.E., Klocke, N.L., Watts, D.G.,Shapiro, C.A., Peterson, T.A., 1994. Extension programming for improving nitro-gen and irrigation management in Nebraska. In: National Symposium onProtecting Rural America's Water Resources: Partnerships for Pollution Solu-tions, Washington, DC.
30. Hanson, B., Bowers, W., Davidoff, B., Kasapligil, D., Carvajal, A., Bendixen, W., 1996.Field performance of microirrigation systems. In: Proceedings of the Fifth Inter-national Microirrigation Congress, Orlando, FL, pp. 769–774.
31. Howell, T.A., 2003. Irrigation efficiency. In: Encyclopedia of Water Science.,http://dx.doi.org/10.1081/E-EWS 120010252, http://www.resear chgate.net /publication/43256707 Irrigation Efficiency
32. Hsiao, T.C., Steduto, P., Fereres, E., 2007. A systematic and quantitative approach toimprove water use efficiency in agriculture. Irrig. Sci. 25, 209–231.
33. Iglesias, E., Blanco, M., 2008. New directions in water resources manage-ment: the role of water pricing policies. Water Resour. Res. 44 (6), 1–11,http://dx.doi.org/10.1029/2006WR005708.
34. Irmak, S., Odhiambo, L.O., Kranz, W.L., Eisenhauer, D.E. 2011. Irrigation Efficiencyand Uniformity, and Crop Water Use Efficiency, Lincoln, NB: University ofNebraska–Lincoln Extension, http://extension.unl.edu/publications
35. IST, 2013. Avaliac, ão Ambiental Estratégica dos novos investimentos EFMA relativosà Rede Secundária de Rega, no âmbito do INALENTEJO (Strategic Environ-mental Assessment of new EFMA's Investments). In: Relatório preliminar:Relatório Ambiental. Instituto Superior Técnico (IST), Lisboa, Available onlineat http://www.inalentejo.qren.pt/images/stories/file980.pdf.
36. Kampas, A., Petsakos, A., Rozakis, S., 2012. Price-induced irrigation water saving:unravelling conflicts and synergies between European agricultural and waterpolicies for a Greek water district. Agric. Syst. 113 (1), 28–38.

37. Knox, J.W., Kay, M.G., Weatherhead, E.K., 2012. Water regulation, crop produc-tion and agricultural water management—understanding farmer perspectiveson irrigation efficiency. Agric. Water Manage. 108 (1), 3–8.
38. Luquet, D., Vidal, A., Smith, M., Dauzatd, J., 2005. More crop per drop: how to makeit acceptable for farmers? Agric. Water Manage. 76 (2), 108–119.
39. Maia, R., 2000. Portuguese–Spanish river basins: bilateral agreements' evolution andcontext. Water Sci. Technol. 42, 227–233.
40. Maneta, L., abril, 2010. Prec¸ o da água de Alqueva não convence agricul-tores. In: Diario de Noticias, http://www.dn.pt/inicio/economia/interior.aspx?content id=1540958.
41. Miller, A., 2007. Environmentally Sustainable Agri-Food Production. EnvironmentalSustainability Knowledge Transfer Network (ESKTN).
42. Molle, F., Berkoff, J., 2007. Water pricing in irrigation: the lifetime of an idea. In:Molle, F., Berkoff, J. (Eds.), Irrigation Water Pricing: The Gap between Theoryand Practice. CABI, Wallingford, pp. 1–20.
43. Molle, F., 2008. Can water pricing policies regulate irrigation use? In: 13th WorldWater Congress, September, Montpellier, France.
44. Molden, D., Oweis, T., Steduto, P., Bindraban, P., Hanjra, M.A., Kijne, J., 2010. Improv-ing agricultural water productivity: between optimism and caution. Agric.Water Manage. 97, 528–535.
45. Nuncio, J., Arranja, C., 2011. Gestão dos aproveitamentos hidroagríco-las de Alqueva. Jornadas Tecnicas APRH, http://sir.dgadr.pt/conteudos/jornadas aph/ apresentacoes /s7/11.pdf.
46. Parker, D., Cohen, D., Osgood, D., Zilberman, D., 2000. Publicly funded weatherdatabase benefits users statewide. Calif. Agric. May–June, 21–25.
47. Pereira, L.S., Cordery, I., Iacovides, I., 2012. Improved indicators of water use perfor-mance and productivity for sustainable water conservation and saving. Agric.Water Manage. 108, 39–51.
48. Perry, C., Steduto, P., Allen, R.G., Burt, C.M., 2009. Increasing productivity in irri-gated agriculture: agronomic constraints and hydrological realities. Agric. WaterManage. 96, 1517–1524.
49. Portoghese, I., D'Agostino, D., Giordano, R., Scardigno, A., Apollonio, C., Vurro, M.,2013. An integrated modelling tool to evaluate the acceptability of irrigationconstraint measures for groundwater protection. Environ. Modell. Softw. 16,90–103.
50. Pulwarty, P., Maia, R., 2013. Adaptation challenges in complex river basins aroundthe world. In: Eighth International Conference of EWRA, Water Resources Man-agement in an Interdisciplinary and Changing Context, 26–29 June, Porto,Portugal.
51 Rita, C., Capelo, O., 2005. Distribuic¸ ão em alta pressão vs. distribuic¸ ão embaixa pressão num bloco de rega do E. F. M. Alqueva (Monte Novo). In:Paper in I Congresso da Rega e Drenagem, Beja, Portugal, Available inwww.cotr.pt/ informacao/web2/Papers/38.pdf.
52. Rosado, L., 2012. A cultura do milho na Fundac¸ ão Eugénio de Almeida. In: VIII Con-gresso Nacional do Milho, Lisboa.
53. Salvador, R., Martínez-Cob, A., Cavero, J., Playán, E., 2011. Seasonal on-farm irrigationperformance in the Ebro basin (Spain): crops and irrigation systems. Agric. WaterManage. 98, 577–587.
54. Santos, J.F., Gomes, J.C., Fialho, I., 2011. Contributo para a discussão dascondic¸ ões de sustentabilidade da component hidroagrícola do EFMA, Jor-nadas Técnicas APRH, Lisboa, LNEC, 12-15 Outubro, http://sir.dgadr.pt/conteudos/jornadas aph/ apresentacoes/s2/32.pdf, Technical Days index athttp://sir.dgadr.pt/jornadasaprh
55. Silva, A.M., 2010. Alqueva: água baixa 70% para agricultores, Correo da Manhã,11 abril, http://www.cmjornal.xl.pt/detalhe/noticias/nacional/economia/agua-baixa-70-para-agricultores, http://boasnoticias.sapo.pt/mobile/noticias.php?id=771

56. Sinclair, T.R., Tanner, C.B., Bennett, J.M., 1984. Water-use efficiency in crop production. BioScience 34 (1), 36–40.Steduto, P., 2007. On the conservative behaviour of biomass water productivity. Irrig.Sci. 25 (3), 189–207.
57. Tollefson, L.C., Wahab, M.N.J., 1994. Better Research-Extension-Farmer Interactioncan Improve the Impact of Irrigation Scheduling Techniques. World Bank, Wash-ington, DC.
58. Vergés, J.C., 2001. The Alqueva Irrigation Agreement (Study subcontracted to Modern Consulter S.L. by Research Contractor WRc plc, for the Environment Agency,UK).
59. WBCSD, 2000. Eco-efficiency. In: Creating More Value with Less Impact.World Business Council for Sustainable Development, http://www.wbcsd.org/web/publications/eco efficiency creating more value.pdf.
60. WssTP, 2012. Irrigated Agriculture Water Saving Options in Irrigation, Lookingfor Efficient Techniques, Irrigation Management and Adapted CroppingPractices. Water Supply and Sanitation Technology Platform (WssTP), http://www.wsstp.eu/files/WSSTPX0001/library/agriculture/SP%20Agriculture%20VI.pdf.
61. WssTP, 2013. Water Reuse: Research and Technology Development Needs forWater Reuse. WssTP, Summary at http://www.wsstp.eu /files/WSSTPX0001/Water%20Reuse%20Ex.Summary%20online.pdf.
62. Zaccaria, D., Oueslati, I., Neale, C.M.U., Lamaddalena, N., Vurro, M., Pereira, L.S.,2010. Flexible delivery schedules to improve farm irrigation and reduce pres-sure on groundwater: a case study in southern Italy. Irrig. Sci. 28, 257–270.
63. Zaccaria, D., Lamaddalena, N., Neale, C.M.U., Merkley, G., 2013. Simulation ofpeak-demand hydrographs in pressurized irrigation delivery systems using adeterministic-stochastic combined model. Part II: Model applications. Irrig. Sci.31, 193–208.

CHAPTER 2

Modelling and Mapping The Economic Value of Supplemental Irrigation in a Humid Climate

D. Rey[a,*], I.P. Holman[a], A. Daccache[b], J. Morris[a], E.K. Weatherhead[a], J.W. Knoxa

[a]Cranfield University, Cranfield, Bedford MK43 0AL, United Kingdom

[b]Mediterranean Agronomic Institute of Bari (IAMB), Via Ceglie, 9, 70010 Valenzano (BA), Italy

ABSTRACT

Irrigation is an essential component of crop production to meet retailer demands for premium quality when rainfall is insufficient. Under drought conditions, irrigation can be constrained by water resources availability, with consequent impacts on yield, quality and revenue. Whilst most agriculture in Europe is rainfed, greater dependence on supplemental irrigation could become more important in humid environments due to a changing climate with greater rainfall uncertainty and higher frequency of droughts. By combining industry and farm level economic data, with geospatial information on agricultural land use, agroclimate, soils and irrigation practices within a GIS, this paper estimates the total financial benefit of outdoor irrigated production in England and Wales assuming no constraints in resource availability and optimal irrigation practices. The analysis suggests that the total net benefits of irrigation in a 'design' dry year are around £665 million, with an average irrigation water productivity in excess of £3.3 per m^3 (close to £1.1 per m^3 excluding soft fruit). Map outputs highlight significant regional differences in water productivity reflecting the composition of land use and the importance of crop mix in determining economic value. A sensitivity analysis to changes in agroclimate, market conditions (crop prices) and water supply (costs) illustrates how the benefits might change under contrasting scenario. The study highlights the importance of supplemental irrigation, even in a humid climate, and the risks that future droughts and/or constraints in water resource availability might have on agricultural systems, livelihoods and the rural economy. The implications for water resources and drought management are discussed.

Keywords: Benefit Crop Drought England and wales Water resources Yield

1. INTRODUCTION

Water is becoming an increasingly scarce resource, not only in arid and drought-prone areas but also in more humid and temperate regions where traditionally rainfall has been abundant (Santos Pereira et al., 2002; Daccache et al., 2012). A drought is normally defined as a natural hazard caused by a period of abnormally low precipitation. Three main types of droughts can be distinguished: a meteorological drought, an agricultural drought, a hydrological drought. A fourth type, socio-economic drought, can also be defined, dealing with drought in terms of supply and demand, taking into account the impact of water shortages on socio-economic systems (Wilhite and Glantz, 1985). A meteorological drought is characterised by a prolonged period of low rainfall; an agricultural drought occurs when a lack of precipitation results in low soil moisture that affects crop growth and development; and a hydrological droughts typically occurs when precipitation deficits lead to below normal water levels in reservoirs and rivers for an extended period (AMS Council, 2013). Over the last three decades, the incidence of droughts in Europe has increased in both intensity and number, mainly in the Mediterranean region (European Commission, 2007). The risk of drought and water scarcity is expected to increase in future in currently dry regions due to a range of factors including climate change and population growth (IPCC, 2014). The agricultural sector is particularly sensitive to drought and water scarcity because of its dependence on water, along with other weather-related factors (Knox et al., 2010a). Agriculture is the dominant user of freshwater in many countries, accounting for 70% of global water withdrawals (FAO, 2004; Calzadilla et al., 2010). Most irrigation occurs in arid and semi-arid areas where there is insufficient rainfall to support crop growth. In these areas, the inter-annual variability in irrigation application is relatively small as irrigation provides the majority of crop water requirements to sustain crop growth. Whilst such areas have tended to be the focus for most research on irrigation demand assessment, water efficiency and economic valuation (Hillel, 1987; Oweis, 1997; Deng et al., 2006; Prasad et al., 2006), supplemental irrigation can also be critical in humid regions where it buffers the effects of rainfall variability so that the adverse effects of low soil moisture content on crop development and particularly quality are reduced (Oweis, 2005).

Given an increasing emphasis on quality assurance, rather than solely yield, supplemental irrigation is essential to ensure the viability and profitability of particular crops in some regions. For example, there has been a marked increase in irrigation of high-value crops such as potatoes and field vegetables in the UK (Morris et al., 2014). However agricultural irrigation is often given the lowest priority for water allocation under drought conditions. This partly reflects a perception that the marginal value of water is relatively low in agriculture compared to its use in other sectors including public water supply, and that there is scope in drought conditions for increasing the 'efficiency in use' of agricultural irrigation. However, in humid regions, the small application depths applied to high-value crops primarily for quality assurance purposes can result in very high financial benefits, and hence the potential economic impacts of any water shortages (abstraction restrictions) can be substantial (Knox et al., 2000).

In order to guarantee public water supply and maintain minimum environmental flows during droughts, the water regulatory agency in England and Wales (Environment Agency, EA) can impose partial or total bans on irrigation abstraction (EA, 2012; Defra, 2014), so water may not be available to farmers when needed (Knox et al., 2010b). This has occurred during recent drought episodes in 1995, 2003, 2006 and 2011–2012 (ADAS, 1999; Marsh, 2004; EA, 2006; EA, 2011a). For instance, formal restrictions were imposed on 600 spray irrigation licences during the 2006 drought, and 300 irrigators were affected by abstraction restrictions in the 2012 drought (EA, 2011b; Vivid Economics, 2013). Hess et al. (2011) estimated that more than half of the total area of irrigated production in England and Wales is currently located in catchments designated as being either 'over-abstracted' and/or 'over-licensed'; these areas are therefore the focus of increasing attention to support more sustainable levels of abstraction through regulatory reform and the revocation of so-called 'time-limited' licences (perpetuity). In order to protect their interests some farmers have formed water abstractor groups (WAGs) to collectively share their risks and knowledge (Leathes et al., 2008), whilst others have taken individual actions including investment in storage reservoirs and precision irrigation.

Drought combined with restrictions on irrigation abstraction can therefore have important agronomic and economic impacts on crop production.

Whilst the impacts of drought on the value of output from UK agriculture as a whole have been assessed (e.g. ADAS (1999) for the 1995 drought; and Anglian Water and University of Cambridge (2013) for the 2010–2012 drought), there is much less understanding of the economic impacts of restrictions on supplemental irrigation in field-scale agriculture and horticulture.

With a changing climate expected to increase irrigation demand (Weatherhead and Knox, 2015; Rodriguez Diaz et al., 2007; Else and Atkinson, 2010) and the frequency and severity of droughts also expected to impact on irrigated agriculture (Fowler and Kilsby, 2004), this study builds on previous research (Morris et al., 1997; Knox et al., 2000) to provide the first comprehensive national spatial assessment of the financial benefits of supplementary irrigation, both in terms of yield and quality assurance in a dry year. This will support policies to better understand the impacts of abstraction restrictions in the sector, and to improve drought management strategies which balance impacts across different economic and environmental sectors. As a result, the farming community will be better equipped to estimate potential losses arising from drought, as will governments and agencies in deciding where and when to impose restrictions on agriculture with minimum economic impact. The analysis is applied to outdoor irrigated cropping in England and Wales, but the approach is applicable in other countries where appropriate datasets are available. The methodology and outputs also have significant implications for the implementation of abstraction controls within the Water Framework Directive (WFD).

2. MATERIAL AND METHODS

In summary, a four staged approach was developed:

1. Deriving and mapping agroclimate variability using potential soil moisture deficit (PSMD) as an aridity indicator;
2. Modelling and mapping theoretical volumetric irrigation water demand for the major irrigated crop categories during a design dry year;
3. Mapping the financial value of supplemental irrigation, by crop category and per unit of water applied (water productivity), and;
4. Conducting a sensitivity analysis to assess how the irrigation benefits might change under contrasting agroclimate, market (price) and water supply (water price) conditions.

A brief description of each stage is given below.

2.1. Mapping agroclimate variability

A dry year in England and Wales in irrigation terms is typically characterised by periods of low rainfall and high evapotranspiration (ET) from June to August (Weatherhead et al., 1997). The combined effects of rainfall and ET on irrigation demand can be reflected via an aridity index using potential soil moisture deficit (PSMD). The advantage of using PSMD over other agroclimatic indicators (e.g. Standardised Precipitation Index, SPI) is that the distribution of rainfall and ET throughout the year is taken into account. It can also be used to identify the dryness or wetness of a specific location for a given year. The index has been widely used internationally to quantify irrigation needs at different scales and assess climate impacts on water demand (Knox et al., 1997, 2010a,b; De Silva et al., 2007; Rodríguez Díaz et al., 2007). It is also used by the regulatory authority in England and Wales for setting licences (permits) for irrigation abstraction (Rees et al., 2003). In this study, the PSMD index for 2010 was calculated using a 5 km × 5 km gridded monthly climatic dataset from the UK Meteorological Office derived from observed historical weather data collected from a range of meteorological stations spatially distributed across the UK (Perry and Hollis, 2004). The rationale for using 2010 data was twofold: Knox et al. (2014) reported that 2010 closely approximated to a 'design' dry year in irrigation terms (statistically defined as the unconstrained water demand with an 80% probability of non-exceedance); secondly, national irrigated area data were available from the most recent 2010 Defra Irrigation Survey (Defra, 2011).

Reference evapotranspiration (ETo) was calculated using the gridded monthly temperature, solar radiation, wind speed and relative humidity data and the FAO Penman-Monteith combination equation (Allen et al., 1998). Using monthly rainfall (Pt) and reference evapotranspiration (ETot) data, the PSMD (mm) for each month (t) is calculated as following:

$$PSMD_t = PSMD_t - 1 + ET_{ot} - Pt \quad (1)$$

In months wherePt > ($PSMD_t - 1 + ET_{ot}$), any initial soil moisture deficit is filled and hence PSMDt=0. In the UK, soil moisture deficits typically start to build up in early spring as ET > P, peak in mid-summer (July–August) and then decline to zero through autumn and winter as P > ET. Therefore in the UK, the estimation of PSMD can start with January as month $t = 1$. The maximum PSMD of the 12 months of each year is the $PSMD_{max}$ value assigned to that year at that location. A map showing the spatial variation in $PSMD_{max}$ for England and Wales at 5 km grid resolution for 2010 was produced and then used to estimate volumetric irrigation demand.

2.2. Mapping theoretical volumetric irrigation demand (m^3)

The spatial volumetric irrigation needs of the main outdoor crop categories in England and Wales were derived by applying the methodology developed by Knox et al. (1997) to spatial datasets of climate, soil type, crop areas and irrigation survey results. In summary, Knox et al. (1997) used a daily time-step water balance model (Hess, 1996) to calculate the theoretical unconstrained annual irrigation need (mm) for each of the major irrigated crop categories (potatoes, field vegetables, sugar beet, soft fruit, orchard fruit, grass, and cereals) grown on a contrasting range of soil types at agroclimatically contrasting locations. These results were then correlated using linear regression analyses against the $PSMD_{max}$. Knox et al. (1997) developed formulae for each crop relating the theoretical irrigation need for any specific year or location to soil type and agroclimate ($PSMD_{max}$).

The spatial volumetric water demand in a grid square was calculated by applying the regression relationships derived by Knox et al. (1997) for each crop type to the gridded agroclimate dataset ($PSMD_{max}$) for 2010, taking account the spatial distribution of soil types from the National Soil Map[1] and estimated irrigated area (ha) in each grid pixel. The irrigated area within each catchment was calculated using a GIS by applying the 2005 and 2010 Defra Irrigation Survey results to the 2 km × 2 km gridded land use dataset derived from the Defra, 2011 June Agricultural Census data. Three different soil types were considered for calculating theoretical water demand, depending on available water capacity (AWC): low AWC (<100 mm/m), medium AWC (100–175 mm/m), and high AWC (>175 mm/m).

2.3. Mapping the financial value of supplemental irrigation (£/m^3)

The positive effect that irrigation has on crops, both in terms of yield and quality (and therefore revenue) (Weatherhead et al., 1997) is crop specific since each crop responds differently to the same volume of water applied. Several studies have estimated the yield response to irrigation (e.g. Morris et al., 1997; based on experimental data and farm records) allowing us to estimate the average net yield increase for each unit (mm) of water applied (Table 1). This is then up-scaled to estimate the net yield increase in each grid pixel according to the theoretical irrigation need (based on the local soil type and $PSMD_{max}$). However, irrigation also has a very important effect on quality which is critical for some crops; for example, on potatoes it is used to limit the incidence of common scab (*Streptomyces scabies* spp.)

which causes skin imperfections or for lettuces where water shortages can negatively impact on size, shape and leaf defects. Irrigation is also essential for soft fruit (strawberries) since the majority of the crop is grown under temporary polytunnels to prevent rain damage and thus attain Grade 1 class fruit, which has a much higher price than the strawberries sold for processing. In this study, we estimated the benefits of irrigation on crop quality, expressed as a percentage increase in crop price (£/t) based on published literature corroborated by farmer interviews. These data have been used to derive estimates of the quality benefits of irrigation for each crop and soil type (Table 2). All costs and prices used in this assessment, including the additional costs due to increased production shown as in Table 1 were checked and confirmed with horticultural and soft fruit growers via a series of face-to-face interviews in 2015.

Table 1. Main economic data and crop yield response to irrigation.

Crop category	Crop price[a] (£/t)	Additional costs due to increased production (% gross benefit)[b]	Average crop yield response per hectare[c] (t/mm)
Early potatoes	155	15%	0.08
Main crop potatoes	104	15%	0.08
Cereals (wheat)	110	3%	0.02
Sugar beet	31	10%	0.13
Vegetables (carrots)	95	15%	0.13
Soft fruit (strawberries)	600	25%	0.03
Grass- graze	95	0%	0.03
Grass- silage	95	22%	0.03
Orchard fruit (dessert apple)	479	25%	0.02

[a] These represent the average national crop price (£/tonne, adjusted for inflation using the Agricultural Price Index for the UK, which measures the monthly price changes in agricultural outputs and inputs for all major crops). Information was collected for all crop categories over the past three drought episodes in the UK (2003, 2004–2006, 2010–2012), from the relevant years of the Farm Business Survey (e.g. Defra 2005), the Agriculture in the English regions (e.g. Defra 2006) and the Farm Management Pocketbook (e.g. Nix, 2010).

[b] Extra costs associated with the additional production due to irrigation. These include: additional harvesting, handling, drying and if relevant, direct packaging and marketing costs (Morris et al., 1997), but not the variable costs of irrigation.

[c] Average yield response to irrigation, normalised by irrigation depth, obtained from previous experimental data and field experience in the UK from Morris et al. (1997). A simplifying assumption of a linear relationship between irrigation depth and yield is used to estimate irrigation benefits, given the impracticability of calibrating biophysical crop models to characterise the non-linear response for all combinations of agroclimate, soils and crop type across the UK.

Table 2. Crop quality benefit arising from irrigation for different available water capacity (AWC) soils, expressed as% increase in the rain-fed crop price.

Quality premium (% increase in price)			
Crop category	**Soil AWC**		
	High	**Medium**	**Low**
Early potatoes	11%	23%	40%
Main crop potatoes	25%	30%	40%
Cereals (wheat)	0%	0%	5%
Sugar beet	0%	0%	5%
Vegetables (carrots)	6%	15%	30%
Soft fruit (strawberries)[a]		429%	429%
Grass	0%	0%	5%
Orchard fruit (dessert apple)	14%	20%	25%

[a] The very large price benefit due to irrigation for strawberries reflects the difference in price between Class 1 strawberries (that can only be achieved

with irrigation, and have an average price of £3150 per tonne) and the lower quality strawberries that result if growers are unable to meet crop water requirements in a dry year (and sold for processing at an average price of £600 per tonne.

Conventionally, it is assumed that crop prices rise during a drought as agricultural production is affected by weather conditions (FAO et al., 2011). However, the final impact on price is influenced by many other factors, such as the timing and onset of the drought, world market prices and the extent to which imported products affect the local domestic food market. Thus, in estimating the benefits of irrigation in financial terms, the average crop price (£/tonne, adjusted for inflation using the government Agricultural Price Index for the UK) over the past three drought episodes in the UK (2003, 2004–2006, 2010–2012) were used (Table 1). Regionally disaggregated price information is available for some crop types, but not all; we therefore assumed national average crop prices for our analysis, having checked that differences were not significant across the country for the available price data (for instance, the price of cereals has a coefficient of variation (CV) across regions <0.05 cereals and <0.16 for potatoes). Price data for these periods were drawn from various sources including Defra (2005, 2006); Nix (2010).[2]

For each crop, the additional costs associated with the increased production (e.g. during harvest, handling, drying) (Table 1) were deducted from the gross benefit. The variable irrigation costs (repairs, fuel, labour and machinery, water charges) were also deducted. The costs of irrigation vary considerably depending on local circumstances, the type of irrigation system, the water source and crop water requirements (Morris et al., 2014). For calculation purposes, we assumed average variable costs of irrigation in the UK (1.50 £/ha mm, based on Nix (2010)).

By combining the yield and quality benefits, the total financial benefit of supplemental irrigation can be estimated. Here, we consider only the extra net benefits derived from irrigation in comparison with the benefits that the farmer would get from the same crops grown under non-irrigated conditions (the counter-factual); i.e. the added value associated with irrigation relative to rainfed production.[3]Fig. 1 shows that the yield and quality benefits are multiplicative, not additive, as the quality premia are relevant to the whole crop not just the yield increase. From the gross benefits illustrated in Fig. 1, the irrigation costs and additional costs associated with higher levels of production must be deducted to obtain the net extra margin (£/ha mm) derived from supplemental irrigation.

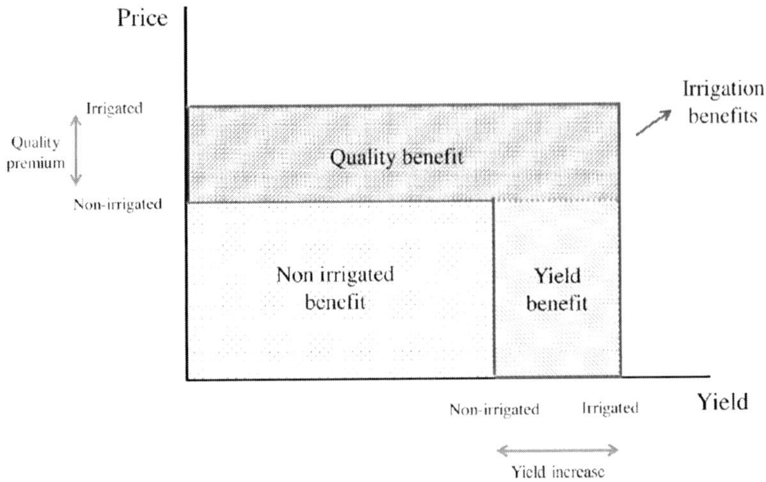

Figure. 1. Conceptual representation of the crop quality and yield benefits attributable to supplemental irrigation.

The following equations were derived and used to estimate the net irrigation benefit (total and per ha) and water productivity (£/m³) for each grid pixel (*i*):

- Total irrigation benefit (£)

$$B_i = \sum_{c=1}^{n} \sum_{s=1}^{m} [((Y_{c,s} + Q_{c,s} - P_c - I) - R_{c,s}) * A_{c,s}] \tag{2}$$

- Benefit per irrigated hectare (£/ha)

$$B_i^{ha} = B_i / A_{c,s} \tag{3}$$

- Water productivity (£/m³)

$$W_i = \sum_{c=1}^{n} \sum_{s=1}^{m} B_i / V_{c,s} \tag{4}$$

where, *B* is the financial irrigation benefit (£). *Y* refers to the yield benefit derived from irrigation (£/ha) and *Q* the quality benefit (£/ha). *P* represents the extra crop costs (£/ha) and *I* the irrigation costs (£/ha). *R* represents the total benefit associated with non-irrigated production (£/ha). Subscript *c* represents each of the eight major irrigated crops (namely, early potatoes, maincrop potatoes, cereals, sugar beet, vegetables, soft fruit, grass, and orchard fruit); subscript *s* denotes the soil available water capacity (AWC) (high AWC, medium AWC and low AWC). *A* is the irrigated area of each crop (ha); *W* is the water average productivity (£/m^3) and *V* is the theoretical irrigation water demand (m^3).

2.4. Sensitivity analysis

For each crop category, agroclimate (PSMD), crop price and water cost were assumed to have the greatest impact on irrigation benefit. Higher PSMD values imply higher irrigation needs (mm), and thus higher benefits derived from fulfilling crop water requirements, as the differences between rainfed and irrigated yield would be higher (Morris et al., 2014). Higher crop prices increase the benefit that irrigators derive from each incremental depth of water applied (mm) to their crops. Costs of water will have a direct impact on water productivity (£/m^3). Using different agroclimate and price scenario, a sensitivity analysis was also undertaken to assess how the irrigation benefits might be impacted by changes in each of these variables.

3. RESULTS

3.1. Irrigated production and water resource availability

Fig. 2a shows the spatial variability in agroclimate across England and Wales for 2010, assumed here to represent a 'design' dry year, defined as a year with an 80% probability of non-exceedance. The areas of highest aridity (high $PSMD_{max}$) are concentrated in eastern and southern England, notably in Norfolk, Suffolk, Essex and Kent. These correspond to parts of the country where irrigated cropping is most concentrated (Fig. 2b) reflecting suitable soils and agroclimatic conditions for potato and field-vegetable production in East Anglia, and soft and orchard fruit production in Kent. The location of irrigated production is also closely correlated with (though not necessarily the cause of) water resources availability with 59% located in either over-licensed and/or over-abstracted catchments (Fig. 2c). In contrast, regions with the lowest aridity extend across much of Wales, the south west and north-west of England, reflecting higher rainfall and less suited land for intensive irrigated vegetable production. In these regions, rainfed production including extensive grassland for livestock is the dominant form of agricultural land use.

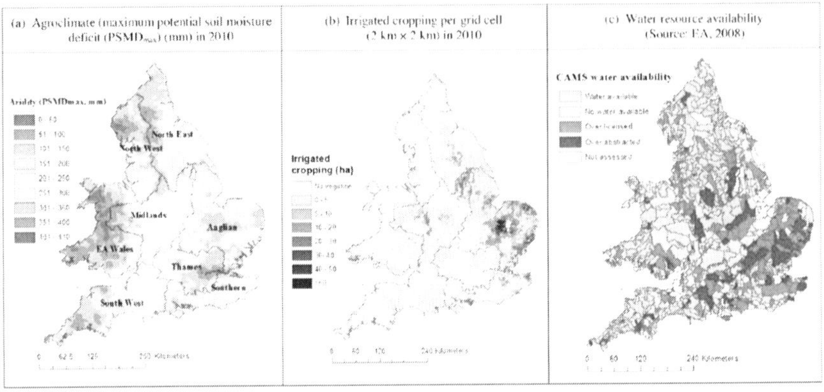

Figure. 2. Maximum potential soil moisture deficit ($PSMD_{max}$) (mm) (a); irrigated cropping (ha) (b); and EA water resource availability(c) for England and Wales, by EA region.

3.2. Irrigation water demand, value and productivity

The calculated spatial distribution of volumetric irrigation demand in England and Wales is shown in Fig. 3a. Nationally, the majority (80%) of irrigation demand is concentrated in central and eastern England, notably in EA Anglian (130×10^6 m^3) and EA Midlands (30×10^6 m^3) regions. The total volumetric demand in 2010 ('design dry year') was estimated to be 200×10^6 m^3. Maincrop potatoes and vegetables have the highest irrigation demand, representing 56% and 23% of total theoretical water demand, respectively.

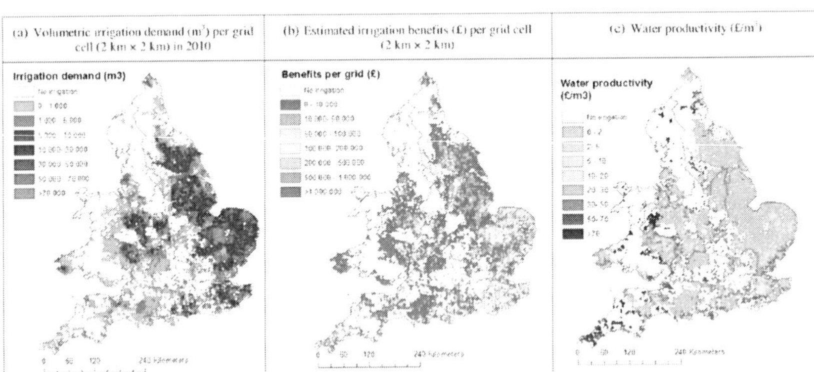

Figure. 3. Volumetric irrigation demand (m^3), estimated irrigation benefit (£) and water productivity (£/m^3) for England and Wales, by EA region.

The financial benefit of irrigated production in 2010 was estimated to be £665 million. As shown in Fig. 3b, the highest benefits accrue in Anglian, Midlands and Southern EA regions, where high-value crops (soft fruit, potatoes and vegetables) are concentrated. Soft and orchard fruit has the highest average benefit per hectare (Table 3) with most production concentrated in EA Anglian and Southern regions. Maincrop potatoes are also an important irrigated crop in EA Anglian region, representing a significant proportion of total benefits in that area (close to 34%). Areas with lower irrigation benefits reflect areas where grassland and cereals dominate the land use mix. Based on the coefficients of variation (*cv*), it can be deduced that the spatial variability in irrigation benefits is very high for cereals and very low for maincrop potatoes and orchard and soft fruit (as these crops benefit from irrigation for quality assurance in almost all areas).

Table 3. Irrigated area and estimated combined yield and quality benefits from irrigation by crop category across England and Wales in the design dry year.

Crop	Irrigated area (ha)	Total benefits (£million)	Average benefit per ha (£/ha)	Standard deviation	Coefficient of variation
Early potatoes	3794	6.36	1,680	388.03	0.23
Maincrop potatoes	31,064	127.47	3919	581.11	0.15
Cereals	9,563	0.14	15	18.94	1.27
Sugar beet	5,898	3.55	598	130.73	0.22
Vegetables	18,413	60.98	3110	1,432.58	0.46
Soft fruit	9297	461.50	54,973	2,236.25	0.04
Grass	3,315	0.87	255	64.45	0.25
Orchard fruit	1,057	4.48	8,834	1,294.41	0.15

As discussed earlier, the total irrigation benefit is a combination of both yield and quality benefits which differ between individual crop types. Table 4 shows the relative contribution of each component to the total irrigation benefit for each crop. The irrigation benefits associated with quality assurance are higher than the yield benefits for soft fruit, maincrop and early potatoes. For example, more than 60% of the total benefit derived from irrigating vegetables in a dry year is derived from an improvement in quality. For other crops, such as grass, sugar beet and cereals, which have a much higher tolerance to drought and are not part of the fresh produce supply chain, most of the irrigation benefits are related to a yield increment with irrigation having a relatively low impact on quality. The proportion of irrigation benefits derived from quality improvements are also higher on low moisture retentive (AWC) soils (Table 2).

Table 4. Estimated average contribution of yield and quality benefits (%) to overall irrigation benefit, by crop category in England and Wales.

Crop category	Yield contribution to total benefits (%)	Quality contribution to total benefits (%)
Early potatoes	47	53
Maincrop potatoes	49	51
Cereals	77	23
Sugar beet	96	4
Vegetables	69	31
Soft fruit	3	97
Grass	96	4
Orchard fruit	65	35

Fig. 3c shows the spatial distribution of benefits per unit of water applied (£/m^3) or water productivity for the composition of land use in 2010. The most productive areas are mainly in the south-east and around the Bristol Channel (south and west) and in the midlands. The data for individual crop types shows soft fruit and early potatoes have the highest average water productivity (£52 and £1.94 per m^3, respectively), with cereals (£0.08 per m^3) and grass (£0.11 per m^3) having the lowest values. Irrigation benefits tend to be highest for high-value (£/ha) crops, where irrigation can make most difference to yield and farm-gate price; and for crops that are associated with relatively high capital and labour production costs, helping to secure their viability. That is the case for strawberry production in the UK. The areas showing the highest water productivity (>£50 per m^3)

correspond to land where soft fruit is the only crop being irrigated. Expressing net benefits per unit of irrigation in this way, however, attributes all net benefits of crop production to the irrigation activity, rather than sharing it amongst the various components of investment, on the understanding that irrigation is an essential, non-substitutable component of production. These high returns to water are indicative of the essential nature of irrigation in England and Wales, where water adds most value in relatively intensive high-value, high-risk production systems.

3.3. Sensitivity analysis

Sensitivity analysis can provide an insight into the impact that changes in agroclimate, crop price or irrigation costs can have on irrigation benefits. As mentioned, PSMD was used in this study as an agroclimate index. Fig. 4 shows the financial value of irrigated agriculture in England and Wales, for different agroclimate and crop price scenario. In a very dry year,[4] with very high crop prices[5] the total economic value of supplemental irrigation could reach almost £1130 million, assuming that crop irrigation needs are fully met. Cereals, grass and sugar beet are most sensitive (in terms of a percentage change in, albeit low, irrigation benefits) to a change in PSMD, with strawberries and early potatoes being least sensitive (Fig. 5). However, it should be highlighted that Figs. 5 and 6 show the 'relative' changes, so for instance for cereals an increase of 148% in average benefit represents only £17/ha. Fig. 6 summarises the effects of changes in farm gate commodity prices on irrigation benefits. The analysis shows, for example, that if the price of cereals were to increase by 10% then the irrigation benefits would increase by nearly 40%.

If water resources become more scarce in future, then the water charges (£/m^3) for agricultural abstraction are likely to increase, with the aim of reducing water use and improving irrigation efficiency. A change in water price will affect the benefits from supplemental irrigation and hence water productivity. Fig. 7 shows the estimated change in water productivity (%) caused by a change in the cost of irrigation water for farmers in different regions. Changes in water costs have an important impact on water productivity, although they may not be significant depending on the crop mix. For example, a 50% increase in water costs would reduce water productivity (£/m^3) by <2%. In those regions with a large proportion of high-value crops such as soft fruit and vegetables, the impact of an increase in water price will be lower, as our results demonstrate. Although current water charges represent a small percentage of the average total crop cost per hectare, if water prices were to increase, then the viability of irrigating some of the least-profitable crops (grass and cereals) would be further reduced.

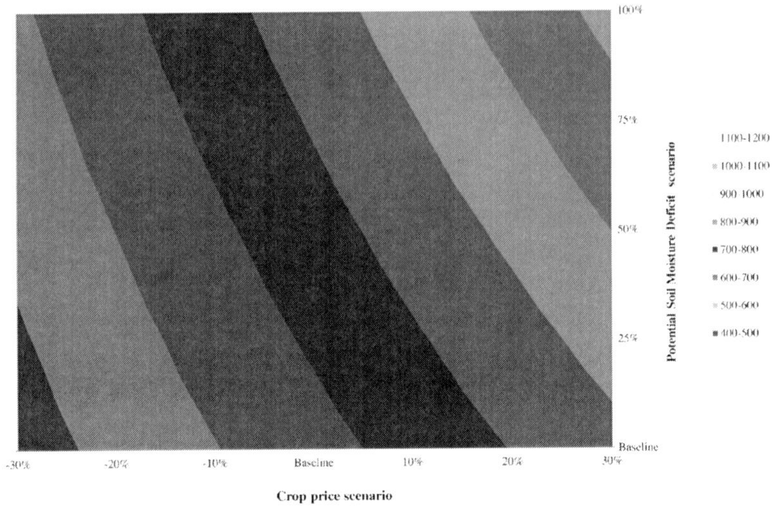

Figure. 4. Estimated financial benefit (£ million) of supplemental irrigation in England and Wales under different agroclimate (PSMD) and crop price scenario.

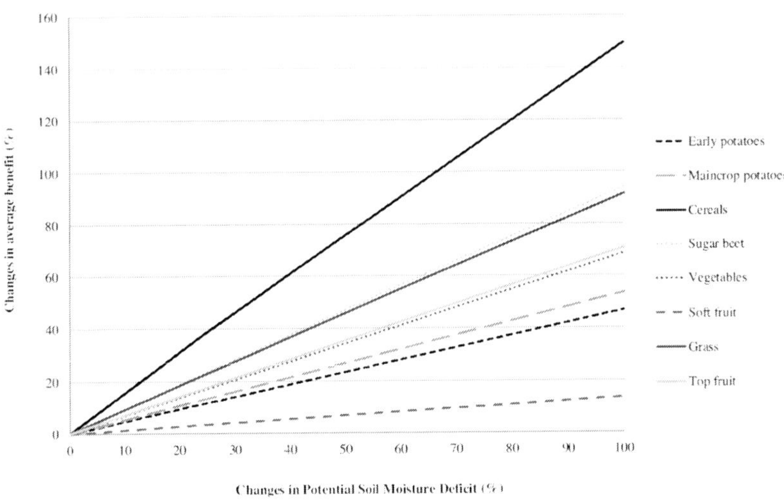

Figure. 5. Estimated change (%) in average irrigation benefit for selected agroclimate ($PSMD_{max}$) scenario.

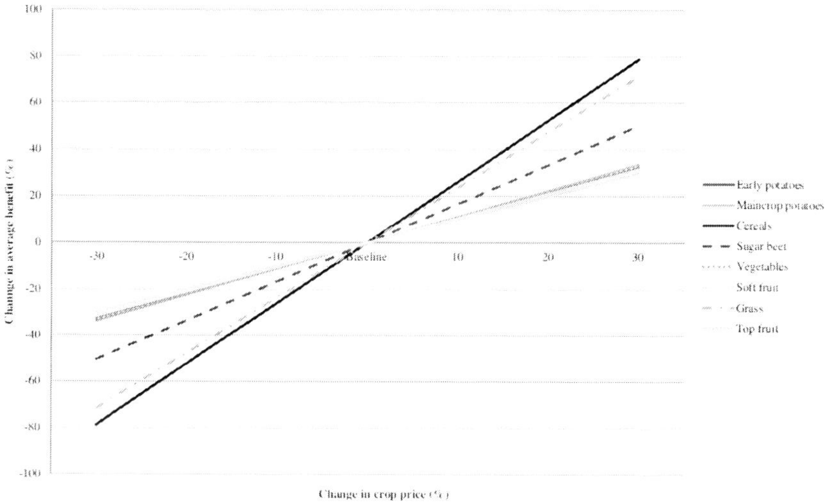

Figure. 6. Estimated changes (%) in average benefit by crop due to a change in crop price.

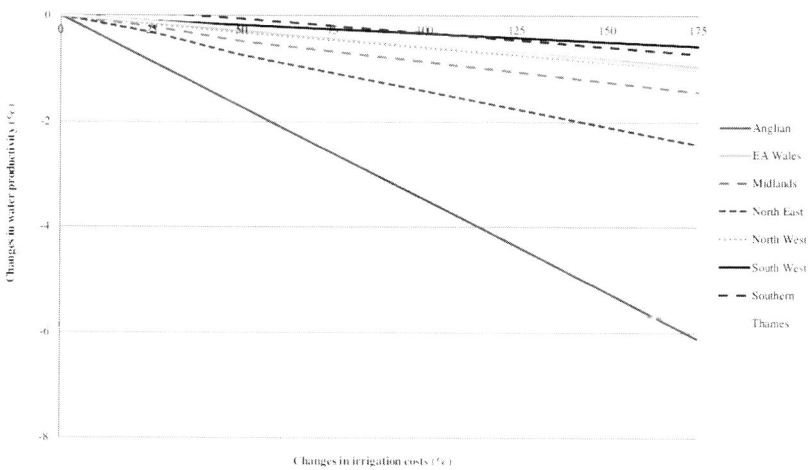

Figure. 7. Estimated change in average water productivity (%) due to an increase in water cost.

4. DISCUSSION

4.1. Implications for water management

This study represents the first international attempt to develop a national assessment of the financial benefits of supplemental irrigation for a defined dry year, compared to rainfed cropping, taking account of the spatial distribution of irrigated crops, and the factors that influence irrigation demand. For the eight crops examined, the highest irrigation benefit is for soft fruit (strawberries) and orchard fruit (dessert apples); and the lowest is for cereals due to the high differential in price between the crops, although differences across regions are also considerable. Our estimate of irrigation demand in EA Anglian Region is marginally higher (130×10^6 m^3) compared to that (104×10^6 m^3) estimated by Knox et al. (2000). However, our results for the average benefit (£/ha) of irrigated maincrop potatoes are consistent with those reported by Morris et al. (2014).

Since the work carried out by Morris et al. (1997); Knox et al. (2000), irrigated agriculture and demand for water has changed markedly (Weatherhead et al., 2015). This is partly due to the changes in cropped area and crop mix over the last decade together with changes in abstraction licensing and water allocation. Irrigation has become increasingly concentrated on fewer high-value crops; coupled with improvements in application technology scheduling and in-field management. The rising real cost of abstracted water and energy for pumping has also led to a reduction in irrigation on some crops where the benefits of irrigation were previously marginal, notably grass and sugar beet. At a national scale, Defra (2011) reported that the total irrigated area in England and Wales has decreased by 43.5% from 2001 to 2010, yet the benefit has increased

In future it is likely that the frequency and severity of droughts in the UK will increase, potentially leading to an increase in crop water requirements. It is likely that irrigation will become more important but also more challenging, as there will be a greater probability of low water availability periods that will affect irrigators, since most agricultural production is concentrated in catchments that are already water stressed. Irrigated agriculture could lose its comparative advantage if constrained, unless the sector adapted to changing water availability. The economic rationale for irrigation is to secure water to add value, particularly through quality assurance, in areas that have comparative advantage for crop production. Many of these areas are in water deficit: thus there is an emerging 'critical irrigation geography' with competing demands between water for irrigation and other uses. During drought conditions, irrigated agriculture is potentially exposed because of its greater dependency on water, with unintended outcomes and consequences. Irrigation may not help in severe drought conditions, unless drought responses for agriculture are explicitly 'designed' into farm irrigation plans and regional drought management strategies.

According to the analyses presented here, the inability to irrigate in future dry years could lead to severe economic impacts for the farming community. Thus, to de-risk their businesses, a combination of irrigation with other drought management strategies will be essential. In recent decades farmers in the UK have invested in irrigation management to cope with a more uncertain water supplies, including for example, building on-farm water storage reservoirs, investing in more efficient technologies or creating water abstractor groups (Holman and Trawick, 2011; NFU 2015). In areas where increased irrigation from direct summer abstraction is not a viable option for coping with drought, other measures will need to be implemented, such as the selection of more drought-resistant varieties (Pidgeon et al., 2006) or changes in planting programmes to avoid late summer maturation (Wreford and Adger, 2010). These adaptation strategies would thus help reduce the potential losses in years when irrigation water was constrained.

In England, irrigation constitutes only 4% of the total cropped area, yet it represents 20% of crop value (EA, 2008). In a dry year, the contribution of irrigated agriculture to the total crop value could be much higher. The benefits derived from irrigation vary spatially in England and Wales. Thus, the magnitude of drought impacts on these benefits will differ depending on which areas of the country are most affected. In a very dry year, as our sensitivity analysis suggests, some crops will be more affected (in terms of financial benefits) than others by an increase in aridity (PSMD). The results from this analysis could help guide water management decisions by regulatory authorities to minimize the overall economic impact of restrictions on water abstractions, changes in the abstraction licensing system and WFD implementation. This includes consideration of the economic impact on incomes, employment and livelihoods in the farm sector, the food and allied industries and other parts of the rural economy where irrigation is important. This wider perspective would also consider regional and national food supply and security issues placed under pressure during drought conditions, as well as the impacts of agricultural water abstraction on the environment. Apart from the potential impacts of abstraction restrictions on the financial benefits of irrigation highlighted in this paper, the water regulatory agency should also take into account the positive and negative externalities (i.e., more water for the environment vs. irrigated agriculture profitability, rural employment and food security), that imposing abstraction restrictions could have on the different water users and the environment.

4.2. Methodological limitations

Finally, this work has some inherent limitations that should be recognised. The study by Morris et al. (1997) was applied to a specific area (Anglian region) where most irrigation is concentrated. Although some of the data used in this study have been updated, our analysis is informed by their results, which might cause errors in the estimation of the financial benefits of irrigation for other regions in England and Wales. Our assumption regarding the relationship between irrigation and crop quality could be improved by developing an irrigation- related quality indicator that could better explain variations in market price. The assumption of a linear relationship between irrigation depth and yield is considered the most feasible option to estimate irrigation benefits for most of the crops in this study, given the difficulty in calibrating biophysical crop models to simulate farm yields for all combinations of agroclimate, soils and crop type. However, it is also probably the greatest source of uncertainty given the assumption of linearity over the 'relevant range' of irrigation water use. Another potential source of error relates to the assumptions on irrigation cost. Although in England and Wales different irrigation systems and water sources (with different associated costs) exist, we assumed the costs associated with direct abstraction from a surface water (river, stream) and overhead application using a hose reel fitted with a gun (93% of the total irrigated area used hose reels in 2010 according to Defra (2011)). However, we recognise that other water sources are used, including groundwater and storage reservoirs which would change the structure of irrigation costs. Finally, we used 'representative' irrigated crops for some crop categories (i.e. carrots for vegetables, wheat for cereals, strawberries for soft fruit, apples for orchard fruit) which could introduce some uncertainty into our estimates. We also assumed that the counterfactual for irrigated crops would be the rainfed equivalent, which is reasonable as we are considering the costs of unexpected droughts or restrictions. However, the counterfactual in areas of highest irrigation needs is likely to be wheat or barley, rather than a rainfed equivalent of the irrigated crop. Finally, the financial assessment has not considered the economic impacts of crop switching due to, for example, changes in future land suitability and/or agroclimate conditions. Further primary data collection could help address the limitations outlined since most are related to constraints in farm level data availability.

5. CONCLUSIONS

Drought episodes can severely impact on agricultural production in the UK, and are projected to become more frequent and extreme. This work presents a methodology to assess the geospatial financial benefits of supplemental irrigation in England and Wales. Despite the fact that many crops grown in the UK are rainfed, our analyses highlight the importance of supplemental irrigation to achieve yield and quality assurance to guarantee profitable crop production in dry years. In a future drier climate with greater aridity, irrigation would only become more crucial to buffer the impacts of the capricious nature of rainfall. Information arising from this work could support water resource planning and decision making to reduce the impacts of low flows on agriculture during drought periods. The procedure developed here could readily be applied to other crop types and/or mixes in other agroclimatic regions worldwide.

ACKNOWLEDGEMENTS

This research forms part of the Natural Environment Research Council (NERC) programme on Droughts and Water Scarcity, funded through the Historic Droughts and MaRIUS projects (grants NE/L010070/1 and NE/L010186/1, respectively). We also acknowledge the UK Meteorological Office for climate data; EDINA for cropping census data and the Environment Agency (EA) for irrigation abstraction data. We also thank Aaron Msowoya, MSc student at Cranfield University, the many farmers that kindly provided valuable farm information to support our research, and the peer reviewers for their insightful feedback. Enquiries for access to the data referred to in this article should be directed to researchdata@cranfield.ac.uk.

REFERENCES

1. ADAS, 1999. A Review of the Impact of the Hot, Dry Summer of 1995 on the MainAgricultural and Horticultural Enterprises in England and Wales. MAFF.
2. AMS Council, 2013. Drought: An information Statement of the AmericanMeteorological Society (accessed on 21.01.15.),https://www.ametsoc.org/POLICY/2013drought amsstatement.html.
3. Allen, R.G., Pereira, L.S., Raes, D., Smith, M., 1998. Cropevapotranspiration—guidelines for computing crop water requirements. FAOIrrigation and Drainage Paper 56.
4. Anglian Water and University of Cambridge, 2013. Water, water everywhere?Encouraging collaborating and building partnerships, (accessed on

15.04.15.),http://www.cisl.cam.ac.uk/business-action/natural-resource-security/natural-capital-leaders-platform/pdfs/water-water-everywhere-scroll.pdf.

5. Calzadilla, A., Rehdanz, K., Tol, R.S.J., 2010. The economic impact of moresustainable water use in agriculture: a computable general equilibriumanalysis. J. Hydrol. 384 (3–4), 292–305, http://dx.doi.org/10.1016/j.jhydrol.2009.12.012.
6. Daccache, A., Keay, C., Jones, R.J.A., Weatherhead, E.K., Stalham, M.A., Knox, J.W.,2012. Climate change and land suitability for potato production in England andWales: impacts and adaptation. J. Agric. Sci. 150 (2), 161–177.
7. De Silva, C., Weatherhead, K., Knox, J., Rodriguez-Diaz, J., 2007. Predicting theimpacts of climate change—a case study of paddy irrigation waterrequirements in Sri Lanka. Agric. Water Manage. 93, 19–29.
8. Defra, 2005. Farm Business Survey, (accessed on 22.04.16.),http://www.farmbusinesssurvey.co.uk/DataBuilder/.Defra, 2006. Agriculture in the English regions (accessed on 22.04.16.).https://www.gov.uk/government/statistics/agriculture-in-the-english-regions.
9. Defra, 2011. Water usage in agriculture and horticulture. In: Results from the FarmBusiness Survey 2009/10 and the Irrigation Survey 2010.
10. Defra.Defra, 2014. Water abstraction reform and drought: supporting information todigital dialogue on drought, September 2014.
11. Deng, X.-P., Shan, L., Zhang, H., Turner, N.C., 2006. Improving agricultural water useefficiency in arid and semiarid areas of China. Agric. Water Manage. 80, 23–40.
12. EA (Environment Agency), 2006. The impact of climate change on severe droughts:major droughts in England and Wales from 1800 and evidence of impact.Science Report SC040068/SR1.
13. EA, 2011. Drought prospects for winter and spring 2011/12. Report–GEHO1111BVBT-E-E.EA, 2011. Impact of long droughts on water resources. Report: SC070079/R5,Environment Agency, UK.EA, 2012. Section 57 Spray Irrigation Restrictions. Environment Agency.
14. EA, 2008. Climate Change Impacts on Water for Horticulture. Environment Agencyand Cranfield University.Else, M., Atkinson, C., 2010. Climate change impacts on UK top and soft fruitproduction. Outlook Agric. 39 (4), http://dx.doi.org/10.5367/oa.2010.0014.
15. European Commission, 2007. Mediterranean water scarcity and drought report,Technical report on water scarcity and drought management in theMediterranean and the Water Framework Directive, Technical Report –009-2007, Mediterranean water scarcity & drought working group.
16. FAO, IFAD, IMF, OECD, UNCTAD, WFP, World Bank, WTO, IFPRI, UN HLTF, 2011.Price Volatility in Food and Agricultural Markets: Policy Responses, Paris,OECD.FAO, 2004. Economic valuation of water resources in agriculture: from the sectoralto a functional perspective of natural resources management.
17. FAO Water Rep.,27.Fowler, H.J., Kilsby, C.G., 2004. Future increases in UK water resource droughtprojected by a regional climate model. Hydrol.: Sci. Pract. 21st Century 1,15–51.
18. Hess, T., Knox, J., Kay, M., Weatherhead, K., 2011. Managing the water footprint ofirrigated food production in England and Wales. In: Hester, R.E., Harrison, R.M.(Eds.), Issues in Environmental Science and Technology 31: Sustainable Water.RSC Publishing, pp. 78–92. 10.1039/9781849732253-00078.
19. Hess, T., 1996. A microcomputer scheduling program for supplementary irrigation.Comput. Electron. Agric. 15, 233–243.
20. Hillel, D., 1987. The efficient use of water in irrigation: principles and practices forimproving irrigation in arid and semiarid regions. World Bank Technical Paper,64.
21. Holman, I.P., Trawick, P., 2011. Developing adaptive capacity within groundwaterabstraction management systems. J. Environ. Manage. 92 (6), 1542–1549.

22. IPCC, 2014. Climate Change 2014. Impacts, Adaptation, and Vulnerability. Contribution of Working Group II to the Fifth Assessment Report of theIntergovernmental Panel on Climate Change, Cambridge, UK: CambridgeUniversity Press, Final Draft.Knox, J., Weatherhead, K., Bradley, R., 1997. Mapping the total volumetric irrigationwater requirements in England and Wales. Agric. Water Manage. 33, 1–18.
23. Knox, J., Morris, J., Weatherhead, K., Turner, A., 2000. Mapping the financialbenefits of sprinkler irrigation and potential financial impact of restrictions onabstraction: a case-study in Anglian Region. J. Environ. Manage. 58, 45–59.
24. Knox, J., Morris, J., Hess, T., 2010a. Identifying future risks to UK agriculturalproduction: putting climate change in context. Outlook Agric. 39 (4), 249–256.
25. Knox, J.W., Rodriguez-Diaz, J.A., Weatherhead, E.K., Kay, M.G., 2010b. Developmentof a water strategy for horticulture in England and Wales. J. Hortic. Sci.Biotechnol. 85 (2), 89–93.
26. Knox, J.W., Daccache, A., Weatherhead, E.K., Groves, S., Hulin A., 2014. Assessmentof the impacts of climate change and changes in land use on future waterrequirement and availability for farming, and opportunities for adaptation.Defra FFG1129 Final Report, March 2014, Cranfield University.
27. Leathes, W., Knox, J.W., Kay, M.G., Trawick, P., Rodriguez-Diaz, J.A., 2008. Developing UK farmers' institutional capacity to defend their water rights andeffectively manage limited water resources. Irrig. Drain. 57 (3), 322–331.
28. Marsh, T.J., 2004. The UK drought of 2003: a hydrological review. Weather 59 (8),224–230.
29. Morris, J., Weatherhead, E.K., Mills, J., Dunderdale, J.A.L., Hess, T., Gowing, D.J.G.,Sanders, C., Knox, J.W., 1997. Spray Irrigation Cost Benefit Study. Final Report,Cranfield University.
30. Morris, J., Ahodo, K., Weatherhead, E.K., Daccache, A., Patel, A., Knox, J.W., 2014.Economics of rainfed and irrigated potato production in a humid environment.In: Bournaris, T., Berbel, J., Manos, B., Viaggi, D. (Eds.), Economics of WaterManagement in Agriculture. CRC Press, pp. 71–97.
31. NFU, 2015. Water Availability in a Changing Climate: A Survey of Irrigated CropGrowers in the River Lark & Wissey Catchments. National Farmers Union, UK.
32. Nix, J., 2010. Farm Management Pocketbook, 41stedition. University of London.
33. Oweis, T., 1997. Supplemental Irrigation: A Highly Efficient Water-use Practice ,International Center for Agricultural Research in the Dry Areas (IRCADA)Aleppo, Syria, p. 16.
34. Oweis, T., 2005. The role of water harvesting and supplemental irrigation in copingwith water scarcity and drought. In. dry, the, year, in:, Wilhite, D. (Eds.),Drought and Water Crisis: Science, Technology, and Management Issues. CRCPress, Boca Raton FL, pp. 191–213.
35. Perry, M., Hollis, D., 2004. The generation of monthly gridded datasets for a rangeof climatic variables over the United Kingdom. Version 2 Meteorological OfficeReport, 22, unpublished.
36. Pidgeon, J.D., Ober, E.S., Qi, A., Clark, C.J.A., Royal, A., Jaggard, K.W., 2006. Usingmulti-environment sugar beet variety trials to screen for drought tolerance.Field Crop Res. 95, 268–279.
37. Prasad, A., Umamahesh, N., Viswanath, G., 2006. Optimal irrigation planning underwater scarcity. J. Irrig. Drain E. 132 (3), 228–237.
38. Rees, B., Cessford, F., Connelly, R., Cowan, J., Bowell, R., Weatherhead, E.K., Knox,J.W., Twite, C.L., Morris, J., 2003. Optimum Use of Water for Industry andAgriculture. Phase III. R&D Technical Report W6-056/TR1. EnvironmentAgency, Bristol.
39. Rodríguez Díaz, J.A., Weatherhead, E.K., Knox, J.W., Camacho, E., 2007. Climatechange impacts on irrigation water requirements in the Guadalquivir RiverBasin in Spain. Reg. Environ. Change 7, 149–159.

40. Santos Pereira, L., Oweis, T., Zairi, A., 2002. Irrigation management under waterscarcity. Rev. Agric. Water Manage. 57, 175–206.
41. Vivid Economics, 2013. The impacts of drought in England, Final report preparedfor Defra, March 2013.
42. Weatherhead, K., Knox, J., Morris, J., Hess, T., Bradley, R., Sanders, C., 1997.Irrigation demand and on-Farm water conservation in england and wales. In:Final Report to MAFF. Cranfield University.
43. Weatherhead, E.K., Knox, J.W., Hess, T.M., Daccache, A., 2015. Exploring irrigationfutures – developments in demand forecasting. Outlook Agric. 44 (2), 119–126.
44. Wilhite, D.A., Glantz, M.H., 1985. Understanding the drought phenomenon: therole of definitions. Water Int. 10 (3), 111–120.
45. Wreford, A., Adger, W.A., 2010. Adaptation in agriculture: historic effects of heatwaves and droughts on UK agriculture. Int. J. Agric. Sustain. 8 (4), 278–289,http://dx.doi.org/10.3763/ijas.2010.0482.

CHAPTER 3

Performance Evaluation of a Prototyped Breadfruit Seed Dehulling Machine

Nnamdi Anosike *, Emmanuel Brown and Chukwnonso Maduka

Department of Mechanical Engineering, Nnamdi Azikiwe University, P.M.B. 5025, Akwa, 420211, Anambra State, Nigeria

ABSTRACT

The drudgery involved in dehulling breadfruit seed by traditional methods has been highlighted as one of the major problems hindering the realization of the full potential of breadfruit as a field to food material. This paper describes a development in an African breadfruit seed dehulling machine with increased throughput of about 70% above reported machines. The machine consists of a 20 mm diameter shaft, carrying a spiral wound around its circumference (feeder). The feeder provides the required rotational motion and turns a circular disk that rotates against a fixed disk. The two disks can be adjusted to maintain a pre-determined gap for dehulling. An inbuilt drying unit reduces the moisture content of the breadfruit for easy separation of the cotyledon from the endosperm immediately after the dehulling process. The sifting unit that separates the shell from the seed is achieved in this design with an electric fan. The machine is design to run at a speed of 250 rpm with an electric motor as the prime mover. The dehulling efficiency up to 86% and breakage of less than 1.3% was obtained at a clearance setting of 12.4 mm between disks. A sifting efficiency of 100% was achieved. Based on the design diameter and clearance between the dehulling disks, the machine throughput was 216 kg/h with an electric power requirement of 1.207 kW.

Keywords: African breadfruit seeds processing; modular design; increased machine throughput; heating element enabled separation unit

1. INTRODUCTION

Dehulling is a process employed to get rid of the outer pericarp and testa (hull) of most cereal grains, grain legumes, nuts and oilseeds using mechanical means [1]. Generally, this process through seed material bulk density reduction offers advantages such as increased temperature distribution management and other rate transfer processes in downstream processing [2].

African breadfruit which belongs to the botanical family Maraceae is a source of highly rated and delicious traditional food in Africa including the eastern and southern regions of Nigeria. It has high nutritional value, containing about 40% carbohydrate and 17% protein with several minerals and vitamins, and is a rich source of vegetable oil [3]. Potential commercial and industrial benefits of breadfruit include raw materials for breakfast cereals and beverages, the production of soap, pharmaceuticals, flour for making biscuits, and paints [4]. Another highlighted potential of breadfruit is the shell which is rich in potash and may be used as filler, bedding, feed cakes and meals for livestock and the polishing of abrasives [5].

African breadfruit is characterized by a spherical shape, and a hard and spongy texture filled with seeds. The seeds are made up of the hull and inner edible endosperm. There are three main varieties of breadfruit: *VarAfricana*, *Var Molis* and *Var Inverse* [6]. There are abundant breadfruit trees both in the savanna region and the eastern part of the country. However, the price of the processed breadfruit seeds is still relatively high compared to other seeds in the local market and there is often not enough quantity to satisfy local market demand.

Many researchers believe that the potential of the breadfruit seed has not been fully harnessed [7,8,9,10]. This is mainly due to the traditional manual means of processing the seed which produces a low yield and increases the cost of processed seeds. The market price of the processed breadfruit can be lowered for local consumption as a staple food and enough quantity can be processed for raw material or industrial feedstock if the manual dehulling approach is replaced with mechanized dehulling and a storage process is adopted.

Currently, the throughput of the existing and reported breadfruit seed dehulling machines is well below other seed shelling machine throughputs. To increase the presence of breadfruit seed to compete with other seeds as a field for food materials, an affordable shelling machine with efficient and reasonable capacity is required as part of the processing processes.

This paper presents a breadfruit seed dehuller machine designed using information from open literatures and existing machines. This is envisaged to increase efficacy of the desired dehuller machine parameters. The performance evaluation of the prototyped machine was done to ascertain efficacy of the machine given the design parameters and results compared with existing machines.

2. DESIGN AND DEVELOPMENT

2.1. Design Approach

The African breadfruit seed dehulling machine was designed and constructed using a modular concept for a low cost, easy adjustability, and easy maintenance. The hopper, dehulling unit, drying unit and aspiration unit are the main units attached to the frame. These major units of the machine can be changed, for instance to increase the throughput. The concept enables individual components to be produced separately and assembled together as a unit with bolts and nuts. This way, the underperforming components are easy to sort out, replace or modify during the testing stage and it is envisaged that this will also be beneficial during the maintenance stage.

2.2. Operational Mode

The electric motor provides the primary motion which rotates the dehulling disk. The breadfruit seeds are introduced through the hopper, while the spiral on the supporting shaft for the disk conveys and compresses the breadfruit seeds into the space between the rotating disk and the fixed disk. An adjustment is provided to determine the length of the gap between the disks.

After the dehulling, the seeds along with the chaff drop into the drying unit where their moisture content is reduced to enhance the separation process in the aspiration unit.

The rubbing action of the disks peels the shell of the breadfruit seed and it moves along with the cracked shell in the direction of rotation to the exit channel, leading to the drying unit. The blower with a heater runs independently and it is to dry the peeled breadfruit seeds and shell of their moisture content, for efficient separation.

The blower in the aspiration unit sifts the hull from the endosperm, thereby delivering a certain quantity of air which is channeled to meet the dropping shells and dehulled seed from the drying unit. The cracked shells are lighter than the dehulled seeds and are carried away by the air in a different direction while the white seeds drop into a receiver where they are collected.

2.3. Prototype of Proposed Dehulling Machine

The overall size of the machine is 500 × 450 × 820 mm^3 (Figure 1). The machine is made up of the frame, the hopper, an electric motor, the dehulling unit, the drying unit and the separation unit. The machine frame is the rigid structure where all the machine components are assembled. The frame is constructed with a hollow, square bar steel material. All the units, including the hopper and hopper under-cover, are fastened with bolts and nuts for ease of disassembling during maintenance and repairs.

1. The Hopper

This is the unit through which the machine is charged with the parboiled breadfruit seeds. The hopper has a rectangular base with the sides inclined at different angles to eliminate the effect of tunneling and arching. At the base of the hopper, a shaft with a spiral around its circumference (power shaft) is located. This shaft is supported with two bearings, one each at both ends. The shaft is driven by an electric motor connected by a belt via a pulley on one end of the shaft.

2. The Dehulling Unit

This unit does the dehulling operation by the attrition method using two disks. It is enclosed by the dehulling unit casing. Within the casing, there are two concentric disks mounted on disk holders and supported on two shafts. One of the disks is mounted on one end of the power shaft with the spiral, while the other disk is clamped on a threaded shaft. Also, the disk mounted on the power shaft with the spiral around its circumference is the rotating disk (it rotates by the motion of the electric motor), while the disk on the threaded shaft is the fixed disk (it does not rotate by the motion of the electric motor), but it has a linear motion which is obtained by adjusting the hand-wheel on other end of the threaded shaft. Both disks can be enclosed together. The gap in between the two disks is adjusted using the threaded shaft supporting the fixed disk. By rotating the hand-wheel on the threaded shaft, the gap in between the two disks is adjusted.

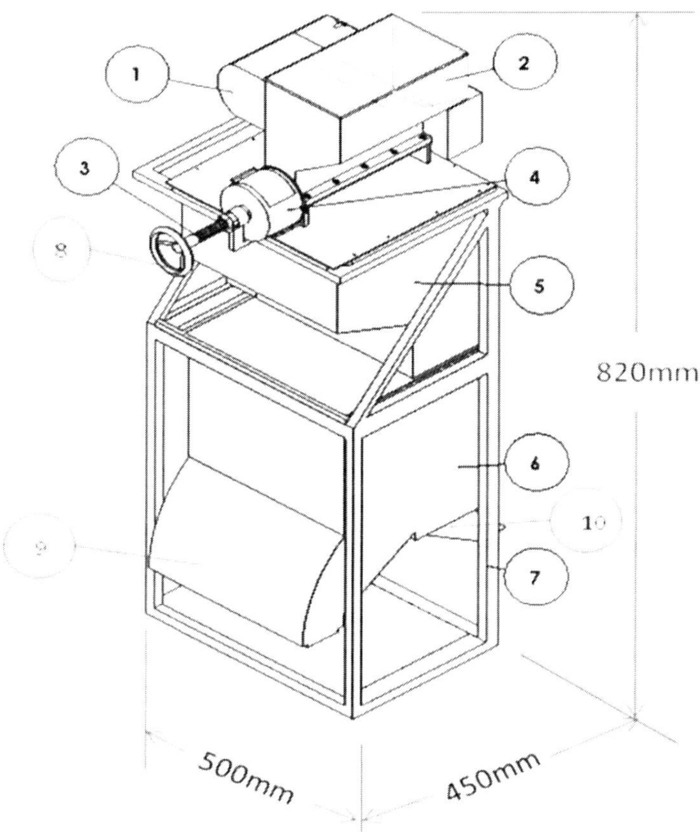

Figure 1. Isometric wireframe view of the machine: ① Electric motor; ② Hopper; ③ Regulating handleshaft; ④ Dehulling unit; ⑤ Drying unit; ⑥ Aspiration unit; ⑦ Frame; ⑧ Regulating handle for fixed disk; ⑨ Discharge for hull; ⑩ Discharge for seeds.

3. The Drying Unit

This unit reduces the moisture content of the breadfruit seeds after dehulling, thereby eliminating the possibility of the cotyledon sticking on the endosperm. This unit has its inlet directly under the discharge hole of the dehulling unit. The unit is comprised of electric fan with a heating element. The heating element supplies the heat required to reduce the moisture content of the cracked seeds. As the dehulled seeds and shells drop, they fall on a part of the drying unit that increases the resident time in which the seeds and shells stay in this unit for proper moisture reduction, and it is stepped and inclined at an angle of 60° to the horizontal.

4. The Separation Unit

This unit completes the separation of the white seeds (endosperm) from the cotyledon. The separation process is achieved by the density differential of the white seeds and the chaff. This unit is comprised of an aspirator (fan) with two discharge holes. The hole directly under the aspirator is for the seeds, while the hole opposite the aspirator (inclined at 30°) is for the chaffs.

2.4. Machine Design Parameters Correlation

The machine parameters affecting the throughput are important for economic appraisal. These parameters include: the space between disks; the effective disk dehulling diameter; and the speed of the rotating dehulling disk. Changes in these parameters affect the power consumption of the dehulling machine. As shown in Figure 2, the prime mover (electric motor) carries more power and is more susceptible to changes in power due to variations in the dehulling force as the dehulling capacity of the machine increases. This relationship between the dehulling force and prime mover power requirement is highlighted in Figure 3a.The relationship betweenthe dehulling disk diameter and the diameter ofthe shaft conecting the dehulling disk and prime mover via a pulley (feeder shaft) is shown in Figure 3b for different breadfruit seed mass at a constant feed with aconstant space between the dehulling disks. Figure 3c,d shows that more breafruit seeds between dehulling disks requires more dehulling power.

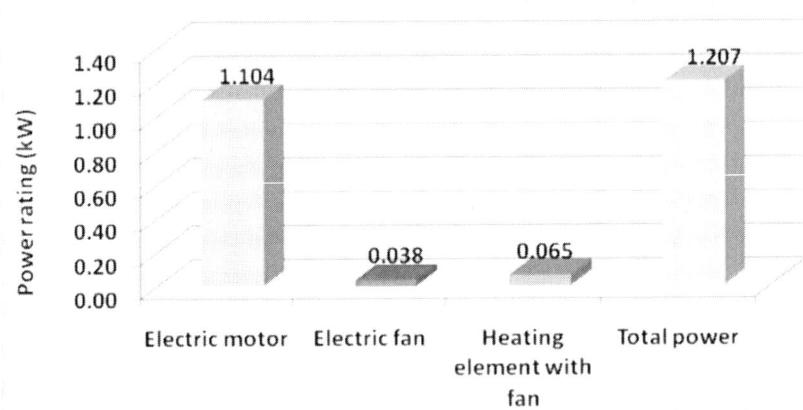

Figure 2. Nominal power rating of the electrical components of the dehulling machine.

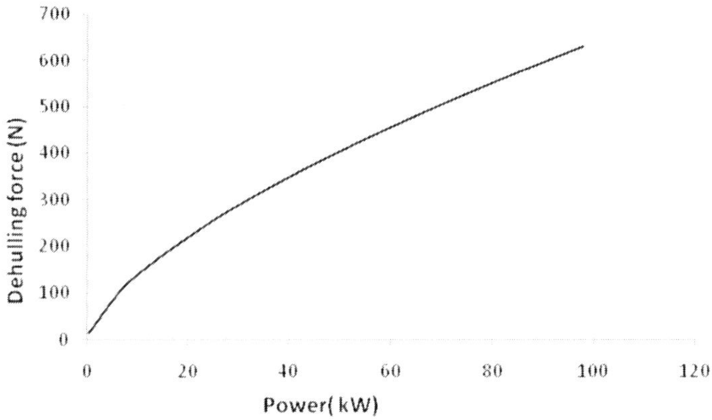

700
600
500
400
300
200
100
0
Dehulling force (N)
0
20
40
60
80
100
120
Power(kW)

(a)

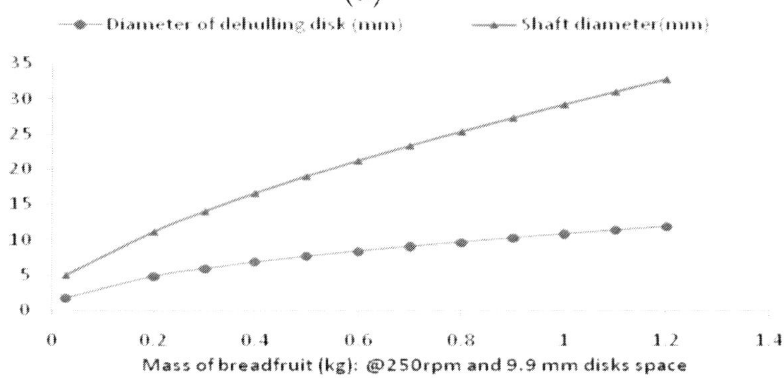

Diameter of dehulling disk (mm)
Shaft diameter(mm)
35
30
25
20
15
10
5
0
0
0.2
0.4
0.6
0.8
1
1.2
1.4
Mass of breadfruit (kg): @250rpm and 9.9 mm disks space

(b)

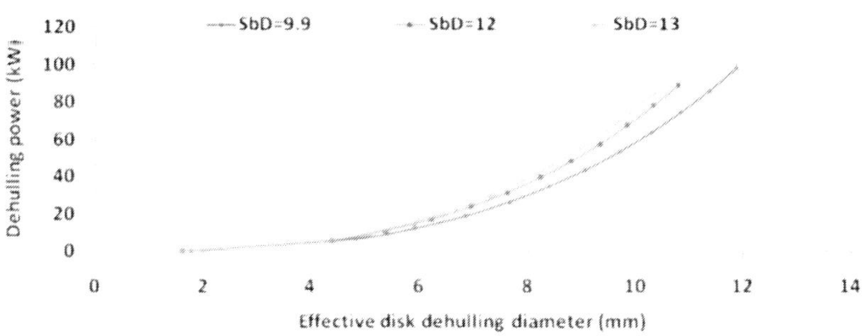

SbD=9.9
SbD=12
SbD=13
120
100
80
60
40
20
0
Dehulling power (kW)
0
2
4
6
8
10
12
14
Effective disk dehulling diameter (mm)

(c)

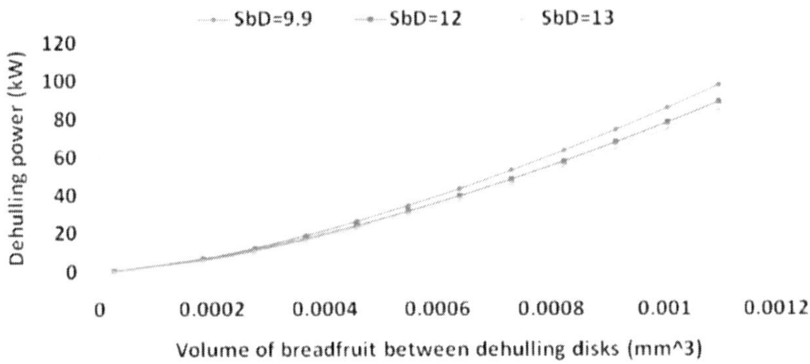

(d)

Figure 3. Engine design parameters correlation at 250 rpm. (**a**) Dehulling force and prime mover power; (**b**) Effective disk dehulling diameter, required shaft (feeder) diameter and machine feed capacity mass; (**c**) Relationship between required dehulling power and effective disk dehulling diameter for different space between disks (SbD); (**d**) Relationship between required dehulling power and volume of breadfruit between dehulling disks for different spaces between disks (SbD).

3. MACHINE PERFORMANCE ANALYSIS

Three variable performance parameters, namely the space between the fixed and rotating dehulling disks (mm), the dehulling speed (rpm), and the moisture content of the breadfruit seeds (%), were used as an off-design performance handle. The machine performance indicators include: throughput (kg/h); electric power consumption (kWh); completely dehulled seeds (%); undehulled seeds; partially dehulled seeds; breakage (%); and sifting efficiency (%).

3.1. Sample Breadfruit Seeds

A commonly available local variety of breadfruit seed bought from Eke Awka market in Anambra State, Nigeria, was used to test the dehuller performance. The seeds were washed with clean warm water and parboiled for 10 min at a temperature of 70 °C. The parboiled sample was divided into three portions to obtain varying moisture content between 10% and 50%. The samples were stored in polythene bags to maintain their moisture content before the dehulling process.

3.2. Test Procedure

The three samples of breadfruit seeds with different moisture contents of approximately 10%, 35% and 50% were tested separately, varying one machine handle at a time. Starting with the space between the fixed and rotating dehulling disks, the regulating handle for the fixed disk is adjusted between 9 mm and 13 mm for a constant speed. This is done for continuous seed feeding for a duration of an hour for each of 9.9 mm, 12 mm and 13 mm spaces between disks. Similarly, the dehulling speed was adjusted from 200 rpm to 400 rpm each time, for a constant space between the fixed and rotating dehulling disks (9.9 mm, 12 mm and 13 mm), lasting an hour for each space setting. The method employed in feeding the breadfruit into the hopper is a continuous feeding approach because the size of the hopper cannot hold all the required quantity for an hour of operation. The performance indicators were calculated using the average of three runs.

4. RESULTS AND DISCUSSION

The results presented in Table 1 are far more promising compared to any other set of parameters for the prototype dehuller at the given breadfruit seed moisture content for this experiment. The results show an increase in throughput compared to existing breadfruit seed dehullers. The 214 kg/h capacity is about a 70% increase compared to 64 kg/h reported by [1] with a 150 rpm dehulling speed. The increment in dehulling speed is 40% with 250 rpm for the present work.

Table 1. Off-design performance of the prototyped breadfruit dehuller [1].

Performance Parameters	Experimental Value
Throughput (kg/h)	214 ± 2.0
Power requirement (kW)	1.3 ± 0.6
Completely dehulled seeds (%)	86 ± 2.0
Partially dehulled seeds (%)	8.0 ± 0.3
Undehulled seeds (%)	2.1 ± 0.3
Breakage (%)	1.0 ± 0.3
Sifting efficiency (%)	99.5 ± 0.2

[1] Moisture content, 36 ± 5.0%; space between dehulling disks, 12.4 ± 0.5 mm; speed of dehulling disks, 250 rpm; continuous feeding.

The increased throughput could be attributed partly due to the increased dehulling speed and probably the dehulling mechanism employed.

Comparing the roller or cam arrangement in [1] with the disks arrangement in this work, it is obvious that disk will have more contact surface area than the roller or cam. Thus, for every revolution of the prime mover disk with more shearing surfaces, there should be arguably more throughput.

There is an increase in breakage, though less than 1%, which could be a result of the increase in the dehulling speed or the variation in moisture content. All other dehuller performance parameters slightly showed recommended improvement.

The regulating handle for the space between the fixed and rotating disks played a vital role in ensuring appropriate space is provided for a particular dehulling speed. In this design, with a clearance between disks of less than 9.9 mm, the breadfruit seeds were crushed for any value of moisture content.

5. CONCLUSIONS

A modular African breadfruit seed dehulling machine with higher throughput has been developed. The result of the performance test showed an improvement compared to reported dehulling machines. The material used for the disks were mild steel; however, disks with lighter weight will save energy and operating costs. Further improvement can be done to reduce the breakage and partially dehulled seeds. A reduction in time taken for the moisture content of the parboiled breadfruit seeds to drop before it is introduced into the machine is necessary.

REFERENCES

1. Omobuwajo, T.O.; Ikegwuoha, H.C.; Koya, O.A.; Ige, M.T. Design, construction and testing of a dehuller for African breadfruit (Treculia africana) seeds. *J. Food Eng.* **1999**, *42*, 173–176
2. Deshpande, S.S.; Sathe, S.K.; Salunke, D.K.; Cornforth, D.P. Effects of Dehulling on Phytic Acid, Polyphenols, and Enzyme Inhibitors of Dry Beans (*Phaseolus vulgaris* L.). *J. Food Sci.* **1982**, *47*, 1846–1850
3. Etoamaihe, U.J.; Ndubueze, K.C. Evaluation of a Dehulling Machine for African Breadfruit (Treculia African). *J. Eng. Appl. Sci.* **2010**, *5*, 312–315
4. Nwaigbo, S.C.; Chinwuko, E.C.; Achebe, C.H.; Tagbo, D.A. Design of breadfruit shelling machine. *Afr. Res. Rev.* **2008**, *2*, 1–16
5. Nwabueze, T.U.; Iwe, M.O.; Akobundu, E.N.T. Physical characteristic and acceptability of extruded African breadfruit-based snacks. *J. Food Qual.* **2008**, *31*, 142–155

6. Okafor, J.C. Delimitation of a new variety of Treculia africana Decaisne subspecie Africana (Moraceae) Bull. *Jard. Bot. Nat. Belg.* **1981**, *51*, 191–199
7. Bassey, M.W.; Schmidt, O.G. Abrasive-disk dehullers in Africa: From research to dissemination. Available online: http://www.idrc.ca/EN/Resources/Publications/Documents/Abrasive-disk-Dehullers-in-Africa-from-Research-to-Dissemination.pdf (accessed on 9 December 2015)
8. Omobuwajo, T.O.; Akande, E.A.; Sanni, L.A. Selected physical, mechanical and aerodynamic properties of African breadfruit (Treculia africana) seeds. *J. Food Eng.* **1999**, *40*, 241–244
9. Reichert, R.D.; Oomah, B.D.; Youngs, C.G. Factors affecting the efficiency of abrasive-type grain legumes investigated with a new intermediate-sized, batch dehuller. *J. Food Sci.* **1984**, *49*, 267–272
10. Reichert, R.D.; Youngs, C.G. Dehulling cereal grains and grain legumes for developing countries Quantitative comparison between attrition and abrasive type mills. *Cereal Chem.* **1976**, *53*, 829–839

CHAPTER 4

The Smart Aerial Release Machine, a Universal System for Applying the Sterile Insect Technique

Ruben Leal Mubarqui[1], Rene Cano Perez[1], Roberto Angulo Kladt[1], Jose Luis Zavala Lopez[2], Andrew Parker[3], Momar Talla Seck[4], Baba Sall[5], Je remy Bouyer[4,6,7*]

[1]*Servicios ae´reos Biologicos y Forestales Mubarqui, Cuidad Victoria, Mexico,*

[2]*Programa Moscamed, Tapachula, Mexico,*

[3]*Insect Pest Control Laboratory, Joint Food and Agriculture Organization/International Atomic Energy Agency (FAO/IAEA) Programme of Nuclear Techniques in Food and Agriculture, Vienna, Austria,*

[4]*Institut Senegalais de Recherches Agricoles, Laboratoire National d'Elevage et de Recherches Veterinaires, Dakar – Hann, Sene gal,*

[5]*Direction des Services Vete rinaires, Dakar, Senegal,*

[6]*Unite´ Mixte de Recherche Contro le des Maladies Animales Exotiques et Emergentes, Centre de Cooperation Internationale en Recherche Agronomique pour le Developpement (CIRAD), Montpellier, France,*

[7]*Unite Mixte de Recherche 1309 Contro le des Maladies Animales Exotiques et Emergentes, Institut national de la recherché agronomique (INRA), Montpellier, France*

ABSTRACT

Background

Beyond insecticides, alternative methods to control insect pests for agriculture and vectors of diseases are needed. Management strategies involving the mass-release of living control agents have been developed, including genetic control with sterile insects and biological control with parasitoids, for which aerial release of insects is often required. Aerial release in genetic control programmes often involves the use of chilled sterile insects, which can improve dispersal, survival and competitiveness of sterile males. Currently available means of aerially releasing chilled fruit flies are however insufficiently precise to ensure homogeneous distribution at low release rates and no device is available for tsetse.

Methodology/Principal Findings

Here we present the smart aerial release machine, a new design by the Mubarqui Company, based on the use of vibrating conveyors. The machine is controlled through Bluetooth by a tablet with Android Operating System including a completely automatic guidance and navigation system (MaxNav software). The tablet is also connected to an online relational database facilitating the preparation of flight schedules and automatic storage of flight reports. The new machine was compared with a conveyor release machine in Mexico using two fruit flies species (*Anastrepha ludens* and *Ceratitis capitata*) and we obtained better dispersal homogeneity (% of positive traps, $p<0.001$) for both species and better recapture rates for *Anastrepha ludens* ($p<0.001$), especially at low release densities (<1500 per ha). We also demonstrated that the machine can replace paper boxes for aerial release of tsetse in Senegal.

Conclusions/Significance

This technology limits damages to insects and allows a large range of release rates from 10 flies/km^2 for tsetse flies up to 600 000 flies/km^2 for fruit flies. The potential of this machine to release other species like mosquitoes is discussed. Plans and operating of the machine are provided to allow its use worldwide.

INTRODUCTION

Area-wide integrated pest management (AW-IPM) has become a prevalent paradigm for insect pest control over the past 50 years [1]. It relies on a good understanding of the ecology of the target populations and optimal integration of both chemical and biological control measures. Only through such an integrated and holistic approach will greater control efficacy and concurrent reduced use of insecticides be achieved. The control of vector-borne diseases in particular is becoming increasingly problematic due to invasions or resurgence of vectors and the necessity to reduce the use of insecticides in view of their environmental and health impacts [2]. Avoiding the negative impacts of reduced insecticide use across large territories necessarily requires an integrative, process-based, preventive and area-wide approach to pest and vector control. Integrated pest management is based on the understanding that each control technique has its advantages and drawbacks (in particular, some techniques are efficient at high densities of the target population whereas others possess an inverse density-dependence efficacy) [3] and that appropriate combination of these techniques (biological, cultural, genetic and chemical) depend on context, taking into consideration the predicted economic, ecological and sociological consequences, with special attention paid to environmental concerns [1].

The sterile insect technique (SIT) and the use of parasitoids are presently among the biological techniques already available for operational programs on a large scale. Both techniques necessitate the use of aerial release of insects if they are to be used in country or continent wide programs. New control techniques based on genetically modified insects or symbionts will also require aerial release of live insects [4].

Examples of recent application of this technique include the combined release of parasitoids (*Diachasmimorpha longicaudata* mainly) and sterile males of *Ceratitis capitata* (1.5 billion pupae produced per week) within the Moscamed regional program (Mexico and Guatemala) [5]. This program eliminated medflies from more than 7.000 km^2 including the state of Chiapas in Mexico [6], and is currently progressing into southern Guatemala.

Similarly, an integrated campaign combining insecticide treatment of cattle, insecticide targets and the release of sterile males eradicated tsetse (*Glossina austeni*) from Zanzibar [7], [8]. This campaign was based on the use of cardboard carton boxes. However, although paper bags (used for fruit flies) or carton boxes are made of biodegradable materials, they still constitute visual pollution. More importantly, they are expensive, require a lot of space in the aircraft, can collapse, and make it very difficult to control the environmental conditions to which the sterile males are exposed, thus reducing their quality [9]. Therefore, aerial release is more and more often achieved through releasing chilled adult sterile insects thanks to automatic aerial release machines on board small single or twin engine aircraft in most current SIT programs. Besides being the fastest method of release, it has been demonstrated to provide uniform distribution over the target areas and to help ensure sterile insect quality and survivability [9]. Aerial release machines are designed with four basic components, a cooling unit, a controler for the release rate, a release mechanism and a navigation system. Early release mechanisms used screw augers [10] which caused mechanical injuries drastically decreasing the survival of insects [11]. This was replaced later by a conveyor belt [12] to adjust and calibrate the rate and desired density by manual and mechanical devices handled by the pilot, and thereafter using automatic navigation systems [9]. The Moscamed program presently uses machines based on conveyor belts and designed by the Mubarqui Company. However, existing machines are not adapted to tsetse release because they can not achieve the very low release rates required (between 10 and 100 per km^2 only).

In this paper, we describe the Mubarqui Smart Release Machine (MSRM), a new design for release machines by the Mubarqui Company, based on the use of vibrating conveyors that avoid damaging insects and allow a very large range of release densities. Moreover, the smaller model is only 64 kg and can be accommodated on board gyrocopters (table 1), which will reduce the costs of aerial release, particularly in small target areas [13]. It is presently in use in the tsetse eradication campaign in Senegal (http://www.fao.org/news/story/en/item/211898/icode/). The machine is calibrated by means of a Geographic Information System (GIS) and environmental parameters are controlled using air conditioning.

Table 1. Types of release machines available as a function of the aircraft and species to be released.

Model	MSRM1	MSRM1	MSRM1	MSRM1	MSRM2
Type of aircraft	CESSNA 401 & 402	CESSNA 206 &207	Maule M7	CESSNA 172	GYROCOPTER
Species	*C. capitata* (7 to 9 mg fly)	*A. ludens* *A. oblique* (19 or 20 mg fly)	*A. ludens*	*C. capitata*	*G. palpalis gambiensis* (21 mg fly)
Capacity	60 million (3 machines)	6 million (1 machine)	6 million (1 machine)	5 million (1 machine)	15.000 (1 machine)
Minimum continuous release rate	400 f/ha	600 f/ha	600 f/ha	600 f/ha	50 f/km^2
Maximum release rate	6000 f/ha	6000 f/ha	6000 f/ha	6000 f/ha	-
Program objectives	600–6000 f/ha	1250–5000 f/ha	1500 f/ha	2000 f/fa	10–100 f/km^2
Swath	500 m	100 and 500 m	100 m	500 m	500 m
Speed	230 km/h	150 km/h	140 km/h	140 km/h	110 km/h
Temperature	4°C	4°C	4°C	4°C	7–8°C
Power needed	90A/24V	60A/12V	60A/12V	60A/12V	40A/12V
Mass (empty)	70 kg * 3	70 kg	70 kg	45 kg	64 kg
Program	MOSCAMED Mexico	MOSCAFRUT Mexico	MOSCAFRUT Mexico	NERETVA Croatia	Tsetse eradication Senegal

In the present paper, we compared the MSRM to the conventional conveyor machine presently in use in the Moscamed and Moscafrut programs in Mexico regarding dispersal and recapture rates of two species of fruit flies. We also provide preliminary data on the tsetse release rates that are used in the eradication program in Senegal.

MATERIALS AND METHODS

Description of the Mubarqui Smart Release Machine (MSRM)

The release device is installed inside an aircraft. It is designed with vibratory feeders, automatic gates and linear actuators with a duct leading the flies outside the aircraft, shaped to avoid Venturi suction in order to increase the accuracy of the release rate (see fig. 1 and fig. 2 for detailed plans of the machines used to release fruit flies (MSRM1) and tsetse flies respectively (MSRM2)). All the components are made of stainless steel. Unlike other machines, such as rotating conveyor or auger machines (previously used in Mexico), the smart machine has no longitudinal moving surface which is a great mechanical advantage that prevents physical damage to the insects. The material moves through micro vibration on a flat horizontal stainless steel tray that vibrates thanks to a powerful electromagnet moving horizontally 0.9 to 1.1 mm at 100 to 300 Hz. The release rate is controlled electronically by varying the frequency, power (acceleration) and displacement (amplitude) of this surface. These variables are operated by a digital controller and a program adapted to the different needs of the project (see Text S1), in this case the calibration of flies per hectare. The machine self-calibrates during the flight without the intervention of the pilot or other operator, thus avoiding human error. The gates are opened and closed automatically when entering or exiting the release area or when conditions are not appropriate for the material to be released.

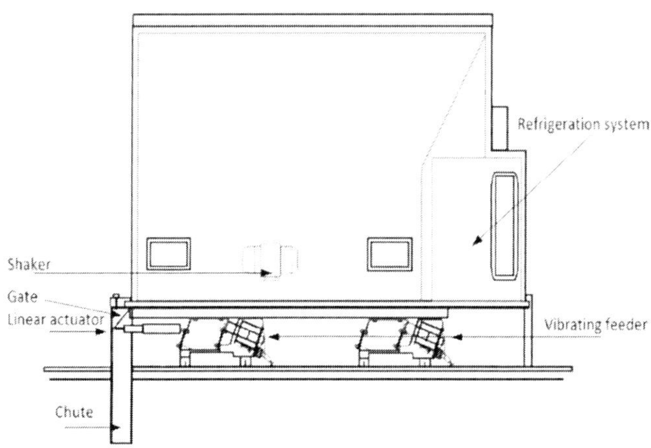

Figure 1. Plan of the version of Mubarqui Smart Release Machine (MSRM1) used to release fruit flies.

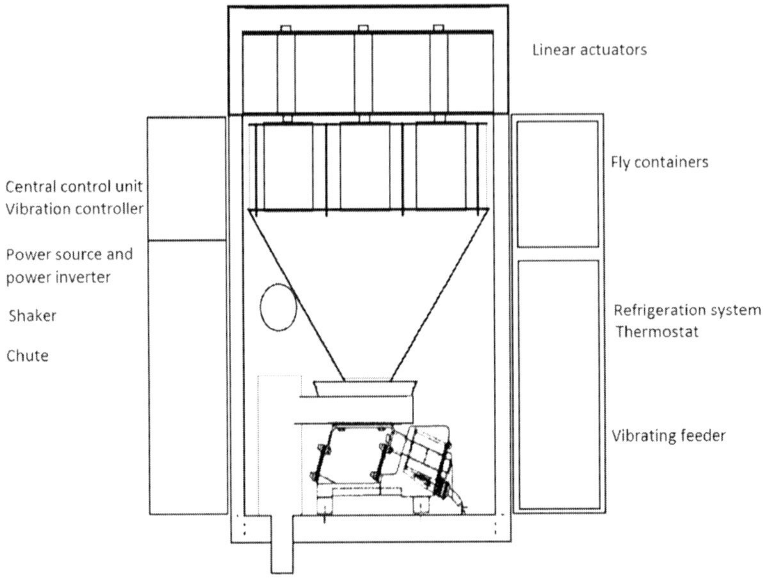

Figure 2. Plan of the version of Mubarqui Smart Release Machine (MSRM2) used to release tsetse flies.

The cooling system and container is a box with a refrigeration system that can be transported from the packing centers with the sterile adults to be released on a given day. This container is made of stainless steel and insulated from outside temperature to keep the biological material in suitable conditions for handling and transport both on the ground and in the air, to the polygon (area with a specific pre-defined release rate) where they will be released. It is equipped with a thermostat and vibrators to avoid air pockets within the mass of flies. Air pockets in the chilled flies must be avoided because it affects the regularity of the release rate [10].

The Central Control Unit receives instructions from the computer and turns them into actions such as opening and closing of the gates, start and stop release mechanism, intensity of the vibrating feeder, and modulation of micro vibrations to maintain accurate rates of release.

The Navigation system is the instrument responsible for guiding the pilot to the release areas (presented as polygons). It also contains the logging, measurement and automation for the release mechanisms.

Finally, the machine is equipped with video cameras connected to the computer, allowing the pilot to confirm that flies are being released.

A full description of operation and calibration of the MSRM is provided in Text S1. Figure S1 presents the installation of MSRM1 on board a Cessna 182 whereas Figure S2 presents its installation on board a Cessna 206 respectively. Figure S3 presents the installation of MSRM2 on board a gyrocopter in Senegal (aerodrome of Kalahari).

Two main types of machine are available according to the aircraft and species to be released (table 1). The tsetse machine is different from the two others in that it contains metallic cylindrical containers that are opened one by one to avoid any physical damage to the flies (fig. 2). The fruit fly machines can have containers of different sizes depending on the aircraft.

Production and handling of the flies used in the trials

The two species of fruit flies (*Anastrepha ludens* and *Ceratitis capitata)* were mass reared at Metapa, Chiapas, Mexico [14], [15]. A genetic sexing strain producing males only was used for *C. capitata* [16], as well as for *A. ludens* (Tapachula 7) [17], reducing by 50% the cost of production, packing, and release.

Flies were packed, held and chilled in the Mexican type towers system [18], composed of 16 trays, each one of 81.77010.3 cm, including one container for pupae (55,000 *C. capitata* or 25,000 *A. ludens*), two trays for food (40 g for *C. capitata* and 20 g for *A. ludens*), one adult resting device and one pillow for water. For both species, mass rearing was achieved using the "Metapa system" [19]. Pupae were irradiated two days before emergence [20]. Pupae were then placed into a hopper and dispensed by means of a pupae dispenser machine which was then placed in trays specific to the tower and including all the other parts (feeding and resting devices). Flies were chilled at the sixth day after packing and then introduced into the release machines.

Regarding *Glossina palpalis gambiensis*, 74% of the pupae originated from Centre International de Rerche-Développement sur l'Elevage en zone Subhumide (CIRDES), Bobo-Dioulasso, Burkina Faso, and the remaining from the Institute of Zoology, Slovak Academy of Sciences, Bratislava, Slovakia. Females were bred in insectaries following standard procedures developed at CIRDES [21], [22]. They were kept at 24°C for female emergence and stored at 10°C when the first males emerged, leaving ~96% of male pupae in this species. They were also transported in isotherm parcels containing phase change gel packs that maintain the temperature around 10°C. Pupae were permitted to emerge at the Institut Sénégalais de Recherche Agricole (ISRA) insectary in Dakar, maintained at 24°C and fed daily with cattle blood containing 10 mg/L of isomethamidium chloride (to avoid transmission of trypanosomes after release) [23]. They were chilled and placed into the MSRM four to six days after emergence.

Field trials to compare MSRM to the conventional conveyor machine

A field trial was organized in 2012 in north-eastern Mexico (San Luis Potosi State, long. 21.937921°, lat. −100.097798) to compare the performance of the smart release machine with a vibrating feeder (MSRM1) to that of the release machine with a flat conveyor (MCRM). Two polygons of 600 ha (MSRM1) and 735 ha (MCRM) separated by 6 km from each other and comprised of citrus tree orchards were used for aerial releases of *A. ludens*. Densities of 2000, 1500, 1000 and 500 flies per ha were released each week for ten weeks, corresponding to a total duration of 39 weeks (only 9 weeks for the last density) in 2012. The swath width was 150 m. Twenty-four and 30 protein baited traps (McPhail) were set in the first and second polygons (corresponding to 0.04 trap per ha) and emptied weekly.

Another trial was organized in Chiapas State (long. 15.263055°, lat. −92.778553), Mexico, using *C. capitata*. The same protocol was used except that the polygons were 3000 ha each and separated by 37 km. The same densities were used but the trial was 20 weeks only (5 weeks for each density). The swath width was 500 m. Thirty-five and 33 traps were set respectively in the MSRM1 and MCRM polygons (corresponding to ~0.01 trap per ha).

The number of flies packed into the device was different between release densities, to mimic a real situation within an eradication project, where the machine load depends on the objective in terms of release densities.

Finally, a third trial was organized in the Niayes area of Senegal (long. −17.1294, lat. 14.7831), using *G. p. gambiensis*, from Dec. 2013 to May 2014. Releases using paper boxes were compared to MSRM2. At the beginning of the trial (7 weeks), paper boxes were kept at ambient temperature (measured with a Hobo thermo-hygrometer station at 13–31°C) and then (8 weeks) within ice-boxes containing phase change gel packs to control the temperature (14–25°C). In MSRM2, we set the temperature inside the machine at 6–10°C for 6 weeks and at 9–12°C for 5 weeks. All the trials were conducted in a polygon of 8100 ha and we used 19 biconical traps to monitor tsetse densities [24]. The swath width was 500 m. Paper boxes were released every 500 m.

All the raw data from these trials are available in Suporting data S1, both at the trap and polygon level.

All trial areas are under SIT control (aerial releases), for more than five years in Mexico, and for two years in Senegal. In San Luis Potosi and Chiapas states, federal government through SENASICA Moscafrut and Moscamed respectively as agriculture authorities and DGAC as aeronautic authorities gave necessary permissions to fly and to release sterile males to implement these control programs. In Senegal, the trials were supervised by the Vet services within the tsetse eradication campaign which received the permit n°0874/MEPN/DE/DEIE/mbf from the Ministry of Environment on 03 April 2012. No specific permissions were required for the research activities presented in this paper, which were implemented within the control areas. The field studies did not involve endangered or protected species.

Statistical methods

R software [25] was used for all statistical analyses.

The rate of positive traps (capturing at least one fly) were analyzed using generalized binomial models, with the release density, the type of release machine and their interaction as fixed effects.

The recapture rates of individual traps were analyzed using generalized linear mixed binomial models [26] using the release density, the type of release machine and their interaction as fixed effects and the trap position as a random effect [27]. In addition, in the case of *C. capitata*, the tree species on which the traps were set (which were all citrus tree in the *A. ludens* trial) was a fixed effect. In the case of tsetse, the release method (paper boxes vs MSRM2 and temperature regimen) and the release density were considered as fixed effects. The lme4 package was used to implement these models [28].

The best model was selected based on the lowest corrected Akaike information criterion (AICc), [29] using the MuMin package [30].

RESULTS

Anastrapha ludens

There was a strong positive impact ($p<0.001$) of release density on the percentage of positive traps (fig. 3 and Table S1). Moreover, the rate of positive traps was higher for MSRM, and the difference was higher at low release densities (1000 per ha and lower, $p<0.001$).

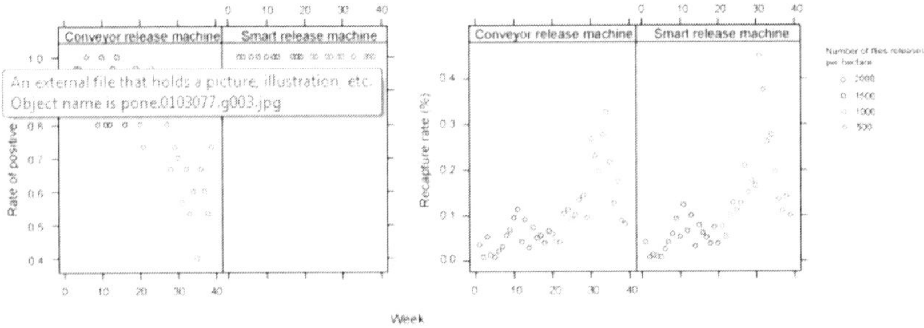

Figure 3. Comparison of conveyor (MCRM) and vibrator (MSRM1) release machines for *Anastrepha ludens* in Mexico. Rate of positive traps (left) and recapture rates (right) are presented for different release densities. A trap is considered positive when at least one fly is captured during a week.

There was also a strong negative impact (p<0.001) of release density on the percentage of recaptured flies (fig. 3 and Table S1). Moreover, the percentage of recapture was higher for MSRM, and the difference was higher at low release densities (p<0.001).

Ceratitis capitata

For *C. capitata* as well as for *A. ludens*, there was a strong positive impact (p<0.001) of release density on the percentage of positive traps (fig. 4 and Table S2). Moreover, the rate of positive traps was higher for MSRM, but the difference was significant only at low release densities (1000 per ha and lower, p<0.001).

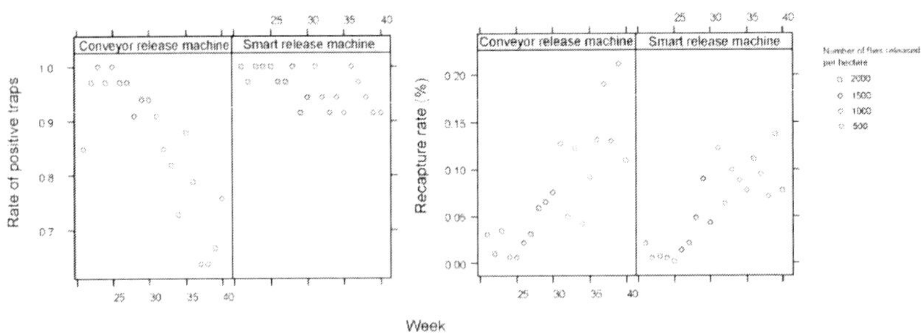

Figure 4. Comparison of conveyor (MCRM) and vibrator (MSRM1) release machines for *Ceratitis capitata* in Mexico. Rate of positive traps (left) and recapture rates (right) are presented for different release densities. A trap is considered positive when at least one fly is captured during a week.

There was a strong negative impact ($p<0.001$) of release density on the percentage of recaptured flies (fig. 4 and Table S2). There was no significant difference between the two release machines on the recapture rate ($p=0.72$). However, this rate of recapture decreased more with density for the MSRM ($p<0.001$).

Glossina palpalis gambiensis

There was a significant negative impact of release density on the recapture rate ($p<0.05$), but not on the percentage of positive traps ($p=0.14$).

The percentage of positive traps was similar ($p>0.14$) between release methods (fig. 5 and Table S3) except that it was lower ($p=0.02$) for MSRM2 used with a temperature range of 6–10°C.

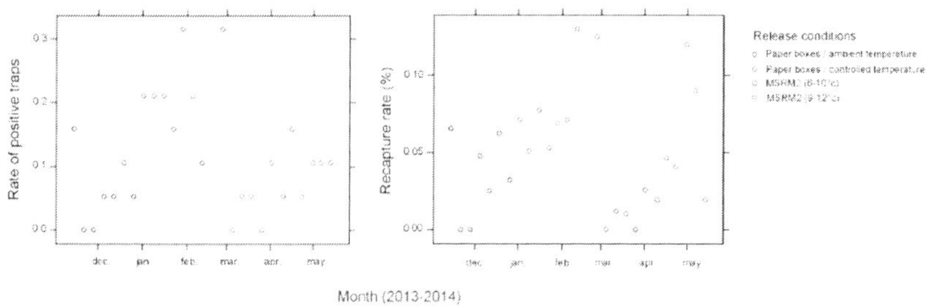

Figure 5. Comparison of paper boxes and vibrator release machine (MSRM2) for *Glossina palpalis gambiensis* in Senegal. Rate of positive traps (left) and recapture rates (right) are presented for different release conditions: paper boxes at ambient temperature (13–31°C), paper boxes at controlled temperature (14–25°C), MSRM2 with the temperature controlled at 6–10°C or 9–12°C. A trap is considered positive when at least one fly is captured during a week.

The recapture rate was better for the paper boxes in controlled temperature conditions than all other methods ($p<0.05$). There was no difference between paper boxes maintained at ambient temperature and the MSRM2 used with a temperature range of 6–10°C ($p=0.34$). However, MSRM2 9–12°C gave better results than paper boxes at ambient temperature ($p=0.05$) and MSRM2 6–10°C ($p=0.005$).

Possible tsetse release rates for MSRM2

In Senegal, the total target area is only about 1000 km^2 [31]. There was a very good correlation ($p<10^{-3}$) between the natural logarithm of the number of tsetse flies released by second and the power of the vibrator (fig. 6). The minimum adjustment rate obtained with MSRM2 was ~50 flies/km² (200 times lower than for fruit flies) with continuous release but the tsetse eradication program requires even less, down to 10 flies/km² (0.1 fly per ha). This was obtained by alternating periods of turning on and off the vibrator (the turning off cannot exceed 18 seconds because the distance between releases must remain lower than the swath width to ensure good dispersal, which is 500 m in this project, whereas the release speed of the gyrocopter is 110 km/h).

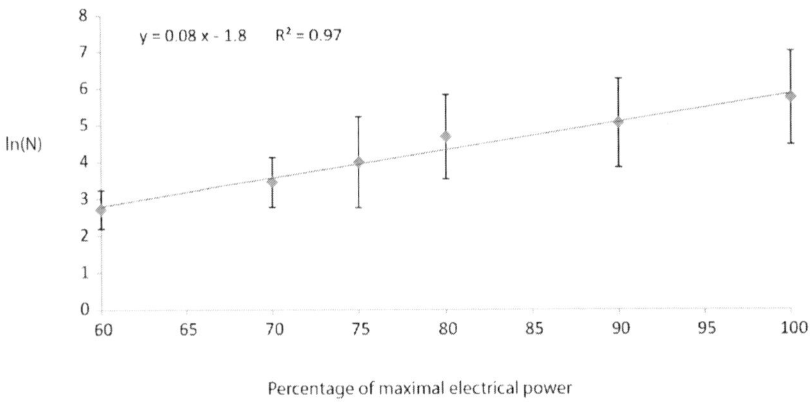

Figure 6. Parameterization of release rates of *Glossina palpalis gambiensis* with the vibrator release machine (MSRM2). The graph presents the linear correlation between the natural logarithm of the number released by second and the power of the vibrator (as a percentage of the maximal electrical power). Vertical bares present the standard errors (three repeats of 1 minute per value). This test was achieved within the tsetse eradication program in Senegal (http://www.fao.org/news /story/en/item/211898/icode/).

Since 28 February 2014, all releases of sterile flies in the Senegal tsetse eradication program have been achieved using MSRM (fig. 7). Overall, 126, 975 flies have been released by the time of writing. The minimal release rates are 0.26, 0.12 and 0.53 flies per ha for RL1, RL2 and RL3 respectively. Release rates are, however, adjusted on a daily basis according to the availability of flies. RL1 is the polygon where the comparison experiments described upon were conducted (fig. 7). Mean number of flies released per week is 14,346 (s.d. 2,397). A video camera allowed confirmation of a very regular release rate of the flies.

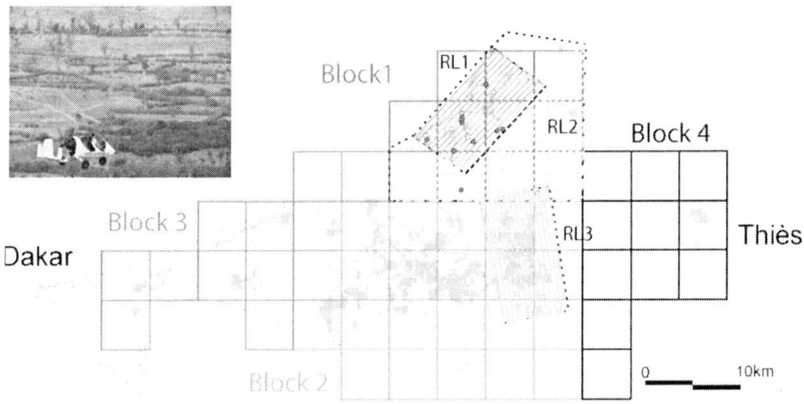

Figure 7. Release patterns of chilled male *Glossina palpalis gambiensis* in Senegal. The control area is subdivided in four blocks targeted sequentially and each block is subdivided in polygons where the release density is adjusted according to the amount of suitable habitats. Since 28 February 2014, all releases of sterile flies are achieved using MSRM2 on board a gyrocopter of the Kalahari aerodrome (picture at the top left of the figure). The minimal release rates are 0.26 flies per ha for RL1, 0.12 flies per ha for RL2 and 0.53 flies per ha for RL3 and actual release rates are adjusted to the availability of flies at the release center (ISRA insectarium, Dakar-Hann, Senegal). The tracks presented on the figures correspond to the flights of 21 March 2014 for RL1, 05 April 2014 for RL2 and 15 April 2014 for RL3. Grey points represent trap positions where no tsetse was captured in April-May 2014 whereas red points present trap positions where sterile males released with MSRM2 were recaptured.

Discussion and perspectives

The first objective of this study was to compare a new release system for chilled adult release, the MSRM using a vibratory feeder instead of a conveyor belt, presently used in MCRM within the Moscamed and Moscafrut programmes in Mexico. The MCRM, provides good results at high release densities for fruit flies [10], but it is relatively inaccurate at low densities. The second objective was to evaluate the possibility of releasing tsetse flies with the same system because no release machine had previously been available for tsetse. MCRM2 appeared to offer an effective means to automatically and aerially release tsetse and we thus compared its efficiency to the standard method (paper boxes at ambient temperature [7]) and an improved method (paper boxes at controlled temperature).

The comparison of MSRM1 and MCRM gave similar results for the two fruit fly species tested. The rate of positive traps gives a good picture of fly distribution due to aerial release [9]. The results show that the MSRM performs better than the conveyor release machine for both fruit fly species, especially at low density, showing a more regular release rate for this machine. Since the same guidance system (MaxNav) was used for both machines during these trials, the better results of MSRM1 can be attributed to its mechanical properties only. This parameter is clearly the most important in eradication programs, since the sterile to wild male ratios must be as homogeneous as possible to induce female sterility over the entire target area [9]. At a release rate of 1500 flies per ha, the two machines also gave much better results (~95% of positive traps) than the auger release systems (70 and 80% of positive traps for the two models tested in [9]).

Tween & Pendon (2007) suggest that the recapture rate is related to longevity or survival and attribute this to fly handling during release activities [9]. Other parameters (trapping intensity and fly release density) were the same in the two release polygons. Mortality (here considered as both true mortality and emigration from the target area), was highly density dependent in this trial. This increased mortality (lower recapture rate) at higher release densities could result from cramped storage in the release device due to the large number of flies used (up to 1.2 million *A. ludens* and 6 million *C. capitata*, corresponding to ~22 kg and ~48 kg respectively). Moreover, the recapture rate was higher for the MSRM for *A. ludens*, especially at low densities (up to 38% higher for 500 flies per ha). For *C. capitata*, however, no significant difference was observed between the two machines. The two machines had similar efficiencies at high densities (more than 1500 per ha).The mean survival for *A. ludens* is ~15 days, causing an accumulation effect which might have created a bias for the first two weeks after a change in release densities. This however did not prevent us from observing a strong signal between series.

In the MOSCAMED project in Mexico, the minimum density of flies to be released is 500 fly/ha (50,000/km^2), whereas MSRM1 could be adjusted down to 100 flies/ha (10,000/km^2) with continuous release for *C. capitata* (table 1). For fruit flies, the intensity of the vibrator has to be adjusted to the stage of the flies available at each daily collect (humid, semi humid or dry flies), because the behavior is different regarding release rates. In the case of tsetse, recapture rates and percentages of positive traps were lower than for fruit flies for all release methods. This was expected because tsetse traps are based on visual stimuli and capture rates are always very low [32]. In the target area, trap efficiency was estimated to be ~0.003 per day per km^2 [33]. MSRM2 with a temperature range of 9–12°C outperformed standard release procedures using paper boxes at ambient temperature (13–31°C), which was used to eradicate *G. austeni* from Zanzibar [7]. However, the recapture rate of MSRM2 with a temperature range of 9–12°C was still slightly lower than with paper boxes in controlled temperature conditions (14–24°C). One possible explanation for this result is that some flies do not become active before reaching the ground using MSRM2. We estimated the maximal speed of cold live flies at 3.2 m/sec (s.d. 0.1) by filming their free fall in front of a white wall. Based on this, it is likely that released tsetse reach the ground within 31 seconds from the release altitude of 100 m in Senegal. However, we also observed using a camera that flies kept at 10°C take 30 seconds to 1 min to wake up at 30°C. Some of them might thus hit the ground and be predated before being able to fly. We will make release trials at higher altitudes (200 m) in the future to try to improve the recapture rate. An alternative hypothesis might be that low temperatures decrease adult tsetse survival, which might explain the better results of the 9–12°C range than the 6–10°C range for MSRM2 (although flies also wake up faster at higher temperatures). It is however impossible to increase temperature furthermore within MSRM2 because upon 12°C, we observed that flies can attach to the stainless steel components thanks to their legs which affects control of release rates.

MSRM2 offers the first opportunity to use the chilled insect release technique in tsetse, where the low release rates needed and the high sensitivity of the flies made it impossible to use auger machines. The use of a gyrocopter reduces flying costs to €320/hour. The speed used with this machine for aerial release was 110 km/h. MSRM2 fulfilled the requirement of the tsetse eradication program in Senegal (http://www.fao.org/news/story/en/item/211898/icode/) for very low release rates (0.1–0.5 flies per ha presently) and has been used routinely since 28 February 2014 to replace in replacement of the boxe. This is also because only ~100 paper boxes can be carried on board a gyrocopter, which limits the release area that can be covered within one flight with this release method. During the field trials in Senegal, no dead flies were observed in the machine after release and the number of flies remaining in the machine was below 50. In the case of tsetse flies, it was demonstrated that when released in a homogeneous density, sterile males are able to aggregate in the same sites preferred by wild flies [34].

The MSRM maintains stable environmental conditions inside the machine thanks to the absence of suction, which limits loss of chilled air. The main drawback is that the machine necessitates a permanent source of power, which can be challenging for small aircraft such as gyrocopters. In single engine aircrafts, the main power source is enough (12V, 100A). In twin engine, the power source is even stronger (24V, 150A). In Senegal however, the use of a gyrocopter was more challenging and we used an auxiliary power unit (APU, 12V, 65A) and an intermediate battery to provide enough power to the cooling system.

We plan to test the machine on parasitoids (*Diachasmimorpha* species), which are more sensitive than other insect species particularly because of long antennae (Montoya et al., 2012). MSRM will also be tested on mosquito species (*Aedes albopictus* in particular) for which the sterile insect technique is still at the R&D stage [35], [36], and for which a new concept called "boosted SIT" has recently been developed, where sterile males are used as specific conveyors of active biocides [37]. Since mosquitoes are very sensitive to mechanical injury [38], the vibrating feeder might provide a technical solution for their release, but this hypothesis remains to be tested.

ACKNOWLEDGMENTS

The authors are very grateful to Udo Feldmann, Jorge Hendricks, Jesus Reyes, Marc Vreysen, Joint Food and Agriculture Organization/International Atomic Energy Agency (FAO/IAEA) Programme of Nuclear Techniques in Food and Agriculture, for their support to the elaboration of the MSRM. They also want to thank Philippe Bouyer for estimating the speed of free fall in live tsetse.

The work of JB is conducted within the project 'Integrated Vector Management: innovating to improve control and reduce environmental impacts' of Institut Carnot Santé Animale (ICSA) excellence network (http://www6.jouy.inra.fr/gabi/Le-Partenariat/L-Institut-Carnot-Sante-Animale-ICSA).

REFERENCES

1. Vreysen M, Robinson AS, Hendrichs J (2007) Area-Wide Control of Insect Pests, From research to field implementation. Dordrecht: Springer. 789 p.
2. McGraw EA, O'Neill SL (2013) Beyond insecticides: new thinking on an ancient problem. Nat Rev Microbiol 11: 181–193.
3. Vreysen MJB, Seck MT, Sall B, Bouyer J (2013) Tsetse flies: their biology and control using area-wide integrated pest management approaches. J Invertebr Pathol 112: S15–S25.
4. McGraw EA, O'Neill SL (2013) Beyond insecticides: new thinking on an ancient problem. Nature Reviews Microbiology 11: 181–193.
5. Villaseñor A, Carrillo J, Zavala JL, Stewart J, Lira C, et al.. (2000) Current progress in the medfly program Mexico-Guatemala. Penerbit Universiti Sains Malaysia. pp. 361–368.
6. Gutiérrez Ruelas JM, Roldán EMC, Hernández DC (2012) Situacion actual de la mosca del Mediterraneo *Ceratitis capitata* (Wiedemann), en Chiapas, Mexico; Aug.; Panama.

7. Vreysen MJB, Saleh KM, Ali MY, Abdulla AM, Zhu Z-R, et al. (2000) *Glossina austeni* (Diptera: Glossinidae) Eradicated on the Island of Unguja, Zanzibar, Using the Sterile Insect Technique. J Econ Entomol 93: 123–135.
8. Vreysen MJB, Saleh K, Mramba F, Parker A, Feldmann U, et al. (2014) Sterile insects to enhance agricultural development: the case of sustainable tsetse eradication on Unguja Island, Zanzibar using an area-wide integrated pest management approach. PloS Negl Trop Dis 8: e2857.
9. Tween G, Rendón P (2007) Current advances in the use of cryogenics and aerial navigation technologies for sterile insect delivery systems. In: Vreysen MJB, Robinson A.S., Hendrichs J, editors. Area-Wide Control of Insect Pests: Springer. pp. 229–238.
10. Enkerlin W (2007) Guidance for packing, shipping, holding and release of sterile flies in area-wide fruit fly control programmes: Food & Agriculture Organisation.
11. Bjeliš M, Radunić D, Bulić P (2011) Pre- and post-release quality of sterile *Ceratitis capitata* males released by an improved automated ground release machine. Journal of Applied Entomology 137: 154–162.
12. Leal Mubarqui R (2005) Manual de operación del método "MUBARQUI". México: Servicios Aéreos Biológicos y Forestales Mubarqui.
13. Tan LT, Tan KH (2011) Alternative air vehicles for sterile insect technique aerial release. Journal of Applied Entomology: 126–141.
14. Liedo P, Salgado S, Oropeza A, Toledo J (2007) Improving mating performance of mass-reared sterile Mediterranean fruit flies (Diptera: Tephritidae) through changes in adult holding conditions: demography and mating competitiveness. Fla Entomol 90: 33–40.
15. Rull J, Reyes J, Enkerlin W (1996) The Mexican national fruit fly eradication campaign: Largest fruit fly industrial complex in the world. In: McPheron BA, Steck GJ, editors. Fruit Fly Pests: A World Assessment of Their Biology and Management. St. Lucie: CRC Press. pp. 561–563.
16. Franz G (2005) Genetic sexing strains in Mediterranean fruit fly, an example for other species amenable to large-scale rearing for the sterile insect technique. In: Dyck VA, Hendrichs J, Robinson AS, editors. Sterile Insect Technique Principles and practice in area-wide integrated pest management. Dordrecht, The Netherlands: Springer. pp. 427–451.
17. Arredondo J, Ruiz L, Hernández E, Montoya P (2013) Efecto del tiempo de hipoxia y de las condiciones de empaque sobre la calidad de *Anastrepha ludens* cepa sexada genéticamente, Tapachula 7: Programa Moscafrut, SAGARPA-IICA.
18. Zavala JL, Gutiérrez JM, Cotoc E, Tirado L (2014) New Mediterranean Fruit Fly Emergence and Release Facility at Tapachula, Chiapas, Mexico. 9th International Symposium on Fruit Flies of Economic Importance (ISFFEI). Bangkok, Thailand.
19. Schwarz AJ, Zambada A, Orozco DHS, Zavala JL, Calkins CO (1985) Mass production of the Mediterranean fruit fly at Metapa, Mexico. Fla Entomol: 467–477.
20. Bakri A, Mehta K, Lance DR (2005) Sterilizing insects with ionizing radiation. In: Dyck VA, Hendrichs J, Robinson AS, editors. Sterile insect technique Principles and practice in area-wide integrated pest management: Springer. pp. 233–268.
21. Bauer B, Filledier J, Kabore I (1984) Large scale rearing of tsetse flies (Diptera, Glossinidae) in the CRTA Bobo-Dioulasso, Burkina based on in vitro feeding techniques. Rev Elev Méd vét Pays trop 37: 9–17.
22. Mutika GN, Kabore I, Seck MT, Sall B, Bouyer J, et al. (2013) Mating performance of *Glossina palpalis gambiensis* strains from Burkina Faso, Mali and Senegal. Entomol Exp Appl 146: 177–185.
23. Bouyer J (2008) Does isomethamidium chloride treatment protect tsetse flies from trypanosome infections during SIT campaigns? Med Vet Entomol 22: 140–144.
24. Challier A, Laveissière C (1973) Un nouveau piège pour la capture des glossines (*Glossina:* Diptera, Muscidae): description et essais sur le terrain. Cah ORSTOM, sér Ent Méd et Parasitol 10: 251–262.

25. R Core Team (2014) R: A language and environment for statistical computing. R Foundation for Statistical Computing, Vienna. Available: http://www.R-project.org.
26. Laird NM, Ware JH (1982) Random-effects models for longitudinal data. Biometrics 38: 963–974.
27. Diggle P, Liang K, Zeger S (1994) Analysis of longitudinal data. Oxford: Clarendon Press.
28. Bates D, Maechler M, Bolker B (2011) lme4: Linear mixed-eects models using S4 classes, R package version 0.999375-40/r1308.
29. Hurvich CM, Tsai C-L (1995) Model selection for extended quasi-likelihood models in small samples. Biometrics 51: 1077–1084.
30. Burnham KP, Anderson DR (2002) Model selection and multimodel inference: a practical information-theoretic approach. New-York: Springer-Verlag.
31. Bouyer J, Seck MT, Sall B, Guerrini L, Vreysen MJB (2010) Stratified entomological sampling in preparation of an area-wide integrated pest management programme: the example of *Glossina palpalis gambiensis* in the Niayes of Senegal. J Med Entomol 47(4): 543–552.
32. Rayaisse JB, Kröber T, McMullin A, Solano P, Mihok S, et al. (2012) Standardizing Visual Control Devices for Tsetse Flies: West African Species *Glossina tachinoides, G. palpalis gambiensis* and *G. morsitans submorsitans*. PloS Negl Trop Dis 6: e1491.
33. Dicko AH, Lancelot R, Seck MT, Guerrini L, Sall B, et al.. (2014) Using species distribution models to optimize vector control: the tsetse eradication campaign in Senegal. Proceedings of the National Academy of Sciences in press.
34. Vreysen MJB, Saleh KM, Lancelot R, Bouyer J (2011) Factory tsetse flies must behave like wild flies: a prerequisite for the sterile insect technique. PLoS Negl Trop Dis 5(2): e907.
35. Boyer S, Toty C, Jacquet M, Lempérière G, Fontenille D (2012) Evidence of Multiple Inseminations in the Field in *Aedes albopictus*. PLoS ONE 7: e42040.
36. Oliva CF, Jacquet M, Gilles J, Lemperiere G, Maquart P-O, et al. (2012) The sterile insect technique for controlling populations of *Aedes albopictus* (Diptera: Culicidae) on Reunion Island: mating vigour of sterilized males. PLoS ONE 7: e49414.
37. Bouyer J, Lefrançois T (2014) Boosting the sterile insect technique to control mosquitoes. Trends Parasitol 30: 271–273.
38. Day MF, Oster II (2012) Physical injuries. Insect Pathology, an Advanced Treatise 1: 29–63.

CHAPTER 5

Sustainable Agricultural Mechanization for Smallholders: What Is It and How Can We Implement It?

Brian Sims [1,*] and Josef Kienzle [2]

[1] 3 Bourneside, Bedford MK41 7EG, UK

[2] Plant Production and Protection Division, FAO, Viale delle Termi di Caracalla, Rome 00153, Italy

ABSTRACT

Smallholder farmers are the main producers of the world's food and they will have to increase production by up to 100 percent by 2050 to feed the growing population. This must be achieved while preserving natural resources and that is why sustainable agricultural mechanization (SAM) will be fundamental to the process. SAM is climate-smart and environmentally benign and essentially means no-till conservation agriculture, which requires specific mechanization inputs. Principally, these are seeders and planters capable of penetrating soil surface vegetative cover to deposit seed and fertilizer at the required depth and spacing; and equipment for management of cover crops and weeds. Mechanization is required not only for crop production, but also for processing and along the entire value chain. Mechanization inputs are usually expensive and so specialist service provision will be the indicated way forward. This will need collaboration from both the private and public sectors and will involve public-private partnerships to be developed in one form or another. Given the poor track record of public sector mechanization provision, the delivery of SAM should be firmly in the hands of the private sector that should be committed to SAM principles or otherwise be incentivized to the concept through smart subsidies. Improved information flows via smallholder farmer-friendly innovation platforms; and continuing development and testing of SAM technologies via regional centres of excellence will both be required—especially for sub-Saharan Africa.

Keywords: climate smart agriculture; environmental protection; mechanization service delivery; public-private partnerships; knowledge creation and dissemination; agricultural transformation

1. INTRODUCTION

The world is facing a crisis originated by a three-pronged attack which threatens the well-being of vast numbers of the planet's people. The crisis—some call it a "perfect storm"—has its origins in the convergence of population growth, natural resource degradation and the impacts of climate change.

The population of the world currently stands at 7.5 billion and is set to rise to around 10 billion by 2050 after that it will continue to rise and may exceed 12 billion by the end of the century [1]. These global figures mask regional variations and it seems that Asia (especially south Asia) and Africa will be home to the vast majority of the global population as their rates of growth are highest. Howard Buffet [2] estimates that the population of Africa is projected to increase from 1.1 billion today to at least 2.4 billion by 2050; while the Asian population will rise from the present 4.5 billion to 5.3 billion over the same time period [3].

The unsustainable advance of the global economy, which seeks to continue to grow year-on-year with a finite natural resource base, means that natural resource capital is everywhere being depleted. Although smallholder farmers are not the main culprits in provoking the widespread global destruction of forests, they are certainly responsible, at least in part, for the degradation of soil, water and woodland resources at the local level. It has been found [4] that globally human-induced degradation of agricultural soils is widespread and varies from light to extreme. It is calculated that we may be losing more than 9 billion hectares of arable land each year through erosion; in Africa it is estimated that 22 percent of productive land is affected by some form of degradation [5]. The term "soil degradation" covers a multiplicity of problems including [5]: loss of organic matter and nutrients; desertification; soil sealing; erosion by wind and water; decline in biodiversity; compaction and; salinization. Additionally, soils can be degraded through contamination resulting from human activities such as mining and road construction, and hydro-geological events such as landslides; frequently several threats may occur simultaneously and so the problems can be multiplied if they are not countered at an early stage. Worldwide agriculture accounts for 70 percent of water extraction (three times more than 50 years ago) [6] and today it is estimated that 40 percent of the world's rural population lives in water-scarce catchments. Competing demands from industry and the growing urban population are exacerbating an already critical situation.

The fifth assessment report (AR5) of the Intergovernmental Panel on Climate Change (IPCC) leaves us in no doubt that the climate changes that are now upon us are mainly due to anthropogenic emissions of greenhouse gasses (GHGs) [7]. The emissions from agriculture, forestry and other land use (AFOLU) now account for 20–22 percent of global CO_2 equivalent GHG emissions (which total around 45 gigatonnes CO_2 eq/year). If emissions for primary production (especially from diesel fuel) and agricultural value chain operations are included, then the AFOLU GHG emissions increase by a further 4.3 gigatonnes CO_2 eq/year [8].

GHG emissions result in global warming and we are now in a desperate race to confine global temperate rise to 2C above pre industrial revolution levels [9]. A warmer atmosphere has more energy and so will result in more extreme weather events. The IPCC has concluded that the Earth can expect more of what we are becoming used to in terms of the increased violence and frequency of such extreme events, including hurricanes, tropical storms, flooding, droughts, heatwaves and wildfires. The corresponding impacts on agricultural production will be huge and widespread from low to high latitudes [10].

Given the importance of family farms for feeding the planet's growing human population, this paper focuses on the issues surrounding sustainable mechanization for this sector. However we include consideration of the sustainable mechanization needs of larger scale farms in Section 2 (Mechanization of conservation agriculture). There are over 500 million family farms in the world and they occupy between 70 and 80 percent of farm land [11]. It is estimated that family farms produce about 80 percent of the world's food in value terms and so it is vital that they are not only supported but supported with technology (especially high quality inputs, including mechanization) that is compatible with sustainable production intensification. Such support should lead to a transformation process that will allow subsistence-oriented smallholders to develop into more market-oriented farmers. In our view this is a principal route to lead smallholders out of poverty through providing training, means and technologies along the value chain for reducing drudgery and increasing labour and land productivity in a sustainable way.

The mechanization of agriculture requires energy and this comes from farm power sources. In large scale agriculture the predominant farm power source is the diesel engine installed in tractors, combines, other self-propelled machines and stationary units where electric motors also play an important role. In the developing world, engine power is generally on the increase as low-cost tractors and other engine-powered technologies become increasingly available, especially from India and China. The use of draught animals for farm power is still important in many developing countries. Although there is a clear trend for animals to be replaced by engine power, this will take some time to achieve, even in rapidly developing countries such as India and China [12]. The Indian tractor manufacturing industry is currently the largest in the world, producing one third of global tractor production [13]. On the other hand, China claims to be the world's leading manufacturer of agricultural machinery in its widest sense, including tractors (both two-wheeled and four-wheeled), implements, other self-propelled machines, pumps and machinery for agricultural product processing along the output value chain [14]. There is a similar trend in Latin America where draught animal power has now declined to constitute about 25 percent of the power source for family farms. In sub-Saharan Africa (SSA) the use of draught animals is still important in many regions, but is constrained by the presence of the tsetse fly (*Glossina* spp.), the vector of trypanosomiasis, regionally in SSA, draught animal power contributes up to 25 percent of farm power [15].

It is still the case, however that human muscles are an extremely important source of farm power throughout the developing world, and especially in SSA where even today humans supply 65 percent of farm power. This constitutes a major limitation for increased land and labour productivity; it impacts heavily on the timeliness of farm operations and is a major source of drudgery. So much so that the African Union has proclaimed that the hand-hoe should be "consigned to the museum" by 2035 [16].

Given the situation that has been described above, we can see that low land and labour productivity can be the direct consequence of a lack of access to, and application of, key production inputs, including mechanization. There is a pressing need to increase global food production to feed the growing, and increasingly urban, global population. FAO [11] estimates that agricultural production globally must increase by 70 percent by 2050, and by 100 percent in developing countries, and so "business as usual" is really not an option if we are to avoid widespread food shortages.

At the same time, we are seeing a continuing degradation of the planet's natural resource base. It is, of course, this natural resource base—our natural capital—that provides the environment for agricultural production. Equally important in this sense are soil and water resources which are prone to abuse from conventional agricultural practices which involve intensive and excessive soil tillage. This situation leads to low productivity on family farms resulting in low family income and an inability to invest in agricultural inputs—including appropriate mechanization inputs. Increased poverty is leading to family disintegration with young, fit males migrating to urban centres both nationally and internationally; leaving women, the elderly and children to take on the burden of food production on the family farm.

Sustainable agricultural mechanization (SAM) can be described as mechanization that is economically feasible, environmentally sensitive and socially acceptable [17]. The United Nations' (UN) Food and Agriculture Organization (FAO) SAM website [18] elaborates on this basic concept as follows:

"Mechanization covers all levels of farming and processing technologies, from simple and basic hand tools to more sophisticated and motorized equipment. It eases and reduces hard labour, relieves labour shortages, improves productivity and timeliness of agricultural operations, improves the efficient use of resources, enhances market access and contributes to mitigating climate related hazards. Sustainable mechanization considers technological, economic, social, environmental and cultural aspects when contributing to the sustainable development of the food and agricultural sector."

The goal of this paper is to analyze the smallholder farm situation, to identify the key elements needed to include the concept of sustainability in future mechanization developments, and to suggest appropriate vehicles for delivering SAM technologies to farm families. The following sections examine more closely the various aspects and implications of SAM, the focus is on mechanization issues, other aspects, such as soil science, are dealt with briefly in order to put mechanization in context. In the next section we reprise the basic pillars of conservation agriculture which underpins the concept of SAM. Then the need to consider mechanization holistically along the agricultural product value chain is debated. Having examined the basic ingredients of SAM, we look at models for getting mechanization inputs delivered to the family farm better. The penultimate section considers the need for SAM development for specific situations (with particular reference to SSA), including R & D and innovation platforms, before summing up with some concluding remarks.

2. THE NEED FOR CONSERVATION AGRICULTURE

The use of soil cultivation implements, including mouldboard, disc and chisel ploughs, is one of the most energy hungry operations in agriculture; it is also the cause of soil structural damage and can create impermeable plough-pans at the depth of penetration. (Using hand-hoes for primary soil cultivation can have a similar effect with the creation of compacted hoe-pans in the soil profile—although it is a main driver of the agricultural mechanization discussion, especially in sub-Saharan Africa, that arduous hand-hoe work should be replaced with mechanization—hence hoe-pans should be a symptom of the past if SAM happens as envisaged. This desire is behind various mechanization initiatives in Africa and one of the aims of this paper is to ensure the application of SAM, and not merely the supply of mechanization inputs with no sustainability strategy in place.)

Tractor-powered ploughing uses vast amounts of energy to invert the top soil, this requires the use of high-powered tractors and can consume many litres of diesel fuel per hectare. In one study, in Nigeria [19] disc ploughing with a 3-disc plough at 30 cm depth consumed around 25 L/ha with a 100 hp tractor. This represents a major production cost and is one of the principal reasons given by farmers for reducing tillage operations. One farmer in Lincolnshire, UK estimates that by switching from mouldboard ploughing to direct seeding his total annual farm fuel consumption has fallen from 92 to 42 L/ha [20].

The physical damage caused by these traditional tillage practices includes the destruction of the natural channel systems in the soil profile created by roots and meso-fauna such as earthworms and termites. These are the channels which allow water to infiltrate and be stored in the profile pores, they also allow gaseous exchange with the atmosphere. Destroying the natural channel system with tillage implements greatly reduces the soil's capacity to absorb rainfall and irrigation water and increases the risk of run-off and soil erosion. If a plough-pan is also created, then the situation is made much worse as the now de-structured "arable layer" is perched on an impermeable layer which impedes the infiltration of water to deeper soil horizons. Soils in this condition suffer from a lack of available water and crops grown in them are increasingly prone to drought.

Tillage not only results in physical soil damage, but has a deleterious impact on soil organic matter and soil life. Soil organic matter (SOM) is the lifeblood of a soil. Formed through the actions of soil fauna and fungi on plant material (crop residues or applied mulch), SOM is the source of natural soil fertility and the principal agent in the formation of soil structure. SOM is oxidized when exposed to the atmosphere and solar radiation, and levels decline in tilled soils. Conversely they increase under no-till, conservation agriculture regimes.

Soil fauna, flora and fungi give a healthy soil its life and constitute the engine driving the production of soil fertility and maintaining a soil structure conducive to healthy crop growth and development. Disturbing the delicate balance and maintaining soil fertility through the application of inorganic fertilizers is not sustainable in the long term, and of course adds to production costs through the enforced use of expensive inputs.

2.1. Climate-Smart Agriculture

Farmers, especially smallholder farmers, have always been dependent on clement weather to raise crops and livestock and be food-secure. Today, with violent and erratic weather events becoming more widespread as the effects of climate change manifest themselves with increasing frequency, smallholder production systems are in danger of becoming ever more precarious. The global rise of interest in climate-smart agriculture (CSA) is a direct consequence of this situation, as is the need to reduce GHG emissions for the AFOLU sector as discussed previously in the light of IPCC analyses and predictions. CSA focuses on three pillars [21]: productivity to provide food security; adaptation both for extreme events and slower, long-term, climate change manifestations affecting both agriculture and the consequent food security implications, and; mitigating the effects of climate change through reduced GHG emissions (especially through carbon sequestration and reduced fuel consumption).

FAO's concept of sustainable crop production intensification (SCPI) [22] recognizes the urgent requirement for increased agricultural productivity to feed a growing world population and the concurrent need to restore and nurture the planet's natural resource base. The essence of SCPI is encapsulated in the "Save and Grow" paradigm which uses an ecosystems approach to raise productivity in an environmentally friendly way. This includes enhancing SOM through the use of organic manures, mulches, cover crops and leguminous crops and trees while minimizing mechanical soil cultivation. Apart from improving soil health, other major aspects of achieving CSA are water management, including precision irrigation and the use of drought-resistant crop varieties; plant protection measures include the employment of crop rotations and effective (especially) biological weed control.

However, care must be taken to select appropriate CSA practices as an analysis of the global situation *vis-à-vis* CSA [23] indicates that there is a risk, because of the misallocation of financial and human resources, of promulgating agricultural systems that do not result in improved food security and increase the risk of accelerated climate change. Many CSA measures, designed to add resilience to agricultural systems to combat the impacts of climate change, are included in the practice of conservation agriculture which is the subject of the following section.

2.2. Conservation Agriculture

Interest in conservation agriculture (CA) has been growing over the last two decades (the seventh world CA congress is taking place in Argentina in 2017). The practice is spreading by an estimated 10 million hectares a year [24] and now covers approximately 11 percent of global arable cropland. CA is far from being a single practice employed over this vast area, instead it is a complementary set of three general principles which are adapted locally to become compatible with the location specific environment. Paraguayan smallholder CA practitioners would have a very different complex of technologies and practices compared to their counterparts in the highlands of East Africa, for example. However the general principles will still apply and will be common to the local adaptations in both regions. These general principles are as follows (see, for example, [24]):

- **Minimal soil disturbance**. The damaging impact of soil tillage has been discussed in the previous section, in CA seed, fertilizer and transplants are placed in the desired position in the soil profile with the least amount possible of disturbance to the soil. Direct seeding and planting is the goal and the mouldboard or disc plough plays no role, although it may be necessary to repair damaged soils prior to embarking on a CA regime. This may entail breaking a plough-pan with a subsoiler or chisel plough, or removing ridges and furrows with one last pass of the mouldboard plough, but these operations will not be necessary once CA has been introduced.

- **Maintaining a permanent organic soil cover**. Keeping the soil surface under organic cover is a very important means of conserving soil moisture, improving soil structure and water infiltration and providing a substrate for conversion into SOM and soil fertility. Soil organic matter cover can also play an important role in weed suppression. Cover will comprise the residues of the previous crop, but this can be augmented by the use of crop associations, rotations and cover crops (especially leguminous cover crops). The maintenance of soil cover can be difficult, especially in semi-arid regions with low dry season precipitation, and where livestock have access to crop residues [25]. In these situations CA farmers may need to provide additional biomass for livestock feed including fodder banks of perennial grass species and trees for dry season browsing [26].

- **Maintaining biodiversity in crop production**. The value of crop rotations to maintain soil fertility, and reduce pest and disease incidence, has long been understood. CA strives to go further than this and encourages the use of crop associations, relay cropping, cover crops and agroforestry. The use of legumes in association with cereal crops like maize is especially useful (Figure 1).

Figure 1. Maize intercropped with *Canavalia ensiformis.* The leguminous Canavalia fixes nitrogen which can be used by the cereal crop.

2.3. Mechanization of CA

2.3.1. Seeders and Planters

Implementing CA requires some specialist equipment [27], especially for sowing seeds through the soil surface mulch of crop residues or cover crops. The process starts with crop residue treatment—residues should be spread evenly across the soil surface behind the combine harvester and this may require a spreader to be fitted at the rear. The seeder (delivering a continuous flow of small seeds) or planter (delivering large seeds at discretely spaced intervals) has two main ways of introducing seed and fertilizer through the surface vegetation. In conditions of light coverage, narrow chisel tined openers can be used (Figure 2). The tine opens a slot behind which seed and fertilizer can be delivered to the required depth.

Figure 2. Narrow chisel-tined slot openers for seed and fertilizer placement are suitable for low surface residue situations. A vertical disc preceding the opener will ease residue flow.

Single discs, inclined to the direction of travel can open a slot in the soil so that seed and fertilizer can be delivered behind the disc in a way similar to a chisel-tined opener (Figure 3). In heavy surface residue conditions, the best approach is the use of double disc openers. These are discs of different diameters, offset to the direction of travel, which can cut through residue, open a 'V'-shaped slot and deliver seed and fertilizer between the discs (Figure 4). In all cases where it is necessary to cut through surface vegetation, a vertical disc preceding the slot opener will usually facilitate smooth residue flow. Keeping soil disturbance to a minimum should always be the aim and the cross-slot opener achieves this by cutting through residue with a vertical disc and delivering seed on one side and fertilizer to the other in horizontal slots produced by inclined lateral wings [28].

Figure 3. Single inclined disc opener. Seed and fertilizer are delivered behind the disc.

Figure 4. Double-disc opener, in this case not only inclined to the direction of travel, but vertically as well. This provides better seed coverage and less surface soil disturbance.

Rice residue is notoriously difficult to cut with vertical discs and one method to deal with heavy residue situations is by lifting the residue with rotating flail-blades before sowing into the cleared surface and depositing the mulch back on top of the seeded soil. This concept, incorporated in the "Happy Seeder" [29] has subsequently been modified to cut and lift the residues with rotating flails either side of the chisel-tined slot opener and seed and fertilizer delivery tubes (Figure 5).

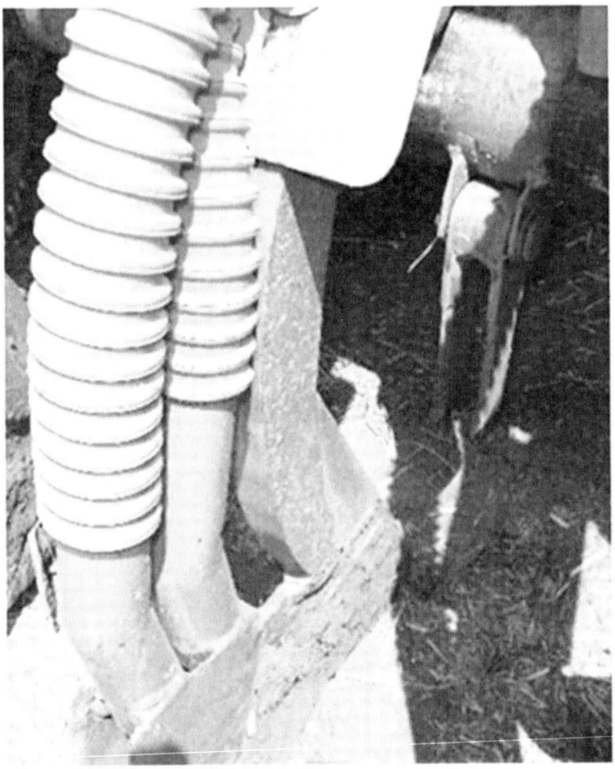

Figure 5. The Turbo Happy Seeder manages rice residue by flailing and lifting it on either side of the chisel-tined no-till slot opener.

Planters for manual use can be as simple as a pointed stick with a small container tied to the farmer's waist for seed and/or fertilizer. An improvement on this basic idea is the hand-operated jab planter which can deliver seed and fertilizer through separate sharp beaks which are closed on entering the soil and opened at planting depth (Figure 6).

Figure 6. A manually operated jab-planter with hoppers for seed and fertilizer.

Reduced tillage, cultivating a strip of soil for crop establishment is an increasingly popular option. Tractor, or draught animal, powered rippers (narrow chisel tines) are used to cultivate a strip of soil into which the crop is planted manually (perhaps with a jab-planter) (Figure 7). Power tillers can be modified by adjusting the lateral placement of blades so that strips are cultivated in the soil. A seed hopper and metering mechanism can be mounted over the rotary cultivator to achieve strip tillage and seed placement in one pass [30].

2.3.2. Power Sources

The power sources used for CA mechanization include tractors (both four-wheel—4 WT, and two-wheel—2 WT); draught animals and humans; the wide range of possible power sources can be seen on the FAO CA website [31]. Generally speaking, and as we have seen above, the power requirement for CA is considerably less than that for tillage-based agriculture—approximately half the energy is expended with no-till CA and so the power of motorized units (4 WTs and 2 WTs) can be halved and this is one of the main attractions of CA for farmers.

Figure 7. A tractor-mounted chisel-tined ripper which can be used to create narrow tilled strips for crop establishment.

2.3.3. Weed Control and Cover Crop Management

Weeds and cover crops can conveniently be managed with the use of herbicides, but this is becoming somewhat controversial as public and environmentalist preferences turn towards a reduction in the use of agro-chemicals. Other options are available and should be employed whenever possible. Cover crops, residues and weeds can be conveniently managed with knife rollers prior to direct drilling—knife rollers are cylinders with blades which crimp the vegetation without chopping it and in many cases this is sufficient to kill the plants (Figure 8).

Weeds can be managed manually by rogueing, mechanically by surface scraping or by removing flower heads above the crop with machines such as a "Weed Surfer" which has a series of rotating blades to decapitate the flowering weeds [32]. The best way to manage weeds is through competition with main crops, associated crops and cover crops; keeping the soil covered with vegetation will suppress weeds and cover crops that include a wide mixture of species are probably the best option. Soil cultivation for weed control is not an option in CA.

Figure 8. An animal-drawn knife roller managing weed growth prior to direct seeding.

If herbicides are to be used, then the conventional range of sprayers can be employed. These range from the ubiquitous back-pack (or knapsack) lever-operated sprayer to animal-drawn and tractor-mounted boom sprayers. One particularly useful innovation for smallholder farmers is the idea of mounting a back-pack sprayer on a small, wheeled, chassis; the pump lever is operated by an eccentric off a drive wheel. The rig, with a 4-nozzle, 2-metre boom is pulled behind the operator ensuring that he is kept well away from possible spray contamination (Figure 9).

3. MECHANIZATION ALONG THE VALUE CHAIN

Increased economic growth in the rural sector with its concomitant improvement in employment possibilities is a key to poverty reduction and so helps with the achievement of the UN's Sustainable Development Goal 1 (An end to poverty). Mechanization in its broadest sense has great potential to increase productivity and improve livelihoods, not only in primary agricultural production (both crop and livestock) but also along the value chain of that production. This will include post-harvest operations such as storage and processing (shelling, milling and husking). It can also include food processing from (fairly) simple packing of fresh produce for supermarket chains (Figure 10) to more industrial type processing, for example of dairy products like yoghurt and ice-cream or juice extraction from aloe (*Aloe vera*) [33].

Figure 9. A back-pack sprayer mounted on a wheeled chassis. The sprayer pump is actuated by an eccentric from one of the wheels, the 2-m boom has four nozzles. A manually operated return valve is required to control pressure and prevent wheel skid.

Figure 10. Small-scale processing and packaging of vegetables for local supermarkets.

Whereas such business ventures, requiring as they do a sound business plan and financial backing, are well within the grasp of larger producers, for smallholders some form of collaborative action will usually be required. Farmer cooperatives can be the solution and there are many examples of farmer cooperatives organized to provide services as varied as crop cleaning, drying and storage, to food processing and manufacturing [34]. The opportunities that arise will be specific to particular locations or regions and their identification can be a useful contribution from NGOs working with rural communities. Cooperatives are also particularly useful for the supply of production inputs (i.e., improving input supply chains) including farm machinery, as they, as bulk buyers, can demand preferential prices from suppliers, something that would not be possible for individual farmers.

The value chain does not end with processing and packaging. There is still a requirement for marketing and transport and it is in this latter category that mechanization input suppliers may find a useful clientele that will allow them to expand their businesses throughout the year. A well-functioning or developing value chain is one key to increasing prosperity and reducing poverty as well as the creation of alternative (and often better paid) jobs compared with subsistence, hand-hoe-powered, agricultural production. Mechanization inputs are the drivers of this transformation process and public-private partnerships are probably the best-bet option for making value chains work in all stakeholders' interests.

A world congress of the International Commission for Agricultural Engineering (CIGR) included a workshop (organized by FAO) which focused on "*creating sustainable and viable machinery supply chains for enhancing mechanization, and of using technology to add value to, and increase the quality of, agricultural products*" [35]. This neatly encapsulates our point on ensuring a well-functioning mechanization input supply chain and output value chain and some examples of success stories (with an SSA focus) discussed in the workshop can be summarized as follows:

Challenges for agricultural mechanization

Access to mechanization by smallholder farmers is a pressing necessity to enable increases in their labour and land productivity. This will require clear guidance from a transparent public sector, free from corruption, to provide an environment conducive to enthusiastic and profitable private sector participation. Local manufacture of locally adapted mechanization inputs should be encouraged and facilitated and demand for mechanization services should be increased through the creation and nurturing of common interest farmer groups (such as CA clubs) and the provision of credit and financial incentives. Mechanization technology transfer is beneficial, especially through south-south collaboration where there is a better understanding of conditions and constraints. At the same time, it is important not to lose sight of the value of participatory adaptation of equipment for local conditions. The conservation of natural capital (especially soil, water and forests) requires a renewed focus on sustainable land management. Supply chains for sustainable mechanization options need to be strengthened to encourage increased demand. Because of rural-urban migration and the impact of pandemic diseases, female-headed farm families are becoming more prevalent, especially in SSA. Improving women's access to farm power through the provision of suitably designed equipment needs to be addressed by the actors in the farm power provision supply chains.

Using technology to add value and quality

The public sector can play a major role in supporting private sector initiatives in the SAM area. Examples of support include funding for market studies, public-private cooperation to innovate in the value chain, and programmes to better link research institutions to the private sector. Consideration of the entire value chain can help to avoid the build-up of constraints (bottlenecks) in the value addition process. That is to say avoiding the situation where the resolution of one problem may give rise to additional problems down the line. Participatory technology development should involve partnerships between manufacturers, researchers and user groups, this will build local capacity for the development of food production and processing technology, rather than dependence on importation. Again, environmental protection is of fundamental importance. Food processing has the potential to be a major polluter—but thinking holistically and considering food processing industry "waste"; products as valuable nutrient inputs for food production is the sustainable solution.

4. MAKING MECHANIZATION AVAILABLE

Smallholder farmers, acting on their own, usually find that investment in mechanization technology is too expensive a proposition for them. As we have discussed above, the application of sustainable agricultural practices requires specific equipment, especially for seeding through vegetative soil cover. Both the availability and affordability of CA-specific equipment are of great importance to scale out CA, but individual smallholders would struggle to acquire the necessary tools. In this case, the possibility of specialist CA service provision should be considered [36]. For this to become reality, appropriate entrepreneurs—willing to provide such services—will need to have access to the correct equipment and this may involve facilitating their access to financial credit. Public sector planners will see that the mechanization "business" can play an influential part in private sector development and so is deserving of some public sector support. This support can take the form of planned strategies to rationalize and promote mechanization [37] which will include partnerships (public-private partnerships or PPPs) to facilitate equipment supply chains (especially through support of local manufacture), supply credit to the rural sector and share financing risks, create improved extension services, demonstrations and farmer-oriented messages with a focus on sustainable agricultural mechanization. Such training would focus specifically on business skills and sustainable agriculture technology practical skills [38]. One example of public sector support for financing is that of the Brazilian government through its More Food (*Mais Alimentos*) programme where concessional loans are offered for the purchase of agricultural machinery. The programme has been extended from Brazil to Mozambique, Ghana, Zimbabwe and Kenya to promote family farming [39]. The idea of making finance available to farmers to encourage vertical integration between input suppliers, farm producers and food processors is not always met with unanimous approval. Brazilian and Japanese investment in the ProSAVANA programme in Mozambique is a case in point [39] where loans are provided to integrate smallholder farmers into agribusiness value chains. ProSAVANA has, however, ignited controversy with civil society organizations arguing that local farmers will be disenfranchised and the beneficiaries will be Brazilian and Japanese multinationals along with the local Mozambican elite. The remainder if this section examines models that are being used to deliver mechanization services to smallholder farmers. Many of them are models discussed during a December, 2016 meeting in Nairobi, Kenya on models for sustainable agricultural mechanization in SSA [40], although not all have an exclusively African focus.

4.1. Machinery Rings in Germany

Machinery rings in Germany are a response to a reduced labour force and inefficient machine use on individual small farms [41], currently there are approximately 230 local machinery rings (MRs) with 192,000 members. MRs are organized by the farmers themselves and constitute a modern centre of a range of services and good technical knowledge. Investments can be made in common or, more usually, by individuals and there is a value chain approach (e.g., planting, weeding, harvesting, loading, cleaning, transport of sugar beet). Machines are hired by farmer ring members and come with a qualified operator.

The experience is that MRs are very successful as the rules are clear and transparent and prices are reasonable. Farmers declare themselves to be in favour of this means of working together to get modern mechanization at an affordable price.

There would appear to be good potential to expand this model to developing country situations and there is good scope for PPPs (e.g., in the design of policy frameworks for facilitating the finance of machinery acquisition) which should be explored further.

4.2. Cooperatives for Promoting Mechanization

Farm machinery cooperatives (*Coopératives d'Utilisation de Matériel Agricole*—CUMAs) are highly participatory (rather than passive) collective investments, managed by farmers in independent groups in the same territory. An illuminating paper [42] gives details of how the CUMA movement in France has contributed to the removal of the major barriers to the economic and social development of family farmers. The history of the CUMA movement shows how social capital is a critical resource in overcoming some incentive problems; the CUMAs have succeeded in creating effective cooperatives through the development of a dense fabric of relationships. The idea of CUMAs has now been established in Benin [43] with three principal aims: to facilitate access to mechanization; to overcome financing difficulties and to increase the profitability of the investment in farm machinery. The result has been an increase in the cultivated area (by 350 percent) and better cultivation to produce increased incomes, enhanced standards of living, and social transformation with the increase in entrepreneurship and involvement of youth. CUMAs promote strong relationships within themselves and they react positively with other unions and cooperatives in the national and international institutional environment.

4.3. International Research Centres

International research centres (especially those of the CGIAR—Consultative Group on International Agricultural Research) are taking an increasing interest in developing technology for sustainable mechanization. These are PPPs as the private sector (farmers and manufacturers) are intimately involved, as is the public sector through the technology development and financing alternatives. One example is

AfricaRice, a member of the CGIAR consortium whose membership comprises 26 countries in West, Central, East and North Africa [44]. Mechanization is a priority area in their strategic plan and there is an Africa-wide mechanization task force to get appropriate technology into the hands of producers and processors; the main activities are: research, fabrication, out-scaling and training. AfricaRice collaborates with public-private partnerships to promote SAM with many technologies along the value chain from transplanters to weeders to threshers to parboilers. The 'ASI' light thresher is particularly suitable for operation by women who are offered agribusiness skills development advice through innovation platforms.

After a lapse in interest, other international research organizations, notably the International Rice Research Institute (IRRI) and the Maize and Wheat Improvement Centre (CIMMYT) are again engaging with the importance of mechanization as a fundamentally necessary agricultural input.

4.4. International Development Organizations

Another example of PPPs is the collaboration between the UN Industrial Development Organization (UNIDO) and private sector heavy equipment manufacturers to establish operator training centres in various African countries. Examples of human capacity development include [45]: the Japan heavy equipment operator school (JHEOS)—Liberia (Komatsu); the Zambia industrial training academy (ZAMITA)—Zambia; the Heavy duty equipment and commercial vehicles academy (HDECoVA)—Ethiopia and; the Académie engins lourds et véhicules commerciaux (AGEVEC)—Morocco.

Public private development partnerships (PPDPs) combine the expertise and resources of the private sector (e.g., Volvo, Festo, Scania) with the public sector and development sector (e.g., UNIDO, SIDA (Swedish International Development Cooperation Agency), USAID (United States Agency for International Development), etc.) to provide a public service by creating a skilled work force.

4.5. CASU Project, Zambia

Creating a sustainable demand for mechanization services is a critical element in out-scaling SAM and the FAO-implemented Conservation Agriculture Scaling-up Project (CASU) in Zambia focuses on raising demand for this input [46]. Challenges with smallholder mechanization, which need to be overcome, include technical problems, from land preparation to planting, weeding and harvest; and postharvest handling. There are also limited financing opportunities due to: the low value of draught animal power (DAP) equipment which is not "economic" for financing institutions; the limited collateral available to smallholder farmers and; the high interest rates required.

The CASU project promotes CA mechanization in a variety of ways including harnessing the potential of existing service providers to focus on CA; linking CA mechanization to on-going social protection programmes (e.g., by using e-vouchers to stimulate demand); the creation of collaborative PPPs for financing and service provision and; the bundling of services (e.g., by offering planting + weeding + harvesting + processing + transport).

4.6. Private Sector Driven Agricultural Mechanization Programme (PSDAMP) in Nigeria

In Nigeria the PSDAMP is a PPP involving the Ministry of Agriculture, financial institutions, agro-machinery manufacturers and service providers [47]. This PPP, initiated in 2012, sets up agricultural equipment hiring enterprises (AEHEs) where there is demand from farmers to hire or lease mechanized equipment for a variety of value chains. AEHEs are needed for a range of reasons including: small and fragmented land holding sizes and poor resource base of smallholder farmers; inadequate skills for operating and maintaining complex equipment; endemic corruption, and; a lack of after sales technical support.

AEHEs are created as safe housing and workshop centres to service a minimum of 1500 ha. A typical list of equipment (costing USD 60,000–123,000) for an AEHE is:

- 3–5 units of tractors + basic implements.
- 5 power tillers.
- 2 mini-harvesters.
- 3 threshers.
- 1 boom sprayer.

Only equipment with a certificate from the National Centre for Agricultural Mechanization (NCAM) is eligible and all foreign equipment must have local technical back-up support. Funding for an AEHE is divided between the public and private sectors and is for four years at an interest rate of 7 percent. The AEHE is owned by service providers and can comprise cooperatives or other farmer groups, individual farmers or private investors. They are run by a 5-person team (leasing manager, accountant, field supervisor, mechanic and security guard); capacity building is built in for technical and managerial staff. The investment is secured against risk by a variety of insurance schemes and machinery tracking. Service back-up is assured and periodic technical training programmes are given in collaboration with manufacturers.

A total of 250 AEHEs is envisaged and to date there are 140 AEHEs with 590 tractors and ancillary equipment.

4.7. Other Examples of Successful Ppps for Mechanization Service Provision

The German Development Institute (*Deutsches Institut für Entwicklungspolitik*—DIE) offers descriptions of a range of PPP models for mechanization service provision [48]. Among the most impressive are the following examples:

MyAgro

MyAgro is a not-for-profit organization active in Mali and Senegal. It is a layaway savings scheme whereby local vendors sell scratch cards for farmers who text the code to deposit their savings in a *myAgro* account; *MyAgro* can then deliver high quality seeds, fertilizer and tools. The result is a secure savings scheme and guaranteed, timely input supply. *MyAgro* provides training and support services on climate-smart and input-saving practices (such as CA). Timely delivery of quality-guaranteed inputs is assured (an especially important aspect for farmers in remote locations). *MyAgro* also adapts tools to local needs, especially the needs of women.

The *myAgro* programme provides a confidential savings method, avoiding the social problems sometimes associated with traditional savings group models. The system is also helpful in ensuring that savings are used for their dedicated purpose (rather than run the risk of being diverted to cover unforeseen emergencies that may crop up) and has resulted in improved access to mechanization inputs.

Hello Tractor

Hello tractor is a social enterprise in Nigeria which provides a machinery hire service. Tractor services are requested by text message and prepayment is made by mobile phone. After satisfactory completion of the task, payment is released to the "smart" tractor owner. Anonymity means that women are more able to supply services.

NWK

NWK in Zambia is a commercial cotton producer and acts as an intermediary between farmers (out-growers) and the bank. Farmers pay according to their cash flow (usually at harvest) and NWK pays for services to be provided when needed. The result is high quality mechanization services provided on time to otherwise vulnerable farmers.

AGCO Future Farm, Zambia

AGCO's Future Farm concept aims to develop a sustainable food production system able to increase farm output through more efficient resource utilization [49]. One ongoing effort is to develop an agricultural machinery hire services model [50] which will indicate the level of income a local agro-dealer and service provider needs to achieve to be viable. This will come from products and services that can be provided to local emerging and small scale farmers, namely contracting services, sales of seed, fertilizer, agro-chemicals and other inputs.

The aim of the model is to provide a tool that enables someone to sit down with an entrepreneur and have an open discussion about what is actually viable from a business perspective. The idea is to figure out what could be possible and start to develop a business case that makes stakeholders comfortable with the proposal.

Several general conclusions can be drawn from the examples outlined above. Firstly only demand-driven and profitable mechanization is viable, service providers and all the other actors in the supply chain need to make a viable living out of the process. Mechanization technology needs to be adapted to farm size and endowment context, low intensity technologies (e.g., 2 WTs and DAP) are feasible at farm level, but higher intensities require shared ownership. The finance provision of all actors needs to be considered in the mechanization supply chain [51].

5. THE NEED FOR THE DEVELOPMENT AND SCREENING OF SAM TECHNOLOGIES AND LOCAL MANUFACTURE

As we have seen, mechanization for smallholder farmers has been somewhat neglected for the past few decades. Efforts to mechanize via public sector service providers in many countries generally failed due principally to unresponsive management structures, soaring costs and subsidized prices. They were generally poorly run and financial losses (Figure 11).

Figure 11. Public sector farm machinery services targeting smallholder farmers have usually been financial failures.

Research and development efforts aimed at mechanization options for smallholder farmers have not always been effectively directed with a focus on identifying needs with the key stakeholders of users and manufacturers; rather inappropriate "solutions" have often been developed in isolation and have subsequently withered on the vine after the fanfare of initial promotion [52] (Figure 12).

Figure 12. The development of animal-drawn wheeled toolcarriers attracted considerable development effort in the 1980s; however farmer adoption was minimal.

During the previously mentioned consultative meeting on mechanization strategies for SAM, held in Nairobi, Kenya in December 2016 [53] two urgent needs were discussed which will put efforts to provide smallholder farmers with appropriate mechanization options, firmly at the forefront of rural development planning. These were support for the establishment of multi-stakeholder innovation platforms, and the creation of a network of centres of excellence for the development and promotion of SAM options for smallholder farmers. An innovation platform is an on-line forum for the dissemination of SAM technology and knowledge; a SAM centre of excellence is a hub for the development of relevant information, training, development and testing. These two proposals will now be discussed.

5.1. Sustainable Mechanization Knowledge Exchange Platform

There is a need for a platform to share knowledge and experience and give visibility to SAM equipment and its application. The platform would bridge the gap and forge a link between machinery users and manufacturers and its outcome would be to demonstrate how to adapt and/or develop technologies for SAM from the knowledge exchange.

The platform would distinguish between short-term and long-term development. Product information would be presented from the users' (usually farmers but also support players such as extension officers) perspective. Participation would be encouraged and opinions on equipment shown. Surveys, statistics and reports would also be included.

An important issue would be technology transfer—matching farmers' needs to manufacturers' offers. There would also be a strong content on capacity building and market information. Longer-term issues to be included would be newly developed equipment, and finance options and models. A country focus would be appropriate for many themes.

The organization could be a collaboration between manufacturers' associations and an agricultural development organization. The platform would need a kick-start from donors and thereafter should be self-sustaining with revenue from advertising and fees from machinery manufacturers wishing to place their products.

The platform should be global as there is a global need for comprehensive information on existing demand (especially for machinery suppliers). Current platforms do not have this information and with this new platform the target would be farmers and farmers' associations as a priority. The platform would offer technology as well as knowledge and include machinery that has the potential to be sustainable (i.e., it would not be a purely commercial website). It is to be hoped that the challenge to investigate further the concept of a SAM platform will soon be taken up and implemented.

5.2. Regional Centres for SAM

Asia is well served by the UN Centre for Sustainable Agricultural Mechanization—UN-CSAM, based in Beijing, China [54]. CSAM grew out of the success of the Regional Network for Agricultural Machinery RNAM and the UN Asian and Pacific Centre for Agricultural Engineering (UNAPCAEM). CSAM became operational in 2012.

A strong case can be made to establish a similar centre with an African focus and possibly also for Latin America which would build on the pioneering work already achieved in Brazil [55]. In the case of SSA, the concept could be a network of centres of excellence. The Network would comprise a central hub (with a secretariat) and regional centres covering West, Central, East and Southern Africa. It would also need to have strong links to North Africa and other networks globally.

- The principal objective of the Network would be to improve smallholder farmers' access to appropriate SAM technologies to help them to improve their livelihoods. It would achieve this through a variety of methods including: collecting and sharing information and knowledge; training and capacity building; testing and evaluation of technologies (although all technologies would be allowed to flourish or fail, there would be no imposition) and; the preparation of catalogues of successful stories (but not prototype development). In addition the Network should be active in the provision of standards, using Organization for Economic Co-operation and Development (OECD), International Organization for Standardization (ISO) and Asian and Pacific Network for Testing of Agricultural Machinery (ANTAM) codes where applicable.

Complementary centres in SSA and globally. In West Africa there are already nine value chain centres of excellence and a tenth is being added (with World Bank funding) on agricultural mechanization. This has the potential to be a West African regional centre of the Network. There is also positive experience with the UN Centre for Sustainable Agricultural Mechanization (UN-CSAM).

In SSA the regional centres can specialize on specific roles, such as testing or capacity building. The Network itself will identify the best choices for regional centres of excellence and this should be undertaken in close collaboration with: the: Governments of the countries involved; the private sector, especially manufacturers of agricultural machinery; farmers' associations seeking to improve access to SAM technology; SAM service provider associations and; National and international institutions research centres, universities, UN organizations (such as FAO and UNIDO).

Funding is likely to emanate from a variety of sources. The private sector is only likely to be interested in joining after initial funding and institutional set-up, once this has happened, they can be invited to become corporate members. Funding for the establishment of the Network could include: the World Bank and UN agencies; foundations (e.g., the Bill and Melinda Gates Foundation); National development agencies; Intergovernmental organizations and; the Private sector.

6. CONCLUSIONS

Sustainable agricultural mechanization involves the application of different forms of power sources in conjunction with appropriate equipment able to do useful work in agricultural production and along the agri-food value chain. Thus, mechanization must meet farmers' needs efficiently and effectively and result in improved farm productivity and reduced drudgery, as well as contributing to the development and competitiveness of the food supply chain. To be sustainable, mechanization must take economic, social, environmental, cultural, and institutional issues into account [56].

As we have seen, SAM holds great potential for improving incomes, livelihoods and employment opportunities for some of the often overlooked segments of rural society: women, children, the elderly and youth. SAM strives to offer life-changing mechanization inputs in a culturally sympathetic way and so these groups can be helped to advance while not upsetting cultural norms, or by forging acceptable new ones. However, it is acknowledged that mechanization is part of a process of transformation from subsistence-oriented farming to market-oriented production. Yet, without having a positive effect on natural capital, mechanization cannot be sustainable and this, perhaps, is the key final message. Sustainable mechanization does not destroy biodiversity nor does it mine natural resources—if mechanization does not result in these impacts then it cannot be called sustainable—despite any process of transformation that it entails.

The focus for SAM should be on smallholder farmers, half a billion of whom account for 80 percent of food production in the developing countries of Asia, Latin America and Africa. Because this group is relatively poor in resources, the best way to give them access to SAM is via professional (trained) private sector SAM service provision. SAM technology is, however, available for and applicable to all farm types and sizes and all power sources (from manual to DAP to engine). The world's farming is turning to climate-smart conservation agriculture at the rate of over 10 million hectares annually. SAM can be provided to the smallholder sector in a number of diverse ways; however, although services should always be provided by the private sector (given the almost universal failure of public sector mechanization provision) the public sector still has an important role to play. This can be through nurturing an attractive environment in which the private sector can do business and includes providing infrastructure, finance and appropriate 360 degree training, entailing all aspects of SAM, in one of the many types of PPP that have been successfully implemented in many countries. Potential trainees should be pre-selected through a rigid process—it should be made clear that they are not just undergoing training but would later on be required to make a business decision to become a SAM service provider with all the entrepreneurial risks implied. Consequently the training will have to cover all aspects from mechanization business type selection to knowledge of environmental sustainability to climate smart, energy efficient ways of machinery application.

To promote the development and scaling out of SAM, two proposals are offered and described. The first is the establishment of a global knowledge exchange platform which would disseminate proven information on SAM technologies, practice and experience in a format readily understandable to, and accessible by, the smallholder farm family community. The second is the establishment of Centres of Excellence for the development and testing of SAM technologies. Based on the positive contribution being made by CSAM in Asia, other regional centres would definitely be recommended for Africa (especially SSA) and, possibly, for Latin America.

REFERENCES

1. Wikipedia. World Population. Available online: https://en.wikipedia.org/wiki/World_population (accessed on 8 March 2017).
2. Buffet, H.G. Foreword. In *Conservation Agriculture for Africa: Building Resilient Farming Systems in a Changing Climate*; Kassam, A.H., Mkomwas, S., Friedrich, T., Eds.; CAB International: Wallingford, UK; Boston, MA, USA, 2017; pp. xi–xiii.
3. Asia Population. Available online: http://www.worldometers.info/world-population/asia-population/ (accessed on 9 March 2017).
4. Global Assessment of Human-Induced Soil Degradation (GLASOD). Available online: http://www.isric.org/projects/global-assessment-human-induced-soil-degradation-glasod (accessed on 9 March 2017).
5. Jones, A.; Breuning-Madsen, H.; Brossard, M.; Dampha, A.; Deckers, J.; Dewitte, O.; Gallali, T.; Hallett, S.; Jones, R.; Kilasara, M.; et al. *Soil Atlas of Africa*; European Commission, Publications Office of the European Commission: Luxembourg, Luxembourg, 2013; p. 176. [Google Scholar]
6. Agriculture at a Crossroads, Findings and Recommendations for Future Farming: Water. Available online: http://www.globalagriculture.org/report-topics/water.html (accessed on 10 March 2017).
7. IPCC. *Climate Change 2014: Synthesis Report. Contributions of Working Groups I, II and III to the Fifth Assessment Report of the Intergovernmental Panel on Climate Change*; IPCC: Geneva, Switzerland, 2014; p. 151.
8. FAO. *The Future of Food and Agriculture—Trends and Challenges*; Food and Agriculture Organization of the United Nations: Rome, Italy, 2017; p. 163.

9. European Commission. Climate Action. Available online: https://ec.europa.eu/clima/policies/international/negotiations/paris_en (accessed on 12 March 2017).
10. Parry, M.L.; Canziani, O.F.; Palutikof, J.P.; Van der Linden, P.J.; Hanson, C.E. *Climate Change 2007: Impacts, Adaptation and Vulnerability. Contribution of Working Group II to the Fourth Assessment of the Intergovernmental Panel on Climate Change*; Cambridge University Press: Cambridge, UK, 2007; p. 976.
11. FAO. *The State of Food and Agriculture. Innovation in Family Farming*; Food and Agriculture Organization of the United Nations: Rome, Italy, 2014; p. 139.
12. Singh, G.; Zhao, B. Agricultural mechanization situation in Asia and the Pacific region. *Agric. Mech. Asia Afr. Lat. Am.* **2016**, *47*, 15–25.
13. Singh, S. Agricultural machinery industry in India. *Agric. Mech. Asia Afr. Lat. Am.* **2016**, *47*, 26–35.
14. Yuanen, G. The current situation and future of agricultural machinery industry in China. *Agric. Mech. Asia Afr. Lat. Am.* **2016**, *47*, 109–114.
15. FAO. Agricultural and Food Engineering Services Technical Report 3. In *Farm Power and Mechanization for Small Farms in Sub-Saharan Africa*; Food and Agriculture Organization of the United Nations: Rome, Italy, 2006; p. 65.
16. Mrema, G. The process of developing a draft framework for SAMS for the transformation of agriculture in Africa: Sending the hand-hoe to the museum. In Proceedings of the Consultative Meeting on Mechanization Strategy: New Models for Sustainable Agricultural Mechanization in Sub-Saharan Africa, Nairobi, Kenya, 1–3 December 2016.
17. Sims, B.; Kienzle, J. Making mechanization accessible to smallholder farmers in sub-Saharan Africa. *Environments* **2016**, *3*, 11.
18. FAO. Sustainable Agricultural Mechanization. Available online: http://www.fao.org /sustainable-agricultural-mechanization (accessed on 16 March 2016).
19. Adewoyin, A.O.; Ajav, E.A. Fuel consumption of some tractor models for ploughing operations in the sandy-loam of Nigeria at various speeds and ploughing depths. *Agric. Eng. Int. CIGR J.* **2013**, *15*, 67–74.
20. Sims, B. A Decade of Conservation Agriculture: Farm Walk at Thurlby Grange Farm, Bourne, Lincolnshire. 2014. Available online: http://www.taa.org.uk/assets/group-articles/Decade%20of%20CA%20at%20Reynolds.pdf (accessed on 8 June 2017).
21. Campbell, B.; Dinesh, D.; Loboguerro-Rodríguez, A.M.; Van Etten, J. Directions for climate action in agriculture. *Agric. Dev.* **2017**, *30*, 2–3.
22. FAO. *Save and Grow: A Policy Maker's Guide to the Sustainable Intensification of Smallholder Crop Production*; Food and Agriculture Organization of the United Nations: Rome, Italy, 2011; p. 102.
23. Lipper, L.; Thornton, P.; Campbell, B.M.; Baedeker, T.; Braimoh, A.; Bwalya, M.; Caron, P.; Cattaneo, A.; Garrity, D.; Henry, K.; et al. Climate-smart agriculture for food security. *Nat. Clim. Chang.* **2014**, *4*, 1068–1072.
24. Mkomwa, S.; Kassam, A.H.; Friedrich, T.; Shula, R.K. Conservation agriculture in Africa: An overview. In *Conservation Agriculture for Africa. Building Resilient Farming Systems in a Changing Climate*; Kassam, A.H., Mkomwa, S., Friedrich, T., Eds.; CAB International: Oxford, UK; Boston, MA, USA, 2017; pp. 1–9.
25. Giller, K.E.; Witter, E.; Corbeels, M.; Tittonell, P. Conservation agriculture and smallholder farming in Africa: The heretics' view. *Field Crops Res.* **2009**, *114*, 23–34.
26. Garrity, D.P. How to make conservation agriculture evergreen. In *Conservation Agriculture for Africa. Building Resilient Farming Systems in a Changing Climate*; Kassam, A.H., Mkomwa, S., Friedrich, T., Eds.; CAB International: Oxford, UK; Boston, MA, USA, 2017; pp. 167–182.
27. Sims, B.G.; Kienzle, J.; Mkomwa, S.; Friedrich, T.; Kassam, A.H. Mechanization of smallholder conservation in Africa: Contributing resilience to precarious systems. In *Conservation Agriculture for Africa. Building Resilient Farming*

Systems in a Changing Climate; Kassam, A.H., Mkomwa, S., Friedrich, T., Eds.; CAB International: Oxford, UK; Boston, MA, USA, 2017; pp. 183–213.
28. Baker, C.J. Seeding openers and slot shape. In *No-Tillage Seeding in Conservation Agriculture*, 2nd ed.; Baker, C.J., Saxton, K.E., Eds.; Food and Agriculture Organization of the United Nations: Rome, Italy; Wallingford, UK; Boston, MA, USA, 2007; pp. 34–59.
29. ACIAR. Fine Tuning the Happy Seeder Technology for Adoption in Northwest India. Australian Centre for International Agricultural Research. Available online: http://aciar.gov.au/project/cse/2006/124 (accessed on 27 March 2017).
30. Sims, B.G. Some advances in mechanization options for conservation agriculture from the fifth world congress of conservation agriculture, Brisbane, Australia, September 2011. *Agric. Dev.* **2012**, *15*, 13–15.
31. FAO. Conservation Agriculture. Available online: http://www.fao.org/ag/ca/ (accessed on 27 March 2017).
32. Sims, B. Organic Farming and Agroforestry in the Fens: A visit to Whitehall Farm, Peterborough. 2014. Available online: https://drive.google.com /file/d/0Bwy IPGne8KZ-WDdLZXl5TGhsQU0/view (accessed on 28 March 2017).
33. Sims, B.G. Small-scale agro-industries for agricultural sector growth: Some practical examples. *Agric. Dev.* **2012**, *16*, 10–13.
34. FAO. Agricultural Cooperatives Are Key to Reducing Hunger and Poverty. 2011. Available online: http://www.fao.org/news/story/en/item /93816/icode/ (accessed on 30 March 2017).
35. FAO. Addressing the challenges facing agricultural mechanization input supply and farm product processing. In Proceedings of the FAO Workshop held at the CIGR World Congress on Agricultural Engineering, Bonn, Germany, 5–6 September 2006; Sims, B.G., Kienzle, J., Cuevas, R., Wall, G., Eds.; Food and Agriculture Organization of the United Nations: Rome, Italy, 2007; p. 71.
36. FAO. *Hire Services by Farmers for Farmers. Diversification Booklet Number 19*; Sims, B., Röttger, A., Mkomwa, S., Eds.; Food and Agriculture Organization of the United Nations: Rome, Italy, 2011; p. 83.
37. Houmy, K.; Clarke, L.J.; Ashburner, J.; Kienzle, J. Agricultural mechanization in sub-Saharan Africa: Guidelines for preparing a strategy. *Integr. Crop Manag.* **2013**, *22*, 93.
38. Sims, B.; Kienzle, J. Mechanization of conservation agriculture for smallholders: Issues and options for sustainable intensification. *Environments* **2015**, *2*, 153–158.
39. Amanor, K.S.; Chichava, S. South-south cooperation, agribusiness and African agricultural development: Brazil and China in Ghana and Mozambique. *World Dev.* **2016**, *81*, 13–23.
40. Consultative Meeting on Mechanization Strategy: New Models for Sustainable Agricultural Mechanization in Sub-Saharan Africa. Available online: http://africamechanize.act-africa.org/ (accessed on 3 April 2017).
41. Hastedt, A. German machinery rings: Experiences and lesson learnt. In Proceedings of the Consultative Meeting on Mechanization Strategy: New Models for Sustainable Agricultural Mechanization in Sub-Saharan Africa, Nairobi, Kenya, 1–3 December 2016.
42. Herbel, D.; Rocchigiani, M.; Ferrier, C. The role of the social and organisational capital in agricultural cooperatives' development: Practical lessons from the CUMA movement. *J. Co-op. Organ. Manag.* **2015**, *3*, 24–31.
43. Nouwogou, K. Promouvoir la mécanisation agricole à travers les coopératives agricoles: Cas de CUMAS au Benin. In Proceedings of the Consultative Meeting on Mechanization Strategy: New Models for Sustainable Agricultural Mechanization in Sub-Saharan Africa, Nairobi, Kenya, 1–3 December 2016.
44. Moreira, J. Sustainable mechanization development strategy options and the role of public-private and private-private partnerships. In Proceedings of the Consultative Meeting on Mechanization Strategy: New Models for Sustainable

Agricultural Mechanization in Sub-Saharan Africa, Nairobi, Kenya, 1–3 December 2016.
45. Papastavrou, S. Mechanization strategies from an industrial perspective. In Proceedings of the Consultative Meeting on Mechanization Strategy: New Models for Sustainable Agricultural Mechanization in Sub-Saharan Africa, Nairobi, Kenya, 1–3 December 2016.
46. Mphatso, M. Facilitating small scale farmers' access to conservation agriculture mechanization. In Proceedings of the Consultative Meeting on Mechanization Strategy: New Models for Sustainable Agricultural Mechanization in Sub-Saharan Africa, Nairobi, Kenya, 1–3 December 2016.
47. Ogirinye, I. Private sector driven agricultural mechanization programme (PSDAMP) and GES application strategy of Nigeria. In Proceedings of the Consultative Meeting on Mechanization Strategy: New Models for Sustainable Agricultural Mechanization in Sub-Saharan Africa, Nairobi, Kenya, 1–3 December 2016.
48. Ströh de Martínez, C.; Feddersen, M.; Speicher, A. *Food Security in Sub-Saharan Africa: A Fresh Look on Agricultural Mechanisation*; Deutsches Institut für Entwicklungspolitik gGmbH: Bonn, Germany, 2016; p. 171.
49. Cultivating Farms of the Future. Available online: http://blog.agcocorp.com/2015/05/cultivating-farms-of-the-future/ (accessed on 6 April 2017).
50. Moore, M.; AGCO. Personal communication, December 2016.
51. Müller, C. Food security in SSA: A fresh look on agricultural mechanization. In Proceedings of the Consultative Meeting on Mechanization Strategy: New Models for Sustainable Agricultural Mechanization in Sub-Saharan Africa, Nairobi, Kenya, 1–3 December 2016.
52. Starkey, P. Deutsches Zentrum für Entwicklungstechnologien—GATE. In *Animal-Drawn Toolcarriers: Perfected yet Rejected: A Cautionary Tale of Development*; Deutsche Gesellschaft für Technische Zusammenarbeit (GTZ) GmbH: Wiesbaden, Germany, 1988; p. 161.
53. FAO; ACT. *Consultative Meeting on Mechanization Strategy: New Models for Sustainable Agricultural Mechanization in Sub-Saharan Africa*; Food and Agriculture Organization of the United Nations (FAO): Rome, Italy; African Conservation Tillage Network: Nairobi, Kenya, 2017; p. 50.
54. United Nations Economic and Social Commission for Asia and the Pacific (ESCAP), CSAM—Centre for Sustainable Agricultural Mechanization. Available online: http://un-csam.org/ (accessed on 11 April 2017).
55. Casão Junior, R.; Guilherme de Araújo, A.; Fuentes Llanillo, R. *No-Till Agriculture in Southern Brazil—Factors that Facilitated the Evolution of the System and the Development of the Mechanization of Conservation Farming*; Food and Agriculture Organization of the United Nations (FAO): Rome, Italy; Agricultural Research Institute of Paraná State (IAPAR): Paraná, Brazil, 2012; p. 77.
56. Kienzle, J. Sustainable agricultural mechanization. In Proceedings of the Consultative Meeting on Mechanization Strategy: New Models for Sustainable Agricultural Mechanization in Sub-Saharan Africa, Nairobi, Kenya, 1–3 December 2016.

CHAPTER 6

Development of Bioelectrochemical Systems to Promote Sustainable Agriculture

Xiaojin Li [1], Ibrahim M. Abu-Reesh [2] and Zhen He [1,*]

[1]*Department of Civil and Environmental Engineering, Virginia Polytechnic Institute and State University, Blacksburg, VA 24061, USA*

[2]*Department of Chemical Engineering, College of Engineering, Qatar University, P.O. Box 2713, Doha, Qatar*

ABSTRACT

Bioelectrochemical systems (BES) are a newly emerged technology for energy-efficient water and wastewater treatment. Much effort as well as significant progress has been made in advancing this technology towards practical applications treating various types of waste. However, BES application for agriculture has not been well explored. Herein, studies of BES related to agriculture are reviewed and the potential applications of BES for promoting sustainable agriculture are discussed. BES may be applied to treat the waste/wastewater from agricultural production, minimizing contaminants, producing bioenergy, and recovering useful nutrients. BES can also be used to supply irrigation water via desalinating brackish water or producing reclaimed water from wastewater. The energy generated in BES can be used as a power source for wireless sensors monitoring the key parameters for agricultural activities. The importance of BES to sustainable agriculture should be recognized, and future development of this technology should identify proper application niches with technological advancement.

Keywords: bioelectrochemical systems; microbial fuel cells; microbial desalination cells; anaerobic digestion; membrane filtration; wastewater reclamation and reuse; wireless sensors

1. INTRODUCTION

Bioelectrochemical system (BES) has drawn great attention in recent years as an emerging technology for energy-efficient wastewater treatment, desalination, sustainable energy generation and value-added chemical production. In principle, BES takes advantage of microbial metabolism with electrodes to generate electricity via extracellular electron transfer (EET) [1] (Figure 1). Exoelectrogens (electrochemically active microorganisms) involved are capable of directly or indirectly transferring electrons to/from electrodes [2], referred to as electrode respiration [3]. Bacterial dissimilatory metal reduction (BDMR) has been regarded as the process closest to electrode respiration [2], with the *Geobacter* and *Shewanella* species identified as the most common BDMR model bacteria used in BES [4]. Development of BES can be classified into the following categories based on their application purposes: microbial fuel cells (MFC) [5], microbial electrolysis cells (MEC) [6], microbial desalination cells (MDC) [7], microbial electrosynthesis cells (MES) [8], *etc.* BES can be applied not only to treat the waste but also to harvest energy and value-added products. For example, electrical power [5] can be captured directly from the oxidation of organic compounds in MFCs while hydrogen [9] and methane [10] can be harvested from MECs.

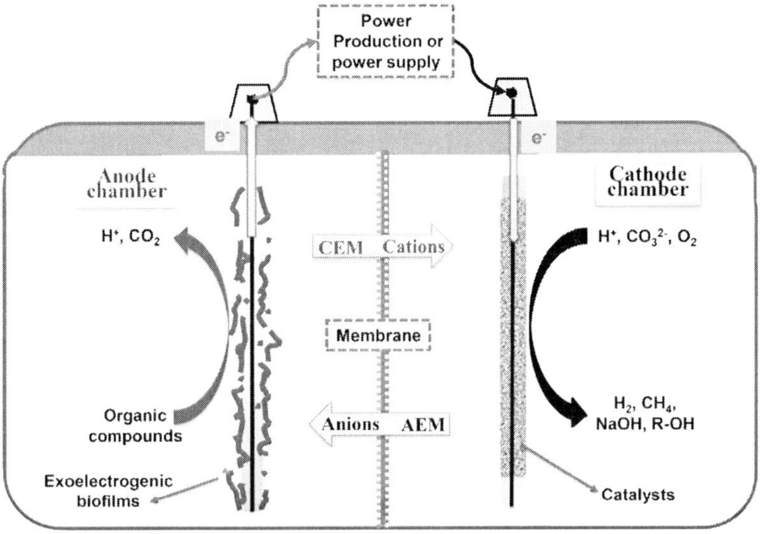

Figure 1. A general schematic of bioelectrochemical systems (BES), CEM—cation exchange membrane, AEM—anion exchange membrane.

Among various forms of BES, MFCs are the most basic one, and extensive efforts have been made towards its development for practical application [11,12,13]. In MFCs, exoelectrogens have the capability of converting chemical energy to electrical energy. Electrons and protons are generated in an anode chamber during the oxidation of organic matters, and then transported through an external electric circuit to terminal electron acceptors (e.g., oxygen, nitrate, *etc.*) in a cathode chamber, incurring reduction reaction; cations such as protons are transferred to the cathode chamber via a separator (e.g., ion exchange membrane) or through the electrolyte. MFCs have the potential for energy efficient wastewater treatment, renewable energy production, water reuse and bioremediation [14]. The substrates used in MFCs include a wide range of organic compounds including digested sludge, municipal sewage, landfill leachate, food wastewater, and marine sediments [15].

Water, energy and nutrient are the key elements for agricultural production that also generates a large amount of waste. The sustainability of agriculture is facing significant challenges [16,17], including an increasing demand for agricultural land and resources due to the rapid growth of population [18], environmental problems caused by excessive consumption of fossil fuels, fertilizers and pesticides, *etc.* Agricultural biomass, such as solid agricultural residues, and wet and dry manure, is considered as a renewable energy source because of its abundance and high organic content. BES appears to be of strong interest to address some of the key issues associated with water, energy and nutrient for sustainable agriculture. This review aims to introduce the past studies of BES related to agriculture, and discuss the critical factors essential for the development of BES for practical applications in sustainable agriculture, including treating agro-industrial waste, providing reclaimed water from saline water and wastewater, and powering the wireless sensors for agricultural monitoring.

2. BES FOR AGRICULTURAL WASTE MANAGEMENT

BES can utilize a wide range of substrates produced from agricultural activities, generating renewable energy (electricity) with simultaneously degrading waste. Previous studies have shown that BES can perform either as a standalone process or as a post-treatment process for treating various types of agricultural waste.

2.1. BES as a Standalone Technology

2.1.1. Animal Waste

Modern livestock agriculture has dramatically increased manure production. Application of manure as fertilizer and soil amendment can result in significant air and water pollution. For example, pollutants such as heavy metals, pathogens, hormones, and antibiotics in agricultural runoff can impair water quality [19]. The emissions of odor, methane, ammonia, and nitrous oxide can also affect air quality [20]. Therefore, animal waste should be treated appropriately to reduce its environmental impact.

Agricultural manure from animal confinements is rich in organic matters, and thus may act as a source of substrate for energy recovery using BES. However, based on the estimate of energy yield per unit mass of feedstock (~10 $kJ \cdot kg^{-1}$ wet manure), manure may have a limited potential for electricity generation via MFC, mostly because of low conversion efficiency and complex substrate composition [21]. Table 1 presents a summary of energy recovery from animal waste in MFCs. In general, the power densities reported in the previous studies are highly diverse, ranging from several milliwatts to several hundred milliwatts per electrode surface area. The power density is largely affected by the substrates, MFC configurations and size [22,23], electrode materials, as well as operating conditions. Cattle manure as a representative of livestock was examined in two different MFC configurations, including a single compartment combined membrane-electrodes (SCME) and a twin compartment brush-type anode electrodes (TBE) without a proton exchange membrane (PEM) [24]. The electricity was produced at the rate of 9.2 mW·kg^{-1} of dry manure in the SCME and 24.3 mW·kg^{-1} in the TBE, suggesting that the brush-type anode design was more efficient than the conventional plate type electrode, probably due to a larger surface area of the electrode. Factors such as moisture content, phosphate buffer solution (PBS), catalyst loading, and electrode area were investigated in a single-chamber, air-cathode MFC fed with cow manure, which showed that a higher moisture content was more suitable for current generation: moisture contents of 80%, 70% and 60% resulted in the maximum power densities of 349 ± 39, 36 ± 9 and 12 ± 2 mW·m^{-2}, respectively [25]. An MFC removed about 84% of BOD (biochemical oxygen demand) from cow slurry, while most of the nitrogen, phosphorus, and potassium were retained (84%, 70%, and 91%, respectively); the maximum power output was only 0.34 mW·m^{-2} probably resulting from the presence of abundant inorganic matter, cellulose and lignin in manure [26].

Table 1. Performance comparison of microbial fuel cells (MFCs) for treating agricultural manure and wastewater.

MFC Type	Feedstock	External Resistance	Max Area Power Density	Max Volume Power Density	Original COD	COD Removal	Ref.
		Ω	$mW \cdot m^{-2}$	$W \cdot m^{-3}$	$mg \cdot L^{-1}$	%	
S-MFC	cattle manure	470	36.6	0.2	1000	-	[24]
T-MFC	cattle manure	470	67	0.3	1000	-	[24]
T-MFC	manure sludge	-	5	-	-	-	[27]
S-MFC	Cow manure	1000	349	-	-	-	[25]
S-MFC	dairy manure	1000	189	4.7	-	-	[28]
T-MFC	cow waste slurry	460	0.34	-	1010	84	[26]
S-MFC	swine wastewater	200	261	1.2	8320	90	[15]
T-MFC	swine wastewater	1000	45	-	8320	-	[15]
S-MFC	swine wastewater	1000	228	-	8270	84	[29]
S-MFC	swine wastewater	10	-	1.08	12980	0.523 kg COD $m^{-3} \cdot day^{-1}$	[30]

S—single-chambered; T—two-chambered.

Swine wastewater is another major animal waste with high strength of organic contaminants, odor problem and pathogenic risk [21]. The studies of swine wastewater treated by MFCs are summarized in Table 1. Typically, swine wastewater was diluted (5–10 times) to prevent inhibition of ammonia on exoelectrogen activities [15,30,31]. An early study used two MFCs to simultaneously generate electricity and treat swine wastewater containing 8320 ± 190 mg·L^{-1} of soluble COD [15]. The maximum power density obtained in a two-chambered MFC was 45 mW·m^{-2}, much lower than 261 mW·m^{-2} in a single-chambered MFC [15]. However, the Coulombic efficiency (CE) was relatively low (8%) in the single-chambered MFC, which was probably due to the diffusion of oxygen into the anodic chamber. In addition, soluble COD removal was increased from 88%–92% when the wastewater was stirred, while CE decreased from 8%–5%. It was found that a maximum power density of 1415.6 mW·m^{-3} could be achieved from swine wastewater at a current density of 3258.5 mA·m^{-3} when using Pt coated graphite felt and CEM; meanwhile, the organic and nitrogen removal rates were 0.523 kg COD m^{-3}·day^{-1} (total anode chamber) and 0.194 kg·N·m^{-3}·day^{-1} (total cathode chamber), respectively [30]. In addition to electricity generation, hydrogen can also be produced in an MEC treating swine wastewater [32]. The overall hydrogen recovery was 28% ± 6% of the COD, and hydrogen gas accounted for 77% ± 11% of total gas volume. In contrast, little hydrogen gas could be recovered by fermentation of the swine wastewater unless it was autoclaved.

Different types of animal waste are rich in nutrients and thus it is of interest to investigate nutrient removal/recovery in BES. An air-cathode single-chamber MFC was used to recover phosphorus in the form of struvite crystal, which precipitated on the surface of the cathode electrode; however, the recovery rate of phosphorus was only 27%, accounting for a small portion of total phosphorus removal (70%–82%) [33]. Ammonia removal was examined in both single- and two-chambered MFCs, and the results suggested that nitrogen losses in the air-cathode system were mainly caused by ammonia volatilization due to elevated pH near the cathode, while nitrogen losses in the two-chambered MFC were primarily due to ammonium ion diffusion through the CEM [34]. In addition, nitrification likely occurred when oxygen was available, as ammonia-oxidizing bacterium *Nitrosomonas europaea* was detected on the cathode electrode.

The results of these previous studies indicate that animal waste has some potential as a renewable feedstock to produce renewable energy by BES. The barriers that can interfere with electricity generation include toxicity of ammonia at high concentrations, volatile fatty acids, as well as methane production [35]. The applications of BES in treating animal waste will depend on many factors such as the cost of the materials, treatment efficiency, and the amount of energy gained and consumed.

2.1.2. Plant Waste

Plant waste generated from agricultural activities is conventionally disposed by landfilling, composting, and incineration, leading to environmental concerns such as greenhouse gas emissions. Plant waste such as cellulose and lignocellulose has been considered as a potential source for renewable energy due to their abundance [36]. For example, biotechnologies have been developed to convert cellulosic biomass to energy products, such as hydrogen and methane [37]. The disadvantages of those bioprocesses include the availability of cellulolytic enzymes, generation of toxic intermediates, disposal of by-products, and high cost of gas separation, purification and storage [38].

BES has been investigated for treating plant waste. However, due to the complex composition of plant waste, the studies about treatment of plant waste by BES are limited, and pretreatment of complex waste to simpler forms appears to be essential. Table 2 shows the performance of the MFCs using cellulose biomass as substrates. It was found that both cellulolytic and exoelectrogenic microorganisms would be required for electricity generation in BES, because no single strain has yet been capable of producing electricity directly from cellulose [39,40]. An early study reported indirect electricity generation from cellulose in an MFC, through *in situ* oxidation of hydrogen that was produced from the anaerobic degradation of cellulose by cellulolytic bacteria (*Clostridium cellulolyticum* and *Clostridium thermocellum*) [41]. A defined coculture of the cellulolytic fermenter *Clostridium cellulolyticum* and the electrochemically active *Geobacter sulfurreducens* was used to generate electricity in a two-chamber MFC fed with cellulose (soluble CMC and insoluble MN301) [42]. The results showed that the coculture achieved maximum power densities of 143 $mW \cdot m^{-2}$ and 59.2 $mW \cdot m^{-2}$ from 1 $g \cdot L^{-1}$ CMC and MN301 cellulose, respectively, while neither pure culture alone could generate electricity from these cellulose sources. Electricity was also produced from cellulose-MFCs using mixed and pure cultures of *Nocardiopsis* sp. KNU and *Streptomyces enissocaesilis* KNU as cellulose-degrading bacteria biocatalysts [43] and mixed cultures with the rumen microbiota containing both strict and facultative anaerobes [44,45]. The low power densities in the MFCs treating cellulose were attributed to the high internal resistance of the two-chamber MFCs related to low conversion rate (Table 2) [36]. Thus, reducing internal resistance of MFCs and developing proper inoculum could increase power density [36]. For example, with a pre-acclimated inoculum from an MEC, the maximum power densities achieved in single- and two-chamber MFCs were 1070 $mW \cdot m^{-2}$ (cathode area) and 880 $mW \cdot m^{-2}$, respectively [36]. As an exception, Rezael *et al.* [39] demonstrated for the first time that electricity can be generated from cellulose in a U-tube MFC using a single bacterial strain (*Enterobacter cloacae*) without exogenous mediators, though a very low power density of 4.9 $mW \cdot m^{-2}$ was obtained.

Table 2. Performance comparison of MFCs for treating cellulose biomass.

MFC Type	Substrate	Strains or Culture	Anode Material	Max. Power Density $mW \cdot m^{-2}$	COD Removal %	Ref.
MFC	3 $g \cdot L^{-1}$ D-0	*Clostridium cellulolyticum* & *Clostridium thermocellum*	Pt-PTFA	130 $A \cdot m^{-3}$ *	-	[41]
T-MFC	1 $g \cdot L^{-1}$ CMC	*Clostridium cellulolyticum* & *Geobacter sulfurreducens*	graphite plates	143	38	[42]
T-MFC	1 $g \cdot L^{-1}$ MN301	*Clostridium cellulolyticum* & *Geobacter sulfurreducens*	graphite plates	59.2	27	[42]
T-MFC	7.5 $g \cdot L^{-1}$ Sigmacell 20	ruman	graphite plates	55		[45]
T-MFC	1.5 $g \cdot L^{-1}$ Sigmacell 20	cellulolytic & exoelectrogenic bacteria	carbon paper	880	50–70	[36]
S-MFC	1.5 $g \cdot L^{-1}$ Sigmacell 20	cellulolytic & exoelectrogenic bacteria	carbon paper	1070	50–70	[36]
3-T-MFC	1 $g \cdot L^{-1}$ rice straw powder	*Nocardiopsis* sp. KNU & *Streptomyces enissocaesilis* KNU	carbon paper	490	-	[46]
U-tube MFC	cellulose	*Enterobacter cloacae*	carbon cloth	4.9	-	[39]

S—single-chambered: T—two-chambered: * current density.

Because of the recalcitrant characteristics of cellulose, pre-treatment processes are necessary to convert cellulose to readily degradable carbohydrates as substrates for BES. Instead of cultivating cellulolytic microorganisms, cellulose hydrolysis can be achieved directly by cellulase, which refers to a group of enzymes involved in cellulose hydrolysis, including endoglucanase, cellobiohydrolase, and β-glucosidase [47]. One drawback of using cellulase is that the reaction can be inhibited by the accumulation of end products (e.g., cellobiose and glucose that can bind active sites or prevent access to substrates) [41,48,49]. Cellulose hydrolysis together with other processes (e.g., fermentation) that simultaneously consume the hydrolysis products will help to address the problem [50]. For example, the combined cellulase of *Novozyme* 188 (β-glucosidase) and *Celluclast* 1.5 L was introduced to increase the power density from 12 ± 0.6 $mW \cdot m^{-2}$ in the absence of the enzymes to 100 ± 7 $mW \cdot m^{-2}$, suggesting that cellulase and exoelectrogens have synergy [40].

In addition to electricity generation, hydrogen gas can also be produced from cellulose in MECs [51,52]. For example, a fermentation-MEC integrated process was used to convert lignocellulose into hydrogen gas [51]. The inoculum of the MEC came from multiple MFCs pre-acclimated to a single substrate, resulting in an improvement in the hydrogen yield and gas production rate. Hydrogen gas production from cellulose was also studied in an integrated system combining dark fermentation and an MFC as a power source for an MEC [52]. A hydrogen production rate of 0.24

$m^3 \cdot H_2 \cdot m^{-3} \cdot day^{-1}$ was achieved at an overall energy recovery efficiency of 23% (based on cellulose removed) [52].

The above studies have demonstrated the technical feasibility of utilizing cellulose for electricity generation or hydrogen production in BES. In general, the power densities produced directly from cellulose are lower than those obtained from soluble substrates, and successful power generation requires specialized cultures and bespoke MFC configuration [36]. With an acclimated culture, reduced cost of enzymes and optimized system design, MFCs and MECs have a potential to be applied to take advantage of those abundant plant wastes from agriculture.

2.2. BES as a Supplementary Technology

For over a century, anaerobic digestion (AD) has been widely used for energy recovery (e.g., methane, ethanol and/or hydrogen) from solid and liquid waste. It has several exceptional advantages including remarkable bioconversion efficiency, low operating costs, and decreased sludge disposal expenses [14,53]. Both AD and MFC are capable of simultaneously treating organic waste and extracting energy from these sources using selected microbial communities [54]. AD systems typically receive a high strength influent ($\geq$1000 mg COD L^{-1}), while BES allows to operate at low COD concentrations ($\leq$1000 mg COD L^{-1}) [55], and perform as effluent polishing processes that convert residuals to electricity (MFC), hydrogen (MEC) or other products such as hydrogen peroxide [56] and caustic solution [57]. In addition, MFCs can directly generate electricity from organic waste without the need for gas purification, and they can perform at low temperatures (<20 °C) while AD does not perform well due to low reaction rates and high solubility of methane under such temperature [58]. Given the advantages and disadvantages of each technology, AD and BES may be integrated to achieve more efficient and thorough bioconversion of waste/wastewater [59]. As shown in Figure 2, MFCs may function as a post-treatment unit for AD, and such a combination could create synergistic effects by taking advantage of the benefits of each process. For example, a thermophilic AD has been coupled with MFCs to evaluate the stability of individual components when operating as a hybrid system [54], leading to an increase of overall energy production and more complete wastewater treatment.

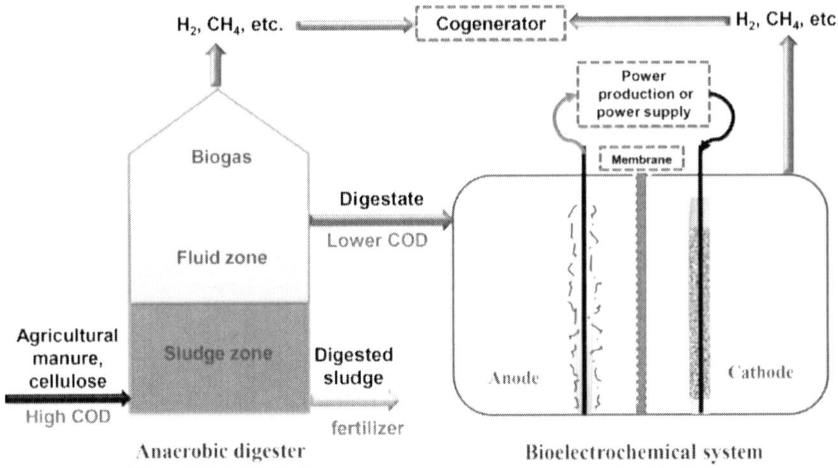

Figure 2. BESs integrated with AD as a post-treatment technology.

BES can help recover nutrients such as ammonia from digester effluent. A high concentration of ammonia from manure and/or produced during the degradation of nitrogenous components (e.g., proteins, urea) will inhibit microorganisms involved in anaerobic digestion [60], thereby affecting the efficiency and stability of the process. Through integrating BES as a side treatment unit that recycles the digester liquid, ammonia can be recovered in either MFCs or MECs. In BES, to maintain charge neutrality, the flux of electrons caused by external power supply needs to be compensated by movement of cations. As a result, ammonium ions in an anode chamber will migrate through a CEM to a cathode [61], where it will be converted to ammonia gas due to the enhanced pH and then ammonia can be recovered by a stripping method. The recovery is affected by the operational parameters, such as current density, pH, ionic strength, and nitrogen concentration. It was showed in an electrochemical system (ES) that NH_4^+ charge transfer efficiency and NH_4^+ flux were achieved 96% and 120 g N $m^{-2}\cdot day^{-1}$ at an energy input of 5 kWh·kg^{-1} N removed, respectively [62]. When being coupled with an upflow anaerobic sludge blanket (UASB) reactor to treat molasses, the ES can effectively control NH_3 toxicity for digester and reduce H_2S emission, due to simultaneous NH_4^+ extraction and oxidation of H_2S in the anode [63]. Oxidation of hydrogen sulfide has also been reported in BES studies [64,65,66,67]. BES in conjunction with anaerobic digestion would achieve similar effects as that of an ES; although BES may have lower performance due to low current generation, it does not require as much energy as the ES, thereby generating energy benefits. Recently, simultaneous ammonia recovery and electricity generation from ammonia-rich wastewater was demonstrated in a hybrid system consisting of a submersible MDC and a continuous stirred tank reactor, which could be applied to counteract ammonia inhibition during AD process [68,69].

A novel wastewater refinery concept has been proposed to recover more resources from waste streams but discharge less into environment [13]. In principle, wastewater with a low loading rate can be directly fed into an MFC, while the high-strength wastewater can be fermented in the AD system before flowing into the MFC system, for biogas production and for providing a suitable wastewater effluent [13]. The concept would also be applicable for treating agricultural waste, such as animal waste and cellulose biomass, achieving more efficient treatment and recovery of energy and other resources.

3. BES FOR FRESHWATER SUPPLY TO AGRICULTURE

Water scarcity has severely affected the agriculture in most countries in the Middle East and North Africa, and many other areas in the world [70]. Agriculture is responsible for the primary water consumption in many regions of the world, accounting for 70% of the total global water demand [71]. In addition, the world population and associated demand for food are expected to increase significantly by 2050 [72]. Therefore, alternative sources of freshwater from seawater or brackish water desalination, and wastewater reclamation and reuse are becoming increasingly important in the future [73,74].

Desalination is an effective approach for producing high quality water, especially in those areas where brackish water and seawater are readily available but freshwater sources are limited [75]. The salt concentration of brackish water is between that of freshwater and seawater. Thus, brackish water desalination is promising as an alternative approach to increase the freshwater supply for drinking, irrigation and other purposes [76]. Mature desalination technologies such as thermal desalination, reverse osmosis (RO), and electrodialysis (ED) are typically energy-intensive and result in significant operating costs, high water prices and potential environmental impacts [73,75,77]. Renewable energy sources such as solar and wind energy have been applied to drive the desalination systems but the capital and operating costs are still high [77]. These drawbacks associated with traditional desalination technologies have implied a need for developing new desalination technologies with economic, energy and environmental benefits [78].

3.1. MDCs for Saline Water Desalination

MDCs have gained great attention as a technology for sustainable wastewater treatment and low-cost desalination [79]. MDCs are derived from MFCs by placing AEM and CEM between anode and cathode, creating a middle chamber for water desalination [80] (Figure 3). To maintain electroneutrality, the electric potential gradient generated by exoelectrogenic bacteria drives cations and anions in the saline solution to migrate through CEM and AEM into the cathode chamber and anode chamber, respectively [81], thereby achieving desalination. The proof-of concept of MDC was firstly proposed by Cao *et al.* [7], and the technology has been advanced through both fundamental research and system development [7,82,83]. Because of the low desalination rate of MDC [84], two potential application niches have been identified. First, MDCs can be applied as a pre-desalination process, resulting in significant energy saving in downstream desalination processes [85]; and second, MDCs will be more suitable for desalinating brackish water rather than seawater, achieving a sound removal efficiency with shortened desalination time [78].

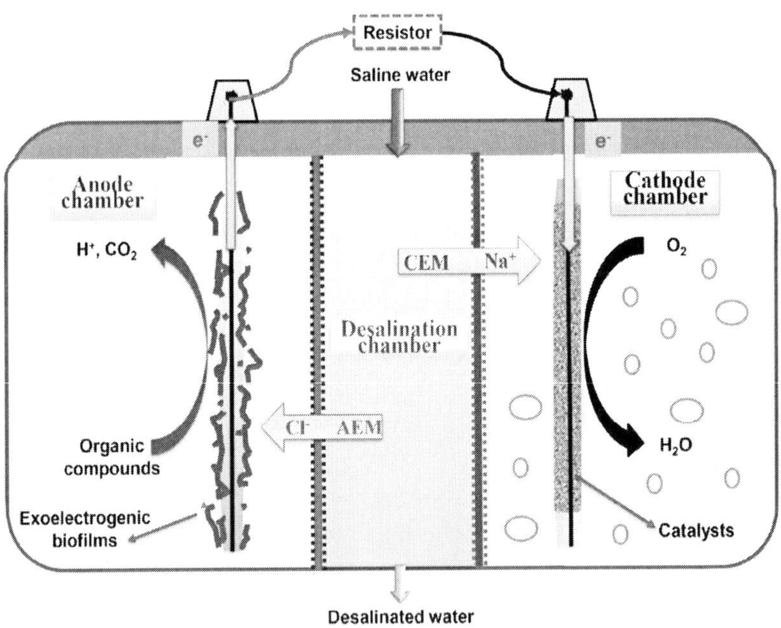

Figure 3. Schematic of a microbial desalination cell.

In many arid regions, brackish water is the main source of water supply [86]. Freshwater scarcity has forced farmers to irrigate crops with brackish water from shallow underground, which may relieve the drought crisis, but can cause the specific ion toxicity to plants and increase the risk of soil salinization [87]. For example, the salt content at different soil depths (upper 1 m soil layer) was significantly increased when brackish water with a salt content of 3.0–5.0 $g \cdot L^{-1}$ was used for irrigation during the two growing seasons [87]. Consequently, high soil water salinity can further affect water uptake by crops due to high osmotic potentials. The recommended salinity for irrigation water has been limited up to 450 $mg \cdot L^{-1}$ of total dissolved solids (TDS) to reduce negative impacts on crops [88].

MDCs have the potential to desalinate brackish water and produce water that meets the irrigation requirement. This is demonstrated in a recent study, in which an MDC fed with three different types of brackish water achieved satisfactory desalination at a suitable hydraulic retention time (HRT) [86]. This MDC decreased the conductivity of the brackish water containing 9.83 $mS \cdot cm^{-1}$ to 0.41 $mS \cdot cm^{-1}$, which met the non-restricted standard for agricultural use [86,89]. The concentration of Na^+ in the desalinated water is a key parameter for assessing the irrigation suitability due to its strong influence on water infiltration and soil aeration [89]. The sodium adsorption ratio (SAR, the ratio of Na^+ content relative to Ca^{2+} and Mg^{2+} contents) has been used to evaluate the potential effects of sodium on crop growth and yield [90]. In the previously mentioned MDC desalination at a HRT of 1.7 d, SAR fell into the range of "slight to moderate restriction on use" for a brackish water sample (TDS = 1.1 $g \cdot L^{-1}$) that had TDS reduced to 110 $mg \cdot L^{-1}$, slightly higher than that of the local tap water (90 $mg \cdot L^{-1}$ TDS) [86]. Furthermore, major ionic species were also effectively removed from this brackish water sample, with final concentrations at comparable levels to those in the tap water (Figure 4). These results have demonstrated that MDCs can reduce the salinity of brackish water by prolonging HRT and generate quality effluent for agricultural irrigation. Future research may focus on scale-up of MDCs and system optimization to further investigate their technical and economical feasibilities for practical application in agriculture.

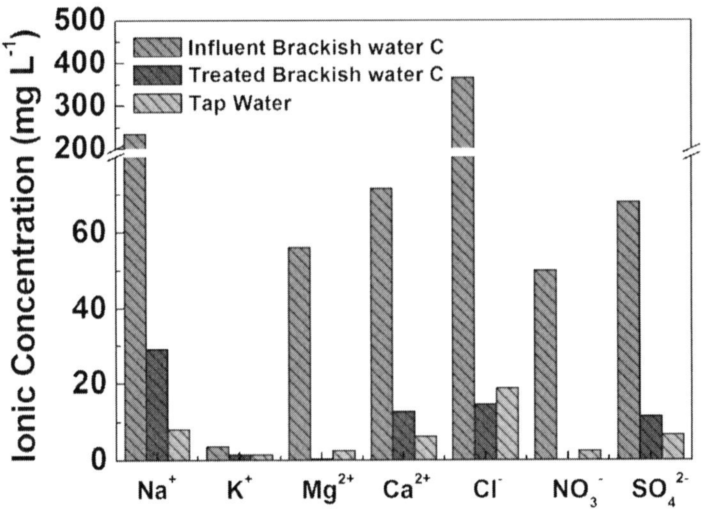

Figure 4. Comparison of individual ion concentrations between the influent/treated brackish water and tap water sampled at Virginia Tech. Reproduced with permission from reference [86].

3.2. BES Integrated with Membrane Filtration for Wastewater Reclamations

Reclaimed wastewater has been widely applied for various purposes [74,91,92]. The application of reclaimed wastewater for agriculture irrigation is a common practice worldwide [93], because of the benefits such as conserving freshwater, saving fertilizers, and eliminating pollutants and nutrients discharging to water bodies [94,95]. However, long-term irrigation with reclaimed wastewater may lead to the changes of soil properties and accumulation of contaminants (e.g., organic matters, heavy metals), which consequently degrade soil quality and impact food safety [96]. Table 3 shows the reclaimed water quality criteria for agricultural irrigation regulated by U.S. EPA.

To achieve a high quality effluent, various membrane separation processes, such as microfiltration (MF), ultrafiltration (UF), nanofiltration (NF), reverse osmosis (RO), *etc.* are adopted in wastewater treatment and reuse, and they are able to eliminate suspended solids (SS), protozoa, bacteria, and even virus [98]. More information regarding membrane technologies for water supply in agriculture can be found in a recent review [99]. Among those membrane processes, membrane bioreactors (MBR) have been applied in wastewater treatment for reuse because of both biological

Table 3. Summary of U.S. EPA guidelines for water reuse for agricultural irrigation [97].

Agricultural Reuse Description	Treatment	Reclaimed Water Quality
• Non-processed food crops • Any crop consumed raw by human	Secondary Filtration Disinfection	• pH = 6–9 • ≤ 10 mg·L^{-1} BOD • ≤ 2 NTU • No detectable fecal coliforms/100 mL • ≥ 1 mg·L^{-1} residual chlorine *
• Processed food crops • Crops which are non-edible by humans, such as fodder, fiber, pasture, *etc.*	Secondary Disinfection	• pH = 6–9 • ≤ 30 mg·L^{-1} BOD • ≤ 30 mg·L^{-1} TSS • ≤ 200 fecal coliforms/100 mL • ≥ 1 mg·L^{-1} residual chlorine *

* A minimum contact time of 30 min.

treatment and physical separation, providing a consistent and high quality effluent for agricultural irrigation to save freshwater resources [99,100]. MBR technology can be integrated with BES to form a new system [101], for example a membrane bioelectrochemical reactor (MBER) can accomplish both direct electricity generation and membrane filtration [102]. To form an MBER, hollow-fiber membranes (HFM) were installed into the anode chamber of a tubular MFC, and membrane fouling was observed to be a key issue especially when operating at high organic loading rates and/or high water flux conditions [103]. To reduce fouling, a fluidized bed MBER was designed by adding granular activated carbon (GAC) in the anode chamber, which significantly reduced membrane fouling and achieved satisfactory removal efficiency of contaminants [104]. This MBER was coupled with an MFC for treating an actual industrial wastewater, showing an exceptional removal performance (Figure 5), and in this system, the MFC was observed as the major process responsible for contaminants removal and energy recovery, while the MBER functioned as post-treatment to obtain a high quality effluent [104]. HFM could also be installed in the cathode of an MBER alleviating membrane fouling by aeration [105]. This modified MBER achieved excellent COD and SS removal (90% and ~2 NTU of turbidity, respectively), while total nitrogen removal was about 69% [105]. Disinfection process may be omitted because the bacteria are retained in the reactor by membranes. In addition, because the treated water is for crop irrigation, nutrients (N/P) do not need to be eliminated, and thus the remaining ammonium, nitrate and/or phosphate could be a valuable nutrient source for crops, which could reach an appropriate level to create a combined benefit of "fertigation" [106]. Therefore, BES integrated with membrane filtration could be an effective approach to supply freshwater for agriculture by wastewater reclamation.

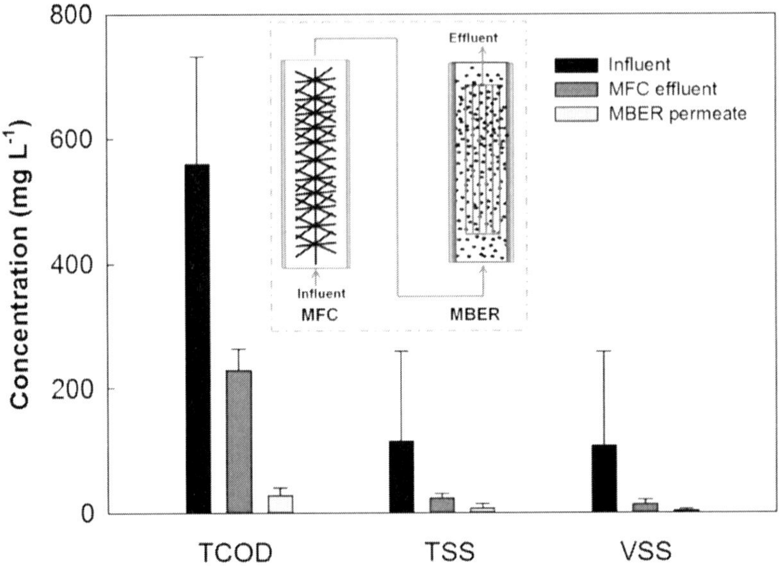

Figure 5. The contaminants removal from cheese wastewater by the coupled MFC–MBER system (see insert). TCOD: total COD; TSS: total suspended solids; VSS: volatile suspended solids. Reproduced with permission from reference [104].

4. BES FOR AGRICULTURAL MONITORING

Wireless sensor network (WSN) represents an important technology used to achieve precision agriculture. WSN can detect and monitor spatial and temporal parameters for decision making in agricultural farm management [107,108,109], thereby increasing efficiency and productivity while minimizing undesirable impacts on environment [110]. WSN devices are mainly powered by either batteries or solar energy [111]. The potential drawbacks associated with these power sources make the sensors unreliable. For example, replacing batteries in a remote location can be very inconvenient and costly, while the solar system is more expensive and highly depends on weather conditions [112].

During the past decade, sediment MFCs (SMFCs) have been extensively studied for contaminant remediation and power generation [113,114,115,116]. SMFCs, consisting of an anode electrode embedded in sediment and a cathode electrode suspended in the water above the anode electrode, can extract bioenergy from aquatic sediments through bioelectrochemical reactions, similar to that in a regular MFC [116,117]

(Figure 6). Unlike traditional MFCs, SMFCs do not require separators or ion exchange membranes because the oxygen gradient along the water column and sediment phases creates potential difference naturally (anaerobic/anoxic/aerobic zones) [117]. The electric power generated from SMFCs depends on the water and sediment conditions, the types of electrode material and cathode catalyst, and the distance between electrodes [117]. Dissolved oxygen (DO) is crucial for the cathodic reaction, and therefore SMFC is typically installed in shallow waters [118]. Previous studies have demonstrated that SMFCs can produce electricity and supply power to wireless sensors in both marine and fresh-water environments [113,119,120]. Capacitors have been adopted to accumulate energy generated from MFCs [121,122,123,124]. For examples, coupled with a power management system (PMS), electric energy extracted by SMFC was stored in ultracapacitors that consistently powered a remote sensor of 2.5 W deployed in the Palouse River, Pullman, WA, USA [125].

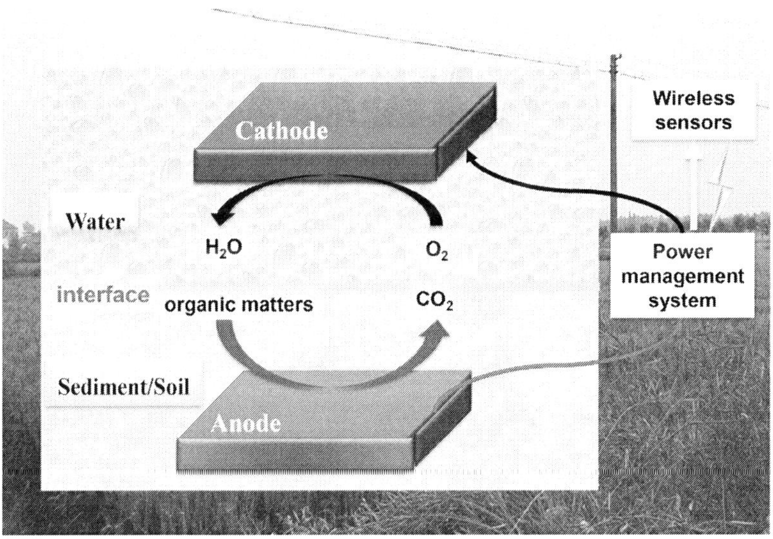

Figure 6. BES for powering wireless sensor for agricultural monitoring.

In precision agriculture, wireless sensors are deployed in fields to acquire micro-climatological data, such as temperature, humidity, sunlight, soil moisture content and wind speed, as well as to manage irrigation, fertilization, and pesticide [126,127,128]. The information obtained from sensors can help develop optimization strategies for crop production and save energy consumption, which is critical for achieving sustainable agriculture [129]. SMFCs may be served as an alternative power source for these wireless sensors, depending on their installation locations. They can be installed in wetlands, rivers or lakes near the farmland. To use the electricity, the output potentials must be boosted and operated by DC–DC converters and a PMS [119,120]. In the area where open water is not available, soil MFCs [130,131,132,133] or plant MFCs [134,135] may be applied. Essentially, they are analogous to SMFCs but oxidize organics in the soil under a low-moisture condition. In addition to the organics in soil/sediment, plants can also excrete organic matters as rhizodeposit, which can be utilized as substrates in MFCs [135]. For example, an MFC was installed in a rice paddy field during the rice-cropping season with graphite felt anode and cathode electrodes placed in the rice rhizosphere and the flooded water above the rhizosphere, respectively [136]. This study found that power generation from the MFC was sunlight dependent, and acetate (one of the major root-derived organic compounds) improved the electricity generation in the dark condition. A maximum power density of 6 $mW \cdot m^{-2}$ (anode area) was achieved in this MFC, with the anode dominant species identified as a specific bacterial population of *Natronocella acetinitrilica*, *Beijerinckiaceae* bacterium and *Rhizobiales* bacterium [136].

In summary, BES might become an effective approach to power wireless sensors used in agriculture for various purposes, such as acquisition of micro-climatological data in the field, management of irrigation, fertilization, and pesticide, monitoring the parameters of agricultural runoff, such as pH, DO, turbidity, conductivity, nutrients (e.g., NO_3^-), *etc.* Further research is needed to improve power generation from two aspects, including the exploration of highly efficient electrodes and the optimization of system design. The choice of adopting SMFCs as a power source in agriculture monitoring will be highly case-specific due to many factors, including the accessibility of water sources, the water level, the characteristics of sediment or soil (e.g., organic/moisture contents, permeability), the abundance and diversity of microorganism communities, and the availability of space for installation.

5. CONCLUSIONS

BES has great potential to be applied for promoting sustainable agriculture in the aspects of waste minimization, resource recovery, water supply, and agricultural monitoring. Despite a large amount of BES literature, the studies related to agriculture are limited. Thus, the interest in agriculture-driven BES research and development should be well recognized. Identification of proper application niches will be critical to BES development. Further studies should explore the BES performance with actual agricultural waste under non-laboratory conditions, system scaling up, and better assessment (e.g., LCA) of BES technology integrated with sustainable agriculture.

ACKNOWLEDGMENTS

This work was made possible by funding from VT College of Engineering Dean's Office Incentive Program, and by NPRP grant # 6-289-2-125 from the Qatar National Research Fund (a member of Qatar Foundation). The statements made herein are solely the responsibility of the authors.

REFERENCES

1. Lu, Z.; Chang, D.; Ma, J.; Huang, G.; Cai, L.; Zhang, L. Behavior of metal ions in bioelectrochemical systems: A review. *J. Power Sources* **2015**, *275*, 243–260
2. Logan, B.E. Exoelectrogenic bacteria that power microbial fuel cells. *Nat. Rev. Microbiol.* **2009**, *7*, 375–381
3. Torres, C.I.; Marcus, A.K.; Rittmann, B.E. Kinetics of consumption of fermentation products by anode-respiring bacteria. *Appl. Microbiol. Biotechnol.* **2007**, *77*, 689–697
4. Yang, Y.; Xu, M.; Guo, J.; Sun, G. Bacterial extracellular electron transfer in bioelectrochemical systems. *Process Biochem.* **2012**, *47*, 1707–1714
5. Liu, H.; Logan, B.E. Electricity generation using an air-cathode single chamber microbial fuel cell in the presence and absence of a proton exchange membrane. *Environ. Sci. Technol.* **2004**, *38*, 4040–4046
6. Call, D.; Logan, B.E. Hydrogen production in a single chamber microbial electrolysis cell lacking a membrane. *Environ. Sci. Technol.* **2008**, *42*, 3401–3406
7. Cao, X.; Huang, X.; Liang, P.; Xiao, K.; Zhou, Y.; Zhang, X.; Logan, B.E. A new method for water desalination using microbial desalination cells. *Environ. Sci. Technol.* **2009**, *43*, 7148–7152
8. Rabaey, K.; Rozendal, R.A. Microbial electrosynthesis—Revisiting the electrical route for microbial production. *Nat. Rev. Microbiol.* **2010**, *8*, 706–716
9. Logan, B.E.; Call, D.; Cheng, S.; Hamelers, H.V.M.; Sleutels, T.; Jeremiasse, A.W.; Rozendal, R.A. Microbial electrolysis cells for high yield hydrogen gas production from organic matter. *Environ. Sci. Technol.* **2008**, *42*, 8630–8640
10. Cheng, S.A.; Xing, D.F.; Call, D.F.; Logan, B.E. Direct biological conversion of electrical current into methane by electromethanogenesis. *Environ. Sci. Technol.* **2009**, *43*, 3953–3958
11. Logan, B.E.; Rabaey, K. Conversion of wastes into bioelectricity and chemicals by using microbial electrochemical technologies. *Science* **2012**, *337*, 686–690
12. Tender, L.M.; Gray, S.A.; Groveman, E.; Lowy, D.A.; Kauffman, P.; Melhado, J.; Tyce, R.C.; Flynn, D.; Petrecca, R.; Dobarro, J. The first demonstration of a microbial fuel cell as a viable power supply: Powering a meteorological buoy. *J. Power Sources* **2008**, *179*, 571–575
13. Li, W.W.; Yu, H.Q.; He, Z. Towards sustainable wastewater treatment by using microbial fuel cells-centered technologies. *Energy Environ. Sci.* **2014**, *7*, 911–924
14. Pant, D.; Singh, A.; van Bogaert, G.; Olsen, S.I.; Nigam, P.S.; Diels, L.; Vanbroekhoven, K. Bioelectrochemical systems (BES) for sustainable energy production and product recovery from organic wastes and industrial wastewaters. *Rsc Adv.* **2012**, *2*, 1248–1263

15. Min, B.; Kim, J.; Oh, S.; Regan, J.M.; Logan, B.E. Electricity generation from swine wastewater using microbial fuel cells. *Water Res.* **2005**, *39*, 4961–4968
16. Hanson, J.; Liebig, M.; Merrill, S.; Tanaka, D.; Krupinsky, J.; Stott, D. Dynamic cropping systems. *Agron. J.* **2007**, *99*, 939–943
17. Öborn, I.; Edwards, A.; Witter, E.; Oenema, O.; Ivarsson, K.; Withers, P.; Nilsson, S.; Stinzing, A.R. Element balances as a tool for sustainable nutrient management: A critical appraisal of their merits and limitations within an agronomic and environmental context. *Eur. J. Agron.* **2003**, *20*, 211–225
18. Dordas, C. Role of nutrients in controlling plant diseases in sustainable agriculture. A review. *Agron. Sustain. Dev.* **2008**, *28*, 33–46
19. Cherry, D.C.; Huggins, B.; Gilmore, K. Children's health in the rural environment. *Pediatr. Clin. N. Am.* **2007**, *54*, 121–133
20. Chadwick, D. Emissions of ammonia, nitrous oxide and methane from cattle manure heaps: Effect of compaction and covering. *Atmos. Environ.* **2005**, *39*, 787–799
21. Lim, S.J.; Park, W.; Kim, T.-H.; Shin, I.H. Swine wastewater treatment using a unique sequence of ion exchange membranes and bioelectrochemical system. *Bioresour. Technol.* **2012**, *118*, 163–169
22. Ieropoulos, I.; Greenman, J.; Melhuish, C. Microbial fuel cells based on carbon veil electrodes: Stack configuration and scalability. *Int. J. Energy Res.* **2008**, *32*, 1228–1240
23. Ringeisen, B.R.; Henderson, E.; Wu, P.K.; Pietron, J.; Ray, R.; Little, B.; Biffinger, J.C.; Jones-Meehan, J.M. High power density from a miniature microbial fuel cell using shewanella oneidensis dsp10. *Environ. Sci. Technol.* **2006**, *40*, 2629–2634
24. Lee, Y.; Nirmalakhandan, N. Electricity production in membrane-less microbial fuel cell fed with livestock organic solid waste. *Bioresour. Technol.* **2011**, *102*, 5831–5835
25. Wang, X.; Tang, J.; Cui, J.; Liu, Q.; Giesy, J.P.; Hecker, M. Synergy of electricity generation and waste disposal in solid-state microbial fuel cell (MFC) of cow manure composting. *Int. J. Electrochem. Sci.* **2014**, *9*, 3144–3157
26. Yokoyama, H.; Ohmori, H.; Ishida, M.; Waki, M.; Tanaka, Y. Treatment of cow-waste slurry by a microbial fuel cell and the properties of the treated slurry as a liquid manure. *Anim. Sci. J.* **2006**, *77*, 634–638
27. Scott, K.; Murano, C. A study of a microbial fuel cell battery using manure sludge waste. *J. Chem. Technol. Biotechnol.* **2007**, *82*, 809–817
28. Kiely, P.D.; Cusick, R.; Call, D.F.; Selembo, P.A.; Regan, J.M.; Logan, B.E. Anode microbial communities produced by changing from microbial fuel cell to microbial electrolysis cell operation using two different wastewaters. *Bioresour. Technol.* **2011**, *102*, 388–394
29. Kim, J.R.; Dec, J.; Bruns, M.A.; Logan, B.E. Removal of odors from swine wastewater by using microbial fuel cells. *Appl. Environ. Microbiol.* **2008**, *74*, 2540–2543
30. Ryu, J.H.; Lee, H.L.; Lee, Y.P.; Kim, T.S.; Kim, M.K.; Anh, D.T.N.; Tran, H.T.; Ahn, D.H. Simultaneous carbon and nitrogen removal from piggery wastewater using loop configuration microbial fuel cell. *Process Biochem.* **2013**, *48*, 1080–1085
31. Nam, J.-Y.; Kim, H.-W.; Shin, H.-S. Ammonia inhibition of electricity generation in single-chambered microbial fuel cells. *J. Power Sources* **2010**, *195*, 6428–643
32. Wagner, R.C.; Regan, J.M.; Oh, S.-E.; Zuo, Y.; Logan, B.E. Hydrogen and methane production from swine wastewater using microbial electrolysis cells. *Water Res.* **2009**, *43*, 1480–1488
33. Ichihashi, O.; Hirooka, K. Removal and recovery of phosphorus as struvite from swine wastewater using microbial fuel cell. *Bioresour. Technol.* **2012**, *114*, 303–307
34. Kim, J.R.; Zuo, Y.; Regan, J.M.; Logan, B.E. Analysis of ammonia loss mechanisms in microbial fuel cells treating animal wastewater. *Biotechnol. Bioeng.* **2008**, *99*, 1120–1127
35. Rittmann, B.E.; McCarty, P.L. *Environmental Biotechnology: Principles and Applications*; McGraw-Hill Education: Boston, MA, USA, 2001

36. Cheng, S.A.; Kiely, P.; Logan, B.E. Pre-acclimation of a wastewater inoculum to cellulose in an aqueous-cathode mec improves power generation in air-cathode mfcs. *Bioresour. Technol.* **2011**, *102*, 367–371
37. Bridgwater, T. Biomass for energy. *J. Sci. Food Agric.* **2006**, *86*, 1755–1768
38. Lynd, L.R. Overview and evaluation of fuel ethanol from cellulosic biomass: Technology, economics, the environment, and policy. *Annu. Rev. Energy Environ.* **1996**, *21*, 403–465
39. Rezaei, F.; Xing, D.; Wagner, R.; Regan, J.M.; Richard, T.L.; Logan, B.E. Simultaneous cellulose degradation and electricity production by enterobacter cloacae in a microbial fuel cell. *Appl. Environ. Microbiol.* **2009**, *75*, 3673–3678
40. Rezaei, F.; Richard, T.L.; Logan, B.E. Enzymatic hydrolysis of cellulose coupled with electricity generation in a microbial fuel cell. *Biotechnol. Bioeng.* **2008**, *101*, 1163–1169
41. Niessen, J.; Schröder, U.; Harnisch, F.; Scholz, F. Gaining electricity from *in situ* oxidation of hydrogen produced by fermentative cellulose degradation. *Lett. Appl. Microbiol.* **2005**, *41*, 286–290
42. Ren, Z.; Ward, T.E.; Regan, J.M. Electricity production from cellulose in a microbial fuel cell using a defined binary culture. *Environ. Sci. Technol.* **2007**, *41*, 4781–4786
43. Hassan, S.H.; Kim, Y.S.; Oh, S.-E. Power generation from cellulose using mixed and pure cultures of cellulose-degrading bacteria in a microbial fuel cell. *Enzym. Microb. Technol.* **2012**, *51*, 269–273
44. Krause, D.O.; Denman, S.E.; Mackie, R.I.; Morrison, M.; Rae, A.L.; Attwood, G.T.; McSweeney, C.S. Opportunities to improve fiber degradation in the rumen: Microbiology, ecology, and genomics. *FEMS Microbiol. Rev.* **2003**, *27*, 663–693
45. Rismani-Yazdi, H.; Christy, A.D.; Dehority, B.A.; Morrison, M.; Yu, Z.; Tuovinen, O.H. Electricity generation from cellulose by rumen microorganisms in microbial fuel cells. *Biotechnol. Bioeng.* **2007**, *97*, 1398–1407
46. Hassan, S.H.; El-Rab, S.M.G.; Rahimnejad, M.; Ghasemi, M.; Joo, J.-H.; Sik-Ok, Y.; Kim, I.S.; Oh, S.-E. Electricity generation from rice straw using a microbial fuel cell. *Int. J. Hydrog. Energy* **2014**, *39*, 9490–9496
47. Reese, E. Polysaccharases and the hydrolysis of insoluble substrates. In *Biological Transformation of Wood by Microorganisms*; Springer-Verlag: Berlin/Heidelberg, Germany, 1975; pp. 165–181
48. Takahashi, M.; Takahashi, H.; Nakano, Y.; Konishi, T.; Terauchi, R.; Takeda, T. Characterization of a cellobiohydrolase (MoCel6A) produced by magnaporthe oryzae. *Appl. Environ. Microbiol.* **2010**, *76*, 6583–6590
49. Sørensen, A.; Lübeck, M.; Lübeck, P.S.; Ahring, B.K. Fungal beta-glucosidases: A bottleneck in industrial use of lignocellulosic materials. *Biomolecules* **2013**, *3*, 612–631
50. Sun, Y.; Cheng, J. Hydrolysis of lignocellulosic materials for ethanol production: A review. *Bioresour. Technol.***2002**, *83*, 1–11
51. Lalaurette, E.; Thammannagowda, S.; Mohagheghi, A.; Maness, P.C.; Logan, B.E. Hydrogen production from cellulose in a two-stage process combining fermentation and electrohydrogenesis. *Int. J. Hydrog. Energy* **2009**, *34*, 6201–6210
52. Wang, A.J.; Sun, D.; Cao, G.L.; Wang, H.Y.; Ren, N.Q.; Wu, W.M.; Logan, B.E. Integrated hydrogen production process from cellulose by combining dark fermentation, microbial fuel cells, and a microbial electrolysis cell. *Bioresour. Technol.* **2011**, *102*, 4137–4143
53. Angenent, L.T.; Karim, K.; Al-Dahhan, M.H.; Wrenn, B.A.; Domíguez-Espinosa, R. Production of bioenergy and biochemicals from industrial and agricultural wastewater. *Trends Biotechnol.* **2004**, *22*, 477–485
54. Weld, R.J.; Singh, R. Functional stability of a hybrid anaerobic digester/microbial fuel cell system treating municipal wastewater. *Bioresour. Technol.* **2011**, *102*, 842–847
55. Kim, J.R.; Premier, G.C.; Hawkes, F.R.; Rodríguez, J.; Dinsdale, R.M.; Guwy, A.J. Modular tubular microbial fuel cells for energy recovery during sucrose wastewater treatment at low organic loading rate. *Bioresour. Technol.* **2010**, *101*, 1190–1198

56. Rozendal, R.A.; Leone, E.; Keller, J.; Rabaey, K. Efficient hydrogen peroxide generation from organic matter in a bioelectrochemical system. *Electrochem. Commun.* **2009**, *11*, 1752–1755
57. Rabaey, K.; Butzer, S.; Brown, S.; Keller, J.; Rozendal, R.A. High current generation coupled to caustic production using a lamellar bioelectrochemical system. *Environ. Sci. Technol.* **2010**, *44*, 4315–4321
58. Verstraete, W.; Morgan-Sagastume, F.; Aiyuk, S.; Waweru, M.; Rabaey, K.; Lissens, G. Anaerobic digestion as a core technology in sustainable management of organic matter. *Water Sci. Technol.* **2005**, *52*, 59–66
59. Pham, T.; Rabaey, K.; Aelterman, P.; Clauwaert, P.; de Schamphelaire, L.; Boon, N.; Verstraete, W. Microbial fuel cells in relation to conventional anaerobic digestion technology. *Eng. Life Sci.* **2006**, *6*, 285–292
60. Sung, S.; Liu, T. Ammonia inhibition on thermophilic anaerobic digestion. *Chemosphere* **2003**, *53*, 43–52
61. Cord-Ruwisch, R.; Law, Y.; Cheng, K.Y. Ammonium as a sustainable proton shuttle in bioelectrochemical systems. *Bioresour. Technol.* **2011**, *102*, 9691–9696
62. Desloover, J.; Abate Woldeyohannis, A.; Verstraete, W.; Boon, N.; Rabaey, K. Electrochemical resource recovery from digestate to prevent ammonia toxicity during anaerobic digestion. *Environ. Sci. Technol.* **2012**, *46*, 12209–12216
63. Desloover, J.; de Vrieze, J.; de Vijver, M.V.; Mortelmans, J.; Rozendal, R.; Rabaey, K. Electrochemical nutrient recovery enables ammonia toxicity control and biogas desulfurization in anaerobic digestion. *Environ. Sci. Technol.* **2015**, *49*, 948–955
64. Rabaey, K.; van de Sompel, K.; Maignien, L.; Boon, N.; Aelterman, P.; Clauwaert, P.; de Schamphelaire, L.; Pham, H.T.; Vermeulen, J.; Verhaege, M. Microbial fuel cells for sulfide removal. *Environ. Sci. Technol.* **2006**, *40*, 5218–5224
65. Sun, M.; Mu, Z.-X.; Chen, Y.-P.; Sheng, G.-P.; Liu, X.-W.; Chen, Y.-Z.; Zhao, Y.; Wang, H.-L.; Yu, H.-Q.; Wei, L.; *et al.* Microbe-assisted sulfide oxidation in the anode of a microbial fuel cell. *Environ. Sci. Technol.* **2009**, *43*, 3372–3377
66. Sun, M.; Tong, Z.-H.; Sheng, G.-P.; Chen, Y.-Z.; Zhang, F.; Mu, Z.-X.; Wang, H.-L.; Zeng, R.J.; Liu, X.-W.; Yu, H.-Q.; *et al.* Microbial communities involved in electricity generation from sulfide oxidation in a microbial fuel cell. *Biosens. Bioelectron.* **2010**, *26*, 470–476
67. Ieropoulos, I.; Greenman, J.; Melhuish, C.; Hart, J. Energy accumulation and improved performance in microbial fuel cells. *J. Power Sources* **2005**, *145*, 253–256
68. Zhang, Y.; Angelidaki, I. Submersible microbial desalination cell for simultaneous ammonia recovery and electricity production from anaerobic reactors containing high levels of ammonia. *Bioresour. Technol.* **2015**, *177*, 233–239
69. Zhang, Y.; Angelidaki, I. Counteracting ammonia inhibition during anaerobic digestion by recovery using submersible microbial desalination cell. *Biotechnol. Bioeng.* **2015**, *112*, 1478–1482
70. Ghermandi, A.; Messalem, R. The advantages of nf desalination of brackish water for sustainable irrigation: The case of the arava valley in israel. *Desalin. Water Treat.* **2009**, *10*, 101–107
71. Assessment, M.E. *Ecosystems and Human Well-Being: Current State and Trends*; Island Press: Washington, DC, USA, 2005
72. UNESCO. *The 3rd United Nations World Water Development Report: Water in a Changing World*; United Nations Educational, Scientific, and Cultural Organization: Paris, France, 2009
73. Shannon, M.A.; Bohn, P.W.; Elimelech, M.; Georgiadis, J.G.; Mariñas, B.J.; Mayes, A.M. Science and technology for water purification in the coming decades. *Nature* **2008**, *452*, 301–310
74. Miller, G.W. Integrated concepts in water reuse: Managing global water needs. *Desalination* **2006**, *187*, 65–75
75. Elimelech, M.; Phillip, W.A. The future of seawater desalination: Energy, technology, and the environment. *Science* **2011**, *333*, 712–717

76. Redondo, J. Brackish-, sea-and wastewater desalination. *Desalination* **2001**, *138*, 29–40
77. Mathioulakis, E.; Belessiotis, V.; Delyannis, E. Desalination by using alternative energy: Review and state-of-the-art. *Desalination* **2007**, *203*, 346–365
78. Zhang, B.; He, Z. Energy production, use and saving in a bioelectrochemical desalination system. *Rsc Adv.***2012**, *2*, 10673–10679
79. Jacobson, K.S.; Drew, D.M.; He, Z. Use of a liter-scale microbial desalination cell as a platform to study bioelectrochemical desalination with salt solution or artificial seawater. *Environ. Sci. Technol.* **2011**, *45*, 4652–4657
80. Wang, H.; Ren, Z.J. A comprehensive review of microbial electrochemical systems as a platform technology. *Biotechnol. Adv.* **2013**, *31*, 1796–1807
81. Zhang, B.; He, Z. Improving water desalination by hydraulically coupling an osmotic microbial fuel cell with a microbial desalination cell. *J. Membr. Sci.* **2013**, *441*, 18–24
82. Jacobson, K.S.; Drew, D.M.; He, Z. Efficient salt removal in a continuously operated upflow microbial desalination cell with an air cathode. *Bioresour. Technol.* **2011**, *102*, 376–380
83. Brastad, K.S.; He, Z. Water softening using microbial desalination cell technology. *Desalination* **2013**, *309*, 32–37
84. Chen, X.; Xia, X.; Liang, P.; Cao, X.; Sun, H.; Huang, X. Stacked microbial desalination cells to enhance water desalination efficiency. *Environ. Sci. Technol.* **2011**, *45*, 2465–2470
85. Mehanna, M.; Saito, T.; Yan, J.; Hickner, M.; Cao, X.; Huang, X.; Logan, B.E. Using microbial desalination cells to reduce water salinity prior to reverse osmosis. *Energy Environ. Sci.* **2010**, *3*, 1114–1120
86. Ping, Q.; Huang, Z.; Dosoretz, C.; He, Z. Integrated experimental investigation and mathematical modeling of brackish water desalination and wastewater treatment in microbial desalination cells. *Water Res.* **2015**, *77*, 13–23
87. Pang, H.-C.; Li, Y.-Y.; Yang, J.-S.; Liang, Y.-S. Effect of brackish water irrigation and straw mulching on soil salinity and crop yields under monsoonal climatic conditions. *Agric. Water Manag.* **2010**, *97*, 1971–1977
88. Ayers, R.S.; Westcot, D.W. *Water Quality for Agriculture*; Fao irrigation and drainage paper 29 rev. 1; Food and Agricultural Organization: Rome, Italy, 1985
89. Scianna, J.; Pick, T.; Logar, R. Plant materials technical note number MT-62: Determining the suitability of salt-affected water and soil for tree and shrub plantings. Available online: http://www.nrcs.usda.gov/wps/portal/nrcs/detail/mt/homeowner/?cid=nrcs2144p2012_057692 (accessed on 28 April 2015)
90. Gardiner, D.T.; Miller, R.W. *Soils in our Environment*; Pearson/Prentice Hall: Upper Saddle River, New Jersey, 2008
91. Pedersen, J.A., Yeager, M.A., Suffet, I. Xenobiotic organic compounds in runoff from fields irrigated with treated wastewater. *J. Agric. Food Chem.* **2003**, *51*, 1360–1372
92. Levine, A.D.; Asano, T. Peer reviewed: Recovering sustainable water from wastewater. *Environ. Sci. Technol.***2004**, *38*, 201A–208A
93. Angelakis, A.; Do Monte, M.M.; Bontoux, L.; Asano, T. The status of wastewater reuse practice in the mediterranean basin: Need for guidelines. *Water Res.* **1999**, *33*, 2201–2217
94. Chen, W.; Lu, S.; Peng, C.; Jiao, W.; Wang, M. Accumulation of cd in agricultural soil under long-term reclaimed water irrigation. *Environ. Pollut.* **2013**, *178*, 294–299
95. Oron, G.; Goemans, M.; Manor, Y.; Feyen, J. Poliovirus distribution in the soil-plant system under reuse of secondary wastewater. *Water Res.* **1995**, *29*, 1069–1078
96. Xu, J.; Wu, L.; Chang, A.C.; Zhang, Y. Impact of long-term reclaimed wastewater irrigation on agricultural soils: A preliminary assessment. *J. Hazard. Mater.* **2010**, *183*, 780–786
97. USEPA. *Guidelines for Water Reuse, 2012*; U.S. Environmental Protection Agency: Washington, DC, USA, 2012. Available online: http://nepis.epa.gov/Adobe/PDF/P2100FS2017K.pdf (accessed on 28 April 2015)

98. Howell, J.A. Future of membranes and membrane reactors in green technologies and for water reuse. *Desalination* **2004**, *162*, 1–11
99. Quist-Jensen, C.A.; Macedonio, F.; Drioli, E. Membrane technology for water production in agriculture: Desalination and wastewater reuse. *Desalination* **2015**, *364*, 17–32
100. Melin, T.; Jefferson, B.; Bixio, D.; Thoeye, C.; de Wilde, W.; de Koning, J.; van der Graaf, J.; Wintgens, T. Membrane bioreactor technology for wastewater treatment and reuse. *Desalination* **2006**, *187*, 271–282
101. Yuan, H.; He, Z. Integrating membrane filtration into bioelectrochemical systems as next generation energy-efficient wastewater treatment technologies for water reclamation: A review. *Bioresour. Technol.* **2015**. in press
102. Wang, Y.-P.; Liu, X.-W.; Li, W.-W.; Li, F.; Wang, Y.-K.; Sheng, G.-P.; Zeng, R.J.; Yu, H.-Q. A microbial fuel cell–membrane bioreactor integrated system for cost-effective wastewater treatment. *Appl. Energy* **2012**, *98*, 230–235
103. Ge, Z.; Ping, Q.Y.; He, Z. Hollow-fiber membrane bioelectrochemical reactor for domestic wastewater treatment. *J. Chem. Technol. Biotechnol.* **2013**, *88*, 1584–1590
104. Li, J.; Ge, Z.; He, Z. A fluidized bed membrane bioelectrochemical reactor for energy-efficient wastewater treatment. *Bioresour. Technol.* **2014**, *167*, 310–315
105. Li, J.; Ge, Z.; He, Z. Advancing membrane bioelectrochemical reactor (MBER) with hollow-fiber membranes installed in the cathode compartment. *J. Chem. Technol. Biotechnol.* **2014**, *89*, 1330–1336
106. Hoover, L.A.; Phillip, W.A.; Tiraferri, A.; Yip, N.Y.; Elimelech, M. Forward with osmosis: Emerging applications for greater sustainability. *Environ. Sci. Technol.* **2011**, *45*, 9824–9830
107. Coates, R.W.; Delwiche, M.J.; Broad, A.; Holler, M. Wireless sensor network with irrigation valve control. *Comput. Electron. Agric.* **2013**, *96*, 13–22
108. Zhang, R.-B.; Guo, J.-J.; Zhang, L.; Zhang, Y.-C.; Wang, L.-H.; Wang, Q. A calibration method of detecting soil water content based on the information-sharing in wireless sensor network. *Comput. Electron. Agric.* **2011**, *76*, 161–168
109. Dong, X.; Vuran, M.C.; Irmak, S. Autonomous precision agriculture through integration of wireless underground sensor networks with center pivot irrigation systems. *Ad Hoc Netw.* **2013**, *11*, 1975–1987
110. Srbinovska, M.; Gavrovski, C.; Dimcev, V.; Krkoleva, A.; Borozan, V. Environmental parameters monitoring in precision agriculture using wireless sensor networks. *J. Clean. Prod.* **2015**, *88*, 297–307
111. Akbari, S. Energy Harvesting for Wireless Sensor Networks Review. Proceedings of the 2014 Federated Conference on Computer Science and Information Systems (FedCSIS), Warsaw, Poland, 7–10 September 2014; Available online: http://ieeexplore.ieee.org/xpl/articleDetails.jsp?reload=true&arnumber=6933124 (access ed on 29 April 2015)
112. Zhang, F.; Tian, L.; He, Z. Powering a wireless temperature sensor using sediment microbial fuel cells with vertical arrangement of electrodes. *J. Power Sources* **2011**, *196*, 9568–9573
113. De Schamphelaire, L.; Rabaey, K.; Boeckx, P.; Boon, N.; Verstraete, W. Outlook for benefits of sediment microbial fuel cells with two bio-electrodes. *Microb. Biotechnol.* **2008**, *1*, 446–462
114. Bond, D.R.; Holmes, D.E.; Tender, L.M.; Lovley, D.R. Electrode-reducing microorganisms that harvest energy from marine sediments. *Science* **2002**, *295*, 483–485
115. Tender, L.M.; Reimers, C.E.; Stecher, H.A.; Holmes, D.E.; Bond, D.R.; Lowy, D.A.; Pilobello, K.; Fertig, S.J.; Lovley, D.R. Harnessing microbially generated power on the seafloor. *Nat. Biotechnol.* **2002**, *20*, 821–825
116. Xu, B.; Ge, Z.; He, Z. Sediment microbial fuel cells for wastewater treatment: Challenges and opportunities. *Environ. Sci.: Water Res. Technol.* **2015**, *1*, 279–284

117. He, Z.; Shao, H.B.; Angenent, L.T. Increased power production from a sediment microbial fuel cell with a rotating cathode. *Biosens. Bioelectron.* **2007**, *22*, 3252–3255
118. Reimers, C.; Girguis, P.; Stecher, H.; Tender, L.; Ryckelynck, N.; Whaling, P. Microbial fuel cell energy from an ocean cold seep. *Geobiology* **2006**, *4*, 123–136
119. Shantaram, A.; Beyenal, H.; Raajan, R.; Veluchamy, A.; Lewandowski, Z. Wireless sensors powered by microbial fuel cells. *Environ. Sci. Technol.* **2005**, *39*, 5037–5042
120. Donovan, C.; Dewan, A.; Heo, D.; Beyenal, H. Batteryless, wireless sensor powered by a sediment microbial fuel cell. *Environ. Sci. Technol.* **2008**, *42*, 8591–8596
121. Ieropoulos, I.; Greenman, J.; Melhuish, C. Imitating metabolism: Energy autonomy in biologically inspired robots. In Proceedings of the 2nd International Symposium on Imitation of Animals and Artifacts, Aberystwyth, UK, 7–11 April 2003; pp. 191–194
122. Ieropoulos, I.; Greenman, J.; Melhuish, C.; Horsfield, I. Ecobot-III-a Robot with Guts. Proceedings of the Alife XII Conference, Odense, Denmark, 19–23 August 2010; pp. 733–740. Available online: https://mitpress.mit.edu/sites/default/files /titles/alife/0262290758chap131.pdf (accessed on 29 April 2015)
123. Ieropoulos, I.; Melhuish, C.; Greenman, J.; Horsfield, I. Ecobot-ii: An artificial agent with a natural metabolism. *J. Adv. Robot. Syst.* **2005**, *2*, 295–300
124. Wilkinson, S. "Gastrobots"—Benefits and challenges of microbial fuel cells in foodpowered robot applications. *Auton. Robot.* **2000**, *9*, 99–111
125. Donovan, C.; Dewan, A.; Peng, H.; Heo, D.; Beyenal, H. Power management system for a 2.5 w remote sensor powered by a sediment microbial fuel cell. *J. Power Sources* **2011**, *196*, 1171–1177
126. Selavo, L.; Wood, A.; Cao, Q.; Sookoor, T.; Liu, H.; Srinivasan, A.; Wu, Y.; Kang, W.; Stankovic, J.; Young, D. Luster: Wireless Sensor Network for Environmental Research. In Proceedings of the 5th International Conference on Embedded Networked Sensor Systems, 6–9 November 2007; ACM: Sydney, Australia; pp. 103–116
127. Baggio, A. Wireless Sensor Networks in Precision Agriculture. In Proceedings of the ACM Workshop on Real-World Wireless Sensor Networks (REALWSN 2005), Stockholm, Sweden, 20–21 June 2005.
128. Vellidis, G.; Tucker, M.; Perry, C.; Kvien, C.; Bednarz, C. A real-time wireless smart sensor array for scheduling irrigation. *Comput. Electron. Agric.* **2008**, *61*, 44–50
129. Hokazono, S.; Hayashi, K. Variability in environmental impacts during conversion from conventional to organic farming: A comparison among three rice production systems in japan. *J. Clean. Prod.* **2012**, *28*, 101–112
130. Huang, D.-Y.; Zhou, S.-G.; Chen, Q.; Zhao, B.; Yuan, Y.; Zhuang, L. Enhanced anaerobic degradation of organic pollutants in a soil microbial fuel cell. *Chem. Eng. J.* **2011**, *172*, 647–653
131. Li, X.; Wang, X.; Zhang, Y.; Cheng, L.; Liu, J.; Li, F.; Gao, B.; Zhou, Q. Extended petroleum hydrocarbon bioremediation in saline soil using pt-free multianodes microbial fuel cells. *RSC Adv.* **2014**, *4*, 59803–59808
132. Rodrigo, J.; Boltes, K.; Esteve-Nuñez, A. Microbial-electrochemical bioremediation and detoxification of dibenzothiophene-polluted soil. *Chemosphere* **2014**, *101*, 61–65
133. Zheng, Y.; Wang, C.; Zheng, Z.-Y.; Che, J.; Xiao, Y.; Yang, Z.-H.; Zhao, F. Ameliorating acidic soil using bioelectrochemistry systems. *RSC Adv.* **2014**, *4*, 62544–62549
134. Moqsud, M.A.; Yoshitake, J.; Bushra, Q.S.; Hyodo, M.; Omine, K.; Strik, D. Compost in plant microbial fuel cell for bioelectricity generation. *Waste Manag. (Oxf.)* **2015**, *36*, 63–69
135. De Schamphelaire, L.; Van den Bossche, L.; Dang, H.S.; Höfte, M.; Boon, N.; Rabaey, K.; Verstraete, W. Microbial fuel cells generating electricity from rhizodeposits of rice plants. *Environ. Sci. Technol.* **2008**, *42*, 3053–3058
136. Kaku, N.; Yonezawa, N.; Kodama, Y.; Watanabe, K. Plant/microbe cooperation for electricity generation in a rice paddy field. *Appl. Microbiol. Biotechnol.* **2008**, *79*, 43–49

CHAPTER 7

Trends and Potential of the Market for Combine Harvesters in Germany

Clemens Fuchs *, Joachim Kasten and Mathias Urbanek

University of Applied Sciences Neubrandenburg, Faculty of Agricultural and Food Sciences, Farm Management, Post Office Box 11 01 21, Neubrandenburg 17033, Germany

ABSTRACT

Combine harvesters used today can achieve throughput rates of over 70 tons of grain per hour; however, their technical performance potential is currently not being exhausted by any means. The global market for combine harvesters, its stocks and recent production is described. According to farm size distribution and regional field sizes, Germany was divided into three combine-harvesting regions. The simulation results show that in Germany about 45,000 units of combine harvesters in three performance categories are necessary. However, the calculations also show that future domestic sales of combine harvesters will depend greatly on the service life of the units currently in use.

Keywords: combine harvesters; technical development; stocks and demand

1. INTRODUCTION

The technical development of the combine machines has been enormous over time. Larger and larger cutting widths, improved performance and advanced electronics in the cockpit lead to the increase of effectiveness of the new machines and as a result fewer harvesters are demanded. This article first briefly describes the technical development and the portfolio development of combines and then forecasts the sales of combines for three sub regions of Germany. The main purpose of the article is to show how much stocks of combine harvesters have changed in the past and what harvester sales figures manufacturers can expect in the future.

2. TECHNICAL DEVELOPMENT AND MANUFACTURERS

Since the invention of the stationary threshing by the Scots in 1785 by Andrew Meikle, revolutionary technological advances have taken place in grain-crop harvesting. The United States and the countries of Western Europe are the pioneers of self-propelled equipment for combine harvesting with ever-higher demands on effectiveness, throughput and precise straw spreading in the swath chopped. Self-propelled and stationary equipment for combine harvesting were used in 2007 worldwide for the harvest of about 695 million hectares of cereals and 153 million hectares of oilseed acreage. Grains account for around two-thirds of cultivated crops planted on the global arable land. In 2007, these planting proportions resulted in a worldwide harvest of 2.35 billion tons of cereals [1] and 150 million tons of oilseeds, which is the most important nutritional basis for a growing world population from the present 7.3 billion people [2].

The technical development of each of the highest performance classes of combine harvester parts has increased steadily over the last 25 years (Table 1). The parameters cutterbar width and grain tank capacity have almost doubled in 25 years in the top performance class of combine harvesters and the engine power tripled approximately. In addition, the high performance potential is complemented by electronic control in critical functions in the workflow. The rapid technological advance in combine harvesting is strongly associated with the world and European leading manufacturers of combine harvesters John Deere, CNH, AGCO and Claas. Combines of the highest performance class use cutters with a working width up to 12 m, equipped with an engine power up to 434 kW and have a grain tank capacity up to 12,500 L for grain crops.

Table 1. Technical development of combine size classes 1985–2010.

Code	Year						
	1985	1989	1995	1999	2004	2005	2010
Engine power (KW/PS = horse power)	147 (200 PS)	206 (280 PS)	243 (330 PS)	265 (360 PS)	316 (430 PS)	390 (530 PS)	434 (590 PS)
Grain tank capacity (L)	6300	9100	9500	9700	11.500	12.000	12.500
Cutting width (m)	6.1	7.3	7.4	9.2	9.2	9.2	12.0

By increasing electronic equipment in the combine, the leading manufacturers pursue the following objectives: facilitating drivers, precise working, maximum capacity use and documentation of numerous performance measures. The large machines, whether combine or self-propelled agricultural machines, are always equipped with the latest technology, because additional costs in relation to the high unit price are less important. After a successful test of new integrated electronic solutions, it will also be installed in the mid-market segment.

Nine types of combine harvesters were offered on the Western European market in the 2010 season by international producers: New Holland and Case I (CHN), John Deere (John Deere), MF, Fendt and Gleaner (AGCO), Claas (Claas), Sampo and Sampo (SDF) and Laverda (ARGO). Technically, the harvester can be divided as follows: The conventional combine has a tangential and straw shaking (four-, five-, six- and eight-shaking versions) are equipped from 85 to 335 kW (115–455 hp (horse power)). Tangential harvesters with one or two separation rotors feature engine outputs from 268 to 431 kW (365–586 hp). In the performance category of axial combine harvesters, the engines range from 205 to 435 kW (279–591 hp). Cutting widths are available from working widths of 3.1 m to 12.0 m. The Austrian provider of cutters, BISO, additionally offers a cutting system with a working width up to 15 m on a modular aluminum construction. From Western European production with international links, there are over 40 combines series, 80 to 100 combine models and numerous additional equipment variants for the Western European market available.

3. STOCKS OF COMBINES AND SALES WORLDWIDE

3.1. Stock of Combines

In 2007, the EU-27, with just under 700,000 machines, possessed the second largest stock of harvesting machines in the world. There are about 2.8 million harvesters, including threshers, on the Asian continent.

In the Member States of the EU-27, the combine stock counted nearly 715,000 machines in 2006 at a threshing area of approximately 70 million hectares of cereals and oilseeds. This corresponds to an average threshing area of 98 hectares per combine in the EU-27, but which varies within the member countries due to different agricultural structures and combine stocks. The largest combine stock within the EU-27, with 124,000 machines, is found in Poland, followed by Germany with 88,500 and France with 78,000 machines. Nearly two-thirds of the 27 member states have a combine stock of less than 20,000 units. Only Malta, the Mediterranean island with a subtropical climate, has no combines. In the agricultural area of Malta, about 10,000 hectares, mainly fruit trees and citrus fruits are grown [4].

3.2. Sales of Combines

Technical progress including new features as well as the structural changes in agriculture have led to a decrease of combine sales from formerly about 48,000 to 7000 units per year in Western Europe within the last 40 years. The average 7000 units per year in the Western European represents 20% a share of the global harvester market in 2004 (Figure 1). In 2004, the world market was slightly more than 35,000 units. The sum of the two separately listed markets in Europe is 14,620 units (41.6%), South America follows with 10,300 units (share 29%) and North America with 8000 units (23% market share) [5].

In the Western European market, the number of combine units of sold is relatively constant with about 7,000 units per year for the last ten years. These numbers include the four-shaker combines up to the central five- and six-shaking combine harvesters with a market share of approximately 43%. In the upper performance category of conventional combines, a market penetration of six- or eight-shaking combine is characteristic with about 37%. The remaining market share of 20% is covered by the hybrid and axial combine harvesters, which show an upward trend [5]. The approximately 6800 sold units in 2007 equals nearly that of 2005; differences were only in the distribution among individual countries. With a total of nearly 4000 units, equivalent to 58.6% of Western Europe combine sales, Germany and France have the largest market in this segment. The high amount of combines in these two countries can be explained by the strong cultivation of cereals; both countries are among the strongest grain producers in the EU-27, along with Poland, Spain and Romania [6].

In contrast to the stable tendency of slightly decreasing volumes in Western Europe, the monetary sales volume has increased because of engine power and upgraded electronic features.

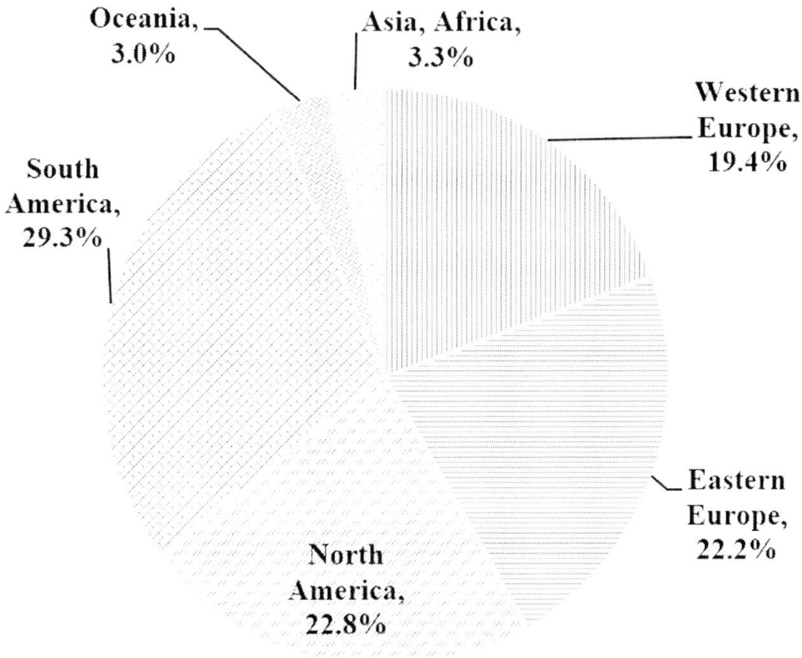

Figure 1. Combine world market in 2004. Source: New Agriculture (adapted from [5]).

Eastern Europe offers good prospects for increasing sales of harvesters, particularly the CIS countries (Commonwealth of Independent States—former Soviet republics, excluding the Baltic States). With the collapse of the Soviet Union, the major combine maker from there has been on the brink of bankruptcy. Before the political changes, 35,000 units were produced annually in the Soviet Bloc. In 1998, in Russia an all-time low of just over 1000 units produced. Furthermore, the current 200,000 unit combine stock is greatly outdated. In addition, 50% of the harvesters are not operational because of technical defects. Furthermore, Russia and Ukraine intend to expand their grain production greatly. Russia alone plans to increase production of cereals to 120 million tons in 2015. Compared to 2004 when 78 million tons of grain was produced, this represents an increase of 65% over a period of 10 years [5]. This means a considerable investment gap and low demand for combine harvesters. This is compounded by faltering production and by increases in import prices after the import tariffs on agricultural machinery have been raised to 15% in the wake of the world economic crisis in 2009. After the trade embargo, which was imposed because of the Crimean crisis in 2014, sales of agricultural machinery to Russia have almost completely collapsed.

The Asian market, with major grain producers China and India that together produced 740 million tons of grain in 2008, is less promising for an increased production of combines from the West. Nevertheless, for example, John Deere Jiamusi in China and Claas in India Faridabad and Chandigarh have production facilities for harvesters to participate in the sales potential of the Southeast Asian market and to show presence as a strategic "global player" [7]. In addition to the leading international manufacturers in India, there are 48 domestic companies (e.g., Preet Agro Industries Private Limited, Nabha, India) manufacturing combine harvesters [8]. With an inventory of 477,000 combine harvesters and threshers, the Indian market, after the Chinese (632,400 units) and Japanese markets (957,000 units), is the most significant in Asia [9].

4. DEVELOPMENT OF DOMESTIC SALES AND THE STOCKS OF COMBINES IN GERMANY

4.1. Production and Domestic Sales

The production of agricultural machinery in Germany covers a wide range of agricultural machinery and supplies for agriculture. In the period from 2006 to 2008, the production increase of combines units was 53%, followed by the plant protection equipment (48%) and planting machines (40%) [10]. In 2008, the 10,692 combines produced predominantly originated from the production sites of Claas in Harsewinkel and John Deere in Zweibrücken. The German agricultural machinery industry is strongly export oriented, especially in the field of combines and forage harvesters. For example, in 2007, 80% of the 6994 combine units manufactured were exported [9].

Domestic sales have more than halved from 2678 in 1998 to 1279 in 2008. Between 2000 and 2008, a decreasing trend can be observed. On average, domestic sales from 1990 to 2008 amounted to 1864 units sold (Figure 2) [11]. The market volume in the harvesters and equipment segment in 2008 amounted to 715 million euros. After the tractor, with a turnover of 1.6 billion euros, the harvesting machinery and equipment market in Germany is the second largest market for the agricultural machinery industry. Totally, sales of tractors and harvesters covered about 50% of the market volume in 2008.

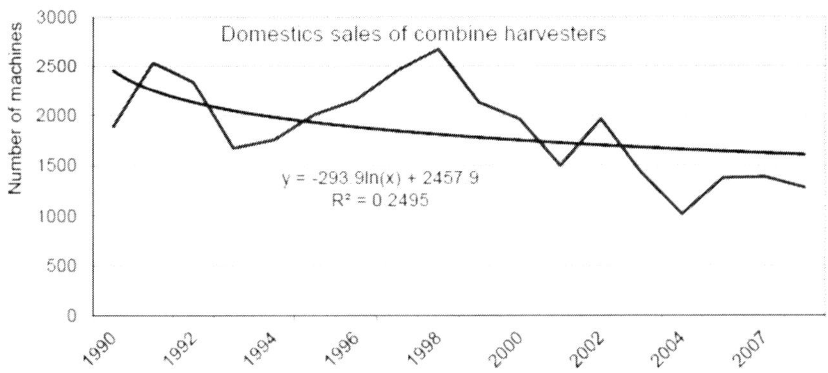

Figure 2. Domestic sales of combine harvesters in Germany (1990–2008) (own calculations based on [11]).

Sales of combine classes are similarly distributed all over Western Europe, where the four-, five- and six-shaking combine harvesters have a market share of about 43%. In the upper performance category of conventional combines, a market penetration of about 37% is typical. The remaining market share of 20% is covered by the hybrid and axial combine harvesters [5].

4.2. Inventory Development of Combine Harvesters

Stocks of combine harvesters rose from the mid-1950s, with the first self-propelled combine harvesters and the highly popular drawn threshing machines, to the year 1975 to total more than 189,000 units. The proportion of self-propelled combines was five times greater in 1976 compared to the drawn combine harvester. In the following years, the proportion of the drawn harvesters has declined sharply. With further technical developments, coupled with increases in performance with increased operational variables as well as due to the structural change in agriculture, the stocks of combines decreased in the following years by half [9].

For the grain crop harvest of 9.9 million hectares with a harvest of 61 million tons, nearly 85,000 combine harvesters were used in 2008. In the average of the years 2003 to 2009, about 9.8 million ha, over 56 million tons, were harvested by 93,200 combines [4]. Although through constructive ways the performance potential of the combines always increased in recent years, the combine's installed capacity is only exploited in the field to half the degree possible, which corresponds to a field efficiency of about 50% [12].

Figure 3 shows the decreasing old stocks of harvesters, beginning in 1990 and, for the period 1990 to 2008, the number of annual harvester sales and for each the fading out due to segregation (colored lines). The accumulated inventory, the sum of the corresponding number of harvesters in a certain year, results in the stock of combines, which has decreased from the previous count of 156,890 in 1990 to 79,153 units in 2008, a drop of approximately 50%.

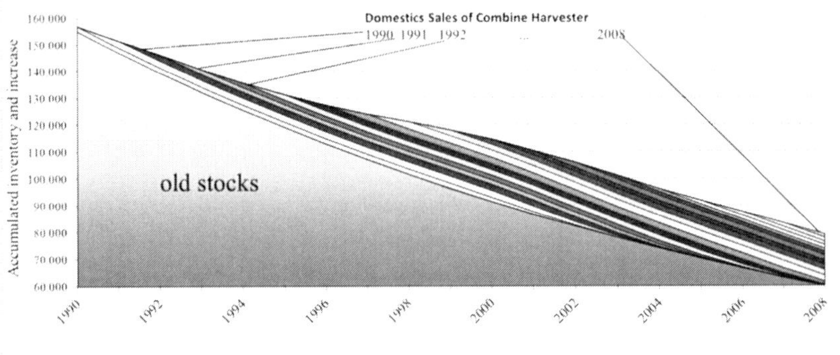

Figure 3. Inventory of combine harvesters (old stocks; domestic sales 1990 to 2008—colored lines; all area together = accumulated inventory) 1990–2008. Source: Own calculations based on [9,11].

According to FAO data, in Germany, in 2007, there were about 85,480 harvesters. Since the period from 1990 to 2008, *i.e.*, the 17 years of the investigation period, 20,248 new combine harvesters were added, which means there are still 63,193 combine harvesters 17 years or older in use [13]. If the old portfolio is updated at a rate of 5%, then in 2022 there would still be 28,632 older harvesters (potentially from before 1990) that remain in the stock. A faster replacement of very old combine harvesters is also conceivable, if newer machines are used to a larger extend. Therefore, the future structure of the combine fleet, in particular with respect to the old stock, is not clearly determined.

4.3. Factors Influencing the Threshing and Deriving the Minimum Capacity

The demand for combine harvesters is determined by several factors. In addition to crop management, other factors affect the extent of the actual performance. Ahead of the other factors, breeding and agronomic measures play an important role how a field is harvested.

Production failures, such as lying grain, even through optimal management efficiency of the harvesters can no longer be sustained. Another factor, which the farmer cannot influence, but determines the threshability decisively, is the weather [14].

Subsequently, the minimum necessary stocking of combines for the region of Germany should be estimated. Here, Germany is divided into three sub-regions, which differ due to local factors (farm size and field size) and where the use of smaller, medium and large harvesters makes sense (Figure 4).

Figure 4. Germany and its three sub-regions for harvester sizes depending on farm size and field size. Source: [15]; own assumptions and changes.

Furthermore, the influence of environmental factors (weather affecting harvest amount, available threshing hours and field efficiency) is simulated in a kind of risk analysis. Unfavorable constellations require the use of a larger number of machines and lead to higher harvesting costs. If there were insufficient harvesters available, higher natural losses and costs could arise; and, in extreme cases, a part of the fields could remain unharvested.

Results of these simulations show the region-specific necessary number of harvesters and the costs of harvest. Based on the average cost of the threshing and the marginal losses in crops, the minimum necessary number of combines is calculated.

4.4. Division into Threshing Regions for the Calculations

The basis for the calculation of the stock of development combines is the distribution of grain crops acreage in Germany in three sub regions. The federal states, with the exception of the city states of Berlin, Hamburg and Bremen, are, in accordance to their average farm size, classified into the following categories: below 40 ha utilized agricultural area (UAA), 40 to 60 ha UUA and 60 ha UAA (Table 2). Accordingly, the relevant crop and areas sown can be assigned. Decision criterion for the allocation of a performance class of combines for a sub-region is the parameter throughput in quantity of grain per hour.

Threshing region 1 (Southern Germany) comprises the federal states Hessen (HE), Nordrhein-Westfalen (NW), Rheinland-Pfalz (RP), Bavaria (BY) and Baden-Württemberg (BW). In these five states, between 18.4 and 24.1 million tons of grain crops were harvested on average in the years 2003 to 2008 on a cultivated area of 3.5 million hectares annually. In threshing region 1, combines with an engine output of 200 kW, a grain tank of 8500 liters and a cutter of 5 m were used for the harvest of grain crops (Table 3). With this combine size, it is possible to achieve a throughput capacity of 15 tons per hour. The four states of Schleswig-Holstein (SH), Sachsen (SN), Niedersachsen (NI) and the Saarland (SL) represent the second largest combine harvest area (North-West Germany), which has a cultivated area of 2.2 million hectare. In the years 2003 to 2008, on average 13.2 million tons of grain were harvested, which corresponds to 27% of the German grain harvest. For the grain crop harvesting in the threshing region 2, a combine with a power size of 275 kW, a grain tank capacity of 10,500 L and a cutting bar width of 7.5 m are used (Table 3). A throughput of 30 tons per hour can be achieved with this harvester size because of the larger farm structures in region 2. In the federal states of Mecklenburg-Vorpommern (MV), Saxony-Anhalt

(ST), Brandenburg (BB) and Thuringia (TH), where on cultivated area of 2.8 million hectares (33.2% of the German grain crops acreage), on average over 16 million tons harvested. In the threshing region 3 (Middle-Eastern Germany), the cultivation of commercial crops is most important. About 47% of the commodities are harvested in these four countries. The utilized agricultural area per farm in Mecklenburg-Vorpommern and Saxony-Anhalt is more than 150 ha, the greatest among the regions. For grain crop harvest in the threshing region 3, combines of the motor power class of over 375 kW, a grain tank capacity of 12,000 L and a cutting bar width of 9 m is common (Table 3). A throughput of 50 tons per hour can be achieved with this harvester size on the large acreage.

Table 2. Classification of threshing regions in Germany on the basis of the average farm size in ha utilized agricultural area (UAA).

Region (Average Farm Size)	Countries	UAA per Farm	Cropping Area	Harvest
		ha	1000 ha	1000 t
Region 1: Southern Germany (<40 ha)	HE	34.9	382.7	2552.6
	NW	31.5	743.0	5836.0
	RP	28.0	302.6	1855.7
	BY	26.4	1443.8	8977.2
	BW	25.1	651.9	4264.4
	Average	**29.2 ha**		
	Sum		***3,524.0 ha***	***23,485.9 t***
	Proportion		***41.0%***	***42.4%***
Region 2: North-West Germany (40 to 60 ha)	SH	57.3	444.7	3360.5
	SN	54.7	574.5	3333.4
	NI	52.3	1161.8	8357.7
	SL	47.6	29.6	164.4
	Average	**53.0 ha**		
	Sum		***2210.6 ha***	***15,216.0 t***
	Proportion		***25.7%***	***27.4%***
Region 3: Middle-Eastern Germany (more than 60 ha)	MV	170.5	834.8	5181.8
	ST	154.2	779.5	5141.8
	BB	96.8	721.9	3277.3
	TH	64.9	518.4	3149.2
	Average	**121.6 ha**		
	Sum		***2854.6 ha***	***16,750.1 t***
	Proportion		***33.2%***	***30.2%***
Germany	**Average**	**56.0 ha**		
	Sum		***8589.2 ha***	***55,452.0 t***
	Proportion		***100.0%***	***100.0%***

Table 3. Combine sizes in relation to the operational structure (farm size) in the regions.

Regions in Germany	Combine Key Figures					
	Motor Power	Grain Tank	Cutting Bar Width	Throughput-Performance	Additional Equipment	Acquisition Costs
Region 1: Southern	200 kW	8500 L	5.0 m	15 t h	Rape cutting table	217.100 €
Region 2: North-West	275 kW	10.500 L	7.5 m	30 t h	--	295.100 €
Region 3: Middle-Eastern	375 kW	12.000 L	9.0 m	50 t h	--	358.400 €

Source: Machinery and equipment [2]; grain harvest [11]; own assumptions.

Applying Monte Carlo method using @RISK [17] with 10,000 simulations for every scenario, the variations on the three external factors, harvest amount (yields), available threshing hours in a certain season and field efficiency, can be considered:

-a small harvest together with high threshing hours and a high field efficiency,

-a great harvest at low threshing hours and low field efficiency, or

-any other possible combinations.

In the first case, a small number of combine harvesters would be sufficient, while a maximum number of harvesting equipment would be needed in the second case. How often unfavorable, average or favorable situations occur is specified with triangular distributions, with extreme values, *i.e.*, unfavorable or favorable situations, occurring less frequently than average values (Figure 5).

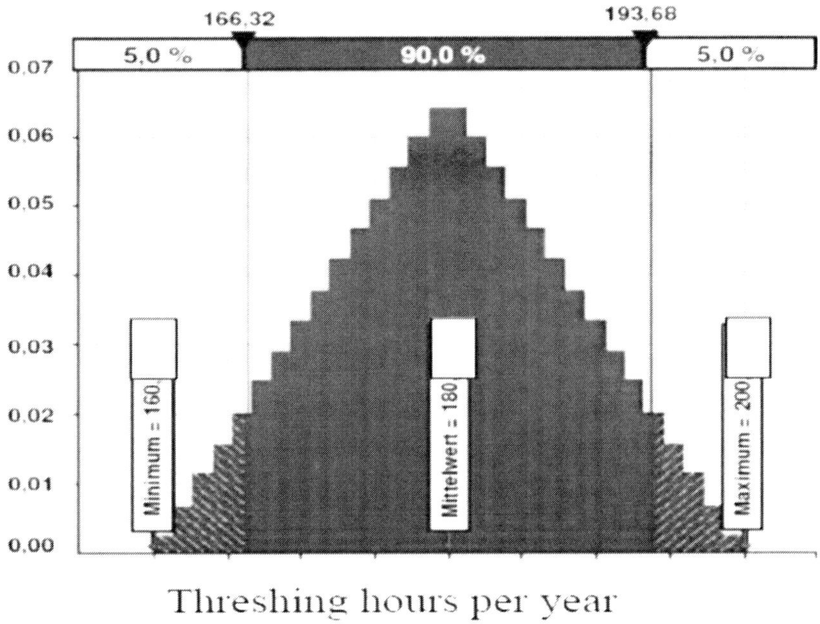

Figure 5. Available threshing hours per year in all regions combine harvesting. Source: [17]; own calculations.

For example, it is assumed that the available threshing hours may vary between 160 h per year as a minimum value (rare), 180 h per year on average and up to 200 h per year as the maximum value (also rarely occurs). Regional differently vary the quantities harvested and the field efficiency, the latter is assumed to be lowest for the small-scale south region (region 1) with an average of 40% (in the span from 30% to 70%) and, at the highest and most expensive in the large structured middle-east area (region 3) with an average of 60% (40%–80%) (Table 3).

The results of the simulations give the number of a necessary minimum set of combines, which is required to harvest the volumes in all simulated years. From these values, the annually sold harvester units can be derived assuming an average service life.

In Table 4, the results of the simulation are summarized. In order to ensure that—with a probability of 100% at 10,000 iterations—the cultivated areas are completely threshed in all years, a combine fleet of 44,385 units would be required. The majority of harvesters (approximately 30,000 units) are for the region 1; this relatively high number is due to the assumed smaller power class of 15 tons per hour. In the other two regions, with approximately 8600 (region 2) and 4900 pieces (region 3), significantly fewer units are required (Figure 6).

Introducing a limit to the fluctuation range to cover for 90% of all cases, *i.e.* cutting off the 5% lowest (best) and 5% highest (worst) values:

- On the minimum cost side, it becomes more expensive, when the harvest could be cheaper due to an "over"-capacity of combines.
- On the other side, it becomes cheaper due to the lower stock of combines; in 5% of all cases, a higher number of combines would be needed. The latter variants could still be realized, namely when other management reserves are developed, for example, additional inter-company co-operation and emergency relief to harvest all under difficult weather conditions.

Table 4. Assumptions (minimum, medium and maximum values of the triangle distributions) and simulation results (in the 90% range).

Region	Key Figure	Unit	Minimum	Average	Maximum
Region 1: Southern Germany	harvest	Mio. t per year	18.43	22.02	24.12
	available threshing hours	h per year	160	180	200
	field efficiency		30%	40%	70%
	threshing costs (90%)	€uro/t	17.83 20.79	26.40	38.63 32.90
	necessary stock of combine harvesters (90%)	number of machines	10.011 12.800	18.068	**30,891** 24.240
Region 2: Northern-West Germany	harvest	Mio. t per year	11.94	13.60	15.63
	available threshing hours	h per year	160	180	200
	field efficiency		35%	50 %	70 %
	threshing costs (90%)	€uro/t	14.26 15.96	18.66	25.57 21.93
	necessary stock of combine harvesters (90%)	number of machines	3190 3830	4980	**8627** 6440
Region 3: Middle-Eastern Germany	harvest	Mio. t per year	13.14	15.70	17.21
	available threshing hours	h per year	160	180	200
	field efficiency		40%	60%	80%
	threshing costs (90%)	€uro/t	12.18 13.18	14.97	20.04 17.29
	necessary stock of combine harvesters (90%)	number of machines	1877 2289	2970	**4867** 3848
Germany	Stock of combine harvesters (90%)	number of machines	15.078 **18,919**	25.737	**44,385** **34,528**

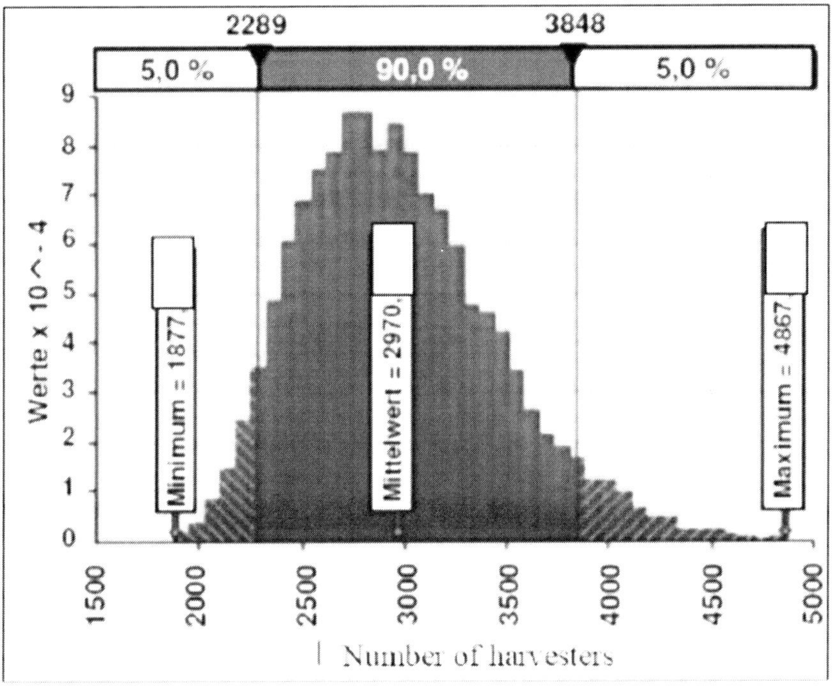

Figure 6. Distribution of the necessary number of harvesters in the threshing region 3. Source: [17]; own calculations (10,000 simulations).

4.5. Future Development of the Number of Combine Harvesters

Historical data for inventory and sales of combines have been analyzed in this study, until 2008 and before. The current stock of combine harvesters in Germany amounting to approximately 80,000 machines is almost twice as high as the calculated necessary inventory with about 44,000 units (Table 4). This explains the relatively low domestic sales in recent years. Due to the overcapacity of combine harvesters, the latest development of destocking, with relatively low replacement, will continue (although powerful modern combine harvesters are not yet in use everywhere) until the necessary stock of about 44,000 combine is reached. For the future, it is assumed that the existing stock of combines will decrease by 5% annually, similar to the period from 1990 to 2008 observed. In addition, the sales of combines per year are estimated (and therefor limited) according the trend line determined during the period from 1990 to 2008 (Figure 2):

$$y = -293.9\text{Ln}(x) + 2457.9 \ (R^2 = 0.2495) \quad (1)$$

where y = number of combines in Germany in units per year, and x = year from 1990 onwards.

Because of the high old stock of harvesters, the annual sales of combines might decrease to 1283 units in 2009 and further to 727 units in 2019 (Figure 7). Thus, on average, over the years 2009 to 2019, a relatively small substitution of an average 1005 combines per year would be sufficient in Germany.

In the calculations for Figure 8, it was assumed that by 2009 for the newly sold combines a 10-year using age could be expected. After this time, annually 20% of the old machines could be eliminated, so after 15 years, none of these machines would be available. In the following years (from 2019), the useful life of the new combines has been limited to a maximum of eight years.

From 2019, the old stock would be so far degraded that a larger replacement demand of approximately 3000 combines a year is expected. In the long term, however, according to the assumptions, about 4400 new combines are annually necessary to replace a total stock of 44,385 units of combine harvester in Germany.

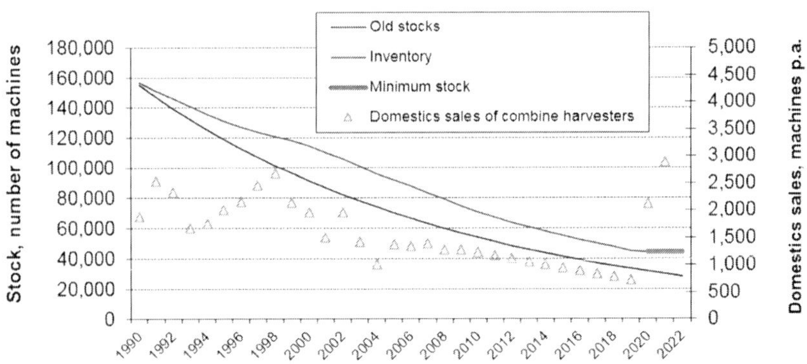

Figure 7. Sales and inventory of combine harvesters (1990–2022). Source: Own calculations based on [9,11].

After adjusting, the stocks of combine harvesters will reach a minimum constant level of at least about 44,000 units from 2019 to 2030; the colored lines in Figure 7 show the initial number of future harvesters sales in a certain year and their path of decrease.

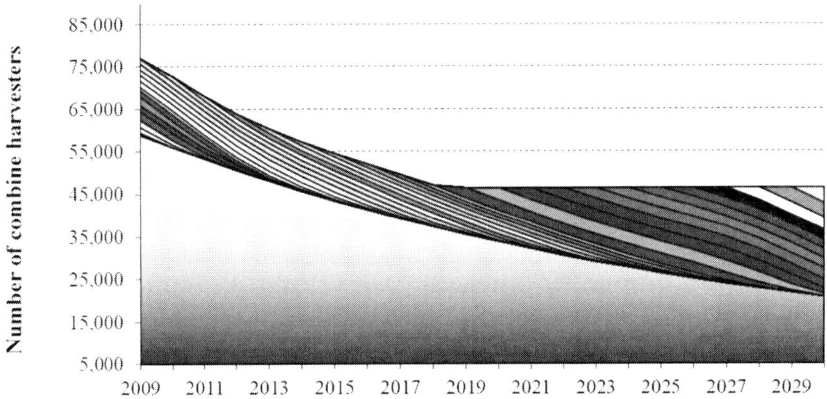

Figure 8. Inventory of combines cumulated from old stock and calculated domestic sales (2009–2030, colored lines). Source: Own calculations based on [9,11].

5. SUMMARY

Development Tendencies and Potential of the Market for Combine Harvesters in Germany

Combine harvesters used today can achieve throughput rates of over 70 tons of grain per hour; however, their technical performance potential is currently not being exhausted by any means. In this paper, a survey of major harvester manufacturers and models is given and factors influencing the combine harvesting performance are presented.

The objective was to reflect the complexity due to factors like size of farm and field size as well as combine performance and environmental factors (weather, with impact on yield and number of hours available for combine harvesting). Therefore, Germany was divided into three combine-harvesting regions. Certain performance categories of combine harvesters were assigned according to farm and regional field sizes. Triangle distributions with assumptions for harvest size, field efficiency, and harvest days are the basis for the applied Mote Carlo simulation, which shows the minimum capacity of combine harvesters necessary to reap the harvest safely.

The simulation results show that, in Germany, 44,385 units of combine harvesters of the three performance categories are necessary. However, the calculations also show that future domestic sales of combine harvesters will depend greatly on the service life of the units currently in use.

REFERENCES

1. Agricultural Production; FAO: 2009. Available online: http://www.fao.org/fileadmin/templates/ess/documents/publications_studies/statistical_yearbook/yearbook2009/b01.xls (accessed on 23 November 2015).
2. Population. Available online: http://www.worldometers.info (accessed on 22 November 2015).
3. Kutschenreiter, W. Mähdrescher als technologische Herausforderung. Available online: http://kutschenreiter.net/index.html (accessed on 22 November 2015).
4. Statistisches Jahrbuch über Ernährung, Landwirtschaft und Forsten der Bundesrepublik Deutschland 2009. Table 104. Anbau, Ertrag und Ernte der Feldfrüchte. Available online: http://www.bmelv-statistik.de/de/statistisches-jahrbuch/ (accessed on 26 November 2015).
5. Kutschenreiter, W. Analyse Mähdreschermarkt: Weniger Maschinen, mehr Leistung. Available online: http://www.dlv.de/grafiken/dlv/NL_11_2005_50-53.pdf (accessed on 23 November 2015).
6. Statistisches Jahrbuch über Ernährung, Landwirtschaft und Forsten der Bundesrepublik Deutschland 2009. Table 654. Bestand an Schleppern und Mähdreschern. Available online: http://www.bmelv-statistik.de/de/statistisches- jahrbuch/ (accessed on 26 November 2015).
7. John Deere Combines. 2009. Available online: http://www.deere.de/de_DE/brochures/downloadcenter/index.html (accessed on 23 November 2015).
8. Singh, G. Agricultural Machinery Industry in India. 2006. Available online: http://agricoop.nic.in/Farm%20Mech.%20PDF/05024-09.pdf (accessed on 23 November 2015).
9. FAOSTAT: Export Quantity of Combines in 2007. Available online: http://faostat.fao.org/site/576/DesktopDefault.aspx?PageID=576#ancor (accessed on 23 November 2015).
10. VDMA: Wirtschaftsbericht 2009. Available online: http://www.vdma.org/wps/portal/Home/de/Branchen/L/LT/Wirtschaft_und_Recht/LT_A_2009 0626_CG_Wirtschaftsbericht_2009 (accessed on 23 November 2015).
11. Statistisches Jahrbuch über Ernährung, Landwirtschaft und Forsten der Bundesrepublik Deutschland 2005 und 2009. Inlandsabsatz der Landmaschinenindustrie. Available online: http://www.bmelv-statistik.de/de/statistisches-jahrbuch/ (accessed on 26 November 2015).
12. Spitzenleistung im HARVEST POOL. Available online: http://www.feiffer-consult.de/dokumente/Prospekt_Harvest_Pool_06.pdf (accessed on 23 November 2015).
13. FAO: Agricultural Production 2009. Available online: http://www.fao.org (accessed on 23 November 2015).
14. Feiffer, A. Drusche ignung als zentrale Führungsgröße im Erntemanagement. Dissertation, Humboldt Universität, Berlin, 2009. Available online: http://edoc.hu-berlin.de/dissertationen/kluessendorf-feiffer-andrea-2009-07-09/PDF/klue ssendorf-feiffer.pdf (accessed on 23 November 2015). [Google Scholar]
15. Bundeslaender. Available online: http://www.weltkarte.com/typo3temp/images/bundeslaender.png (accessed on 23 November 2015).

16. Statistical Offices of the Federal and State Governments. Germany 2010. Available online: http://www.destatis.de/jetspeed/portal/cms/Sites/destatis/Internet/DE/Content/Statistiken/LandForstwirtschaft/StrukturenLandwirtschaftlicherBetriebe/Aktuell,templateId=renderPrint.psml (accessed on 23 November 2015).
17. PALISADE. The DecisionTools. @RISK 5.5. Available online: www.palisade.com (accessed on 23 November 2015).
18. Urbanek, M. Entwicklungstendenzen am Mähdreschermarkt. Master's Thesis, University of Applied Sciences Neubrandenburg, Neubrandenburg, Germany, 2010.

CHAPTER 8

Double Lighting Machine Vision System to Monitor Harvested Paddy Grain Quality during Head-Feeding Combine Harvester Operation

Mahirah Jahari [1,2,*], Kazuya Yamamoto [1], Munenori Miyamoto [3], Naoshi Kondo [1], Yuichi Ogawa [1], Tetsuhito Suzuki [1], Harshana Habaragamuwa [1] and Usman Ahmad [4]

[1] *Graduate School of Agriculture, Kyoto University, Kitashirakawa-Oiwakecho, Sakyo-ku, Kyoto 606-8502, Japan*

[2] *Department of Biological and Agricultural Engineering, Faculty of Engineering, University Putra Malaysia, 43400 Serdang, Selangor, Malaysia*

[3] *Agricultural Research Center, Agricultural Operations Business, Yanmar Co., Ltd., 2481, Umegahara Maibara, Shiga 521-8511, Japan*

[4] *Department of Mechanical and Biosystem Engineering, Bogor Agricultural University (IPB), Jawa Barat 16680, Indonesia*

ABSTRACT

A machine vision system to evaluate harvested paddy grain quality during harvesting using double lighting was developed. The prototype consisted of a low-cost web camera and two lighting systems: a ring white LED for front lighting, and a flat dome white LED light for backlighting. Both lighting systems were arranged in a coaxial axis, making the system simple, compact and easy to handle. The aim of the system is to analyse the captured images and determine the amount of unwanted materials (rachis branch, grass and leaves, and stems) and damaged grain (brown and crack rice) present in the paddy as it is being harvested. In this paper, we introduce the first step in the development of the system: the design and selection of components to optimize the performance of the system to monitor harvested paddy grain quality. The idea would be to mount the system on top of the inlet channel of the grain tank of a combine harvester to provide real-time assessment of harvesting operational parameters.

Keywords: machine vision system; grain monitoring; combine harvester

1. INTRODUCTION

Paddy grain quality is an important parameter that is continually changing from field to consumption. More precisely, paddy grain quality can be defined in terms of the sum of all the attributes that lead to consumer product acceptability. Although this quality is often subjectively assessed by humans, including such attributes as appearance, smell, texture, and flavor, there are six main physical characteristics which objectively determine the material quality of paddy: moisture content level, grain maturity, varietal purity, present of dockage (such as straw, leaves and rachis branch), and present of discolored and cracked grains.

For these material paddy grain quality characteristics, most research has either focused on pre-harvest attributes or on post-harvest attributes, particularly the milling process. Research on pre-harvest attributes has demonstrated the significant within-field variability of paddy grain quality (protein and moisture content) [1,2,3], while in the post-harvest milling process, quality assessment has focused on physical characteristics, such as immature grain, varietal purity, and discolored grain (identification of damaged grain kernels) [4], inspecting and identifying rice varieties [5], degree of milling [6], yield prediction, and percentage of whole kernels [7], *etc.* To date, though, no research has been done on quality assessment of the material attributes of the paddy during harvesting. In this study, we focus on a methodology to assess, in real-time, the material quality attributes of the harvested paddy, such as dockage and cracked grain.

The rationale for this is that farmers need to monitor the harvested paddy grain and adjust their combine harvester operation settings (such as travelling speed, threshing depth, sieves, wind blower, *etc.*) to optimize grain yield, minimize grain damage and quality deterioration, and minimize dockage (see Table 1). Currently, in order to do this, though, they need to continually stop the harvester and inspect the harvested paddy grain in the grain tank during harvesting. Improper adjustment of operating parameters can lead to grain damage, increased grain loss, excessive unwanted material (straw, grass and *etc.*) and clogging [8,9]. Moreover, the farmers can only qualitatively assess the damaged grain and amount of dockage in the harvested grain, leading to inconsistent assessment, and an assessment that is highly dependent on the experience and expertise of the farmer.

Recent developments in precision and computer-assisted automated agriculture using sensors and machine vision [10] promise to help the farmer to make better management decisions. Besides the precision and efficiency offered, a machine vision system offers farmers the opportunity to build value-added characteristics and brand awareness of their product through consistency, minimizing the need for further downstream processing, and the robustness and ease of use of their product [11]. Such advantages could be obtained if the current time-consuming and inconsistent manual inspection of paddy grain quality during harvest could be replaced by a machine vision system that aids in real-time decision making for combine harvester operations.

In order to achieve this, we aim to develop a paddy grain monitoring system which can be mounted in the grain tank of a head-feeding combine harvester. The monitoring system can conduct real-time image analysis of harvested paddy when it enters the grain tank. The results of image analysis will then be used to inform the operator what parameters should be adjusted, or used as input parameters for automatic adjustment. This paper reports the first step in the development of a machine vision system to evaluate harvested paddy grain quality in a combine harvester grain tank.

Table 1. Harvesting grain conditions and necessary adjustment of the combine harvester parameters.

Type of Conditions	Harvesting Grain Condition	Necessary Adjustment of Combine Harvester Setting
Unwanted material	Many long rachis branches	Increased travelling speed Stronger wind blower Narrower opening sieve Narrower dust ejection
	Large amount of grass and leaves	Stronger wind blower Narrower opening sieve Shallower threshing depth
	Many stem	Narrower opening sieve
Damage Grain	Large amount of brown and cracked rice	Decreased travelling speed Weaker wind blower Wider opening sieve

To construct such a machine vision system it will be important, as a first step, to select an appropriate illumination system which optimizes the detection of damaged grain and unwanted material in the harvested grain. Moreover, the system will need to be simple, compact, and relatively low cost, as well as physically robust as it will be used in harsh field conditions. To date, most machine vision grading of agriculture products has used front lighting systems for color related information. Brosnan *et al.* (2004) reported that the quality of rice is based on variety of properties such as size, shape, color, chalkiness and number of broken rice kernels, color and chalkiness [11]. Shantaiya *et al.* (2010) used the morphological and color features to identify different varieties of rice using feed forward neural network [12]. Verma (2010) used the back propagation through time neural to sort the rice into chalky, sound, and broken kernels [13]. Yao *et al.* (2010) develop an inspection system of rice exterior quality (head rice rate, chalk rice, crackle rice) based on computer vision [14].

Novini (1990) reported that most illumination system arrangements could be grouped as one of the following: front lighting, back lighting, and structured lighting [15]. Normally front lighting and structured lighting gives color information, while back lighting generates instant contrast as it creates dark silhouettes against a bright background. By enhancing image contrast, a well-designed illumination system can improve accuracy and lead to successful image analysis [16]. Therefore, this paper focuses on the design and selection of lighting components, and the construction of a robust, prototype machine vision system to monitor harvested paddy grain quality with the expectation that the system will then need to be further developed to enable farmers to operate the combine harvester at full capacity and maximum efficiency through real-time measurement of and adjustments to harvesting operations.

2. THE PROPOSED MACHINE VISION SYSTEM FOR MONITORING HARVESTED GRAIN QUALITY

2.1. Design of Double Lighting System

In this study, a double lighting system was developed for monitoring harvested paddy grain: a frontlight system and a backlight system. Each will in turn be discussed.

2.1.1. Frontlight System

Selection of an appropriate illumination system is the most important component of the machine vision system. In terms of selection critieria, uniformity of light distribution brightness is of primary concern. The Ring-type white LED (Hayashi Watch-works Co., Ltd., type: HDR55W, Tokyo, Japan) and the Dome-type white LED (CCS Inc. Co., Ltd., type: HPD-100SW, with a temperature of 5500 K, Kyoto, Japan) illumination systems were compared using a whiteboard to evaluate their distribution brightness (Figure 1).

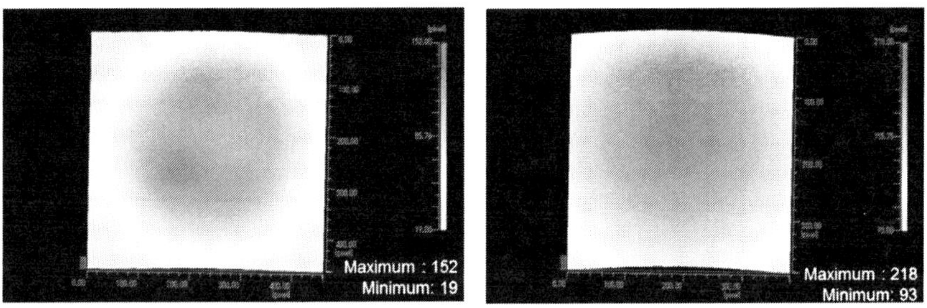

(a) (b)

Figure 1. Brightness distribution by using whiteboard for different types of lighting system. (**a**) Ring light; (**b**) Dome light.

These figures show a distribution brightness in the range of 0 to 255 (from minimum to maximum pixel value) for the two lighting systems. For example, the Ring light has a maximum brightness of 152 and a minimum of 19 (Figure 1a), while the Dome light ranged from 218 to 93 (Figure 1b); there is not such a big difference between them. In order to make the light system compact and for ease of mounting in the grain tank, the Ring LED light was a better choice for the frontlight system. This is because the camera can be placed between the Ring lights and thus maintain a compact system. In addition, a Ring light is not as heavy as a Dome light.

These Ring lights are a white LED type (CCS Co., Ltd., type: LDR2-90-30SW with a temperature of 5500 K) with Polarizing (PL) filter. The Ring LED lights were set at an angle of 45° to the object plate to concentrate the light at the center of the target objects [17]. As the targeted illumination area was set at 50 mm × 50 mm, the Ring LED lights were set up at a distance of 47 mm in front of the object plate.

2.1.2. Backlight System

The distribution brightness on a whiteboard of a Flat dome light (CCS Inc. Co., Ltd., type: LFX2-100SW with a temperature of 6600 K) and a Coaxial light (CCS Inc. Co., Ltd., type: LFV-35SW with a temperature of 6500 K) were compared for use in a backlight system (Figure 2).

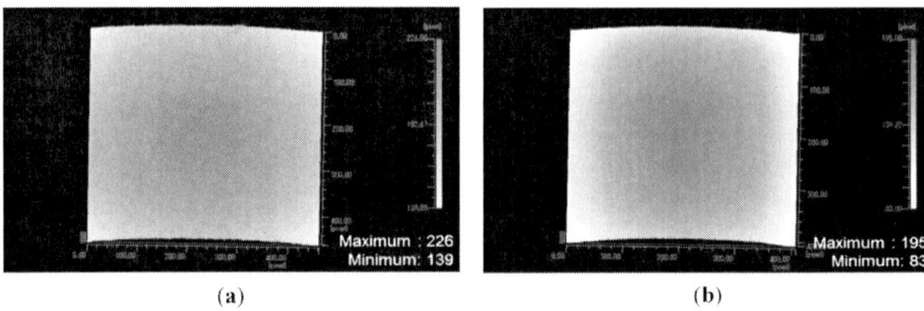

Figure 2. Brightness distribution by using whiteboard for different types of lighting system. (**a**) Flat dome light; (**b**) Coaxial light.

These figures show the distribution brightness in the range of 0 to 255 (from minimum to maximum pixel value) for the two lighting systems. For example, the Flat dome light has a maximum brightness of 226 and a minimum of 139 (Figure 2a), a small variation in in brightness denoting a relatively uniform light distribution, while the Coaxil light ranged from 195 to 83 (Figure 2b). Therefore, we chose the Flat dome light, with its more uniform light distribution, to be used in the backlight system.

Normally, a critical problem for image acquisition when using both a front and back lighting set up on the same axis with conventional (diffusion plates) and LEDs (for example: CCS Co., Ltd., type: TH-63X60SW) is that the frontlight image of the objects will be poorly contrasted from the background color; making it difficult to distinguish between the target objects and the background.

In our system, we used a square white Flat dome light unit (CCS Co., Ltd., type: LFX2-75SW with a temperature of 6600 K) with a dotted pattern for indirect backlighting to resolve this problem during frontlit image acquisition. This Flat dome LED light has dots printed on the surface of the light diffusion plate (Figure 3). The dot pattern on the surface of the light diffusion plate controls illumination diffusion and transmission [18]. It can illuminate objects with a more uniform and diffused light. Besides that, when using this Flat dome light in the backlighting system, a PL filter does not need to be used in front of the backlight system.

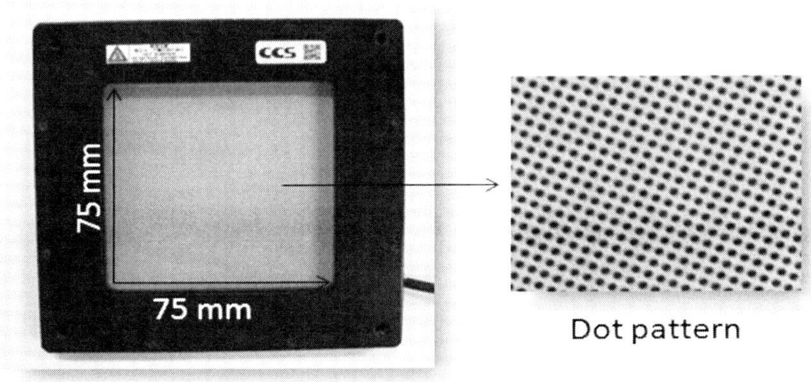

Figure 3. Flat dome light units (indirect lighting).

As a result, the background of a frontlight image will be dark in color and thereby ease differentiation between target objects and the background. The conventional method for taking a frontlight image is to put the sample on a black plate for high contrast between the background and objects. However, this makes it difficult to take the backlight image with the same orientation and conditions.

2.2. Overall Setup of Double Lighting System

2.2.1. Design of Whole System

A schematic of the proposed double lighting system is shown in Figure 4. The front and backlight systems were aligned along the same axis.

The advantage of using a double lighting system aligned along the same axis is that both frontlit and backlit images can be sequentially acquired simply by turning the backlighting off and on. Moreover the same orientation allows the system to be more compact and easily mounted in the a grain tank of a head-feeding combine harvester. This double lighting system has been patented in Japan (Patent No. 5780642) [19].

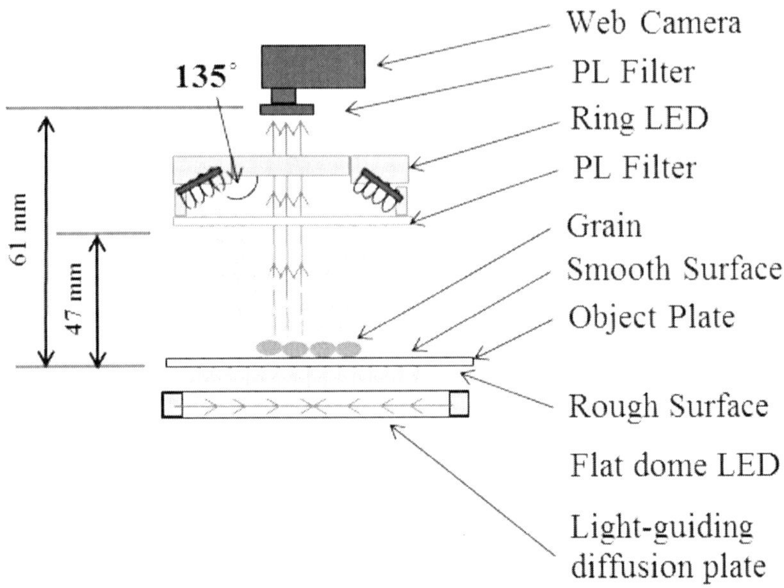

Figure 4. Layout of schematic diagram for double lighting system.

In order to acquire high-quality images, camera operating parameters were first optimized. Logicool Webcam Software version 1.1 (Logitech. Com., Romanel-sur-Morges, Switzerland) was used to adjust camera parameter settings and image capture specifications, including combinations of shutter speed and gain for ease of background removal and object classification. Once these camera parameters were optimized, images were captured at a resolution of 22 pixel/mm and a size of 1600 by 1200 pixels. The color images were 24-bit depth in the RGB color model. The data were stored in JPEG format on a computer. Control of brightness for both lighting systems was performed by a power unit (CCS Co., Ltd., type: PD2-3024-2).

The first experimental prototype was very simple and acquired sequential front- and backlight images via on-off control of the backlight source alone (Figure 5).

Figure 5. Appearance of the experimental equipment.

2.2.2. Object Plate

As shown in Figure 4, a glass plate: one side smooth and the other rough, was used as the object plate to produce a robust imaging system. The rough surface of the object plate performs the same function as a diffuser panel: distributing light uniformly. Samples were placed on the smooth surface of the object plate and the backlight LEDs attached to the rough surface side. Normally, after paddy grain has been removed from the sampling area, dust and water (for high moisture content grain) will remain on the surface of the object plate. By ensuring the surface of the sampling area on the object plate is smooth, the cleaning process should be easier and to lead to a rapid image capture of the next sample.

2.2.3. Camera

To maintain the compactness of the system and restrain costs, we used a WEB camera with a resolution of 2 megapixels (Logicool Co., Ltd., type: C905M, UXGA class) and PL filter.

3. MATERIALS AND METHOD

3.1. Target Materials

Paddy (*Kinuhikari*) harvested in mid-September 2011 in Shiga Prefecture, Japan was used as representative samples. Within this sample, 4 major objects will be detected: paddy with a long rachis branch (Figure 6a); grass (Figure 6b); stems (Figure 6c); and brown and cracked rice (Figure 6d).

Normally, the harvested grain will contain many paddy with a long rachis branch if the combine harvester is operated at too slow a travelling speed, the wind blower is too weak, the sieve opening is too wide, or dust ejection is set too wide during harvest. Too much grass in the harvested paddy grain reduces the quality of the harvested grain and can create problems later in storage. Excessive amounts of grass in the harvested paddy grain is caused by setting the wind blower too weak, the sieve opening too wide, or the threshing depth too deep during harvest. Setting the sieve opening too wide will also lead to the inclusion of too much stem material in the harvested paddy grain. Figure 6d shows the brown and crack rice. The presence of abundant brown or cracked rice indicates the combine harvester was being operated at too high a travelling speed, or the wind blower was set too strong or the sieve opening too narrow.

3.2. Image Acquisition Method

3.2.1. Frontlight System Performance

Two frontlight images of the paddy, as well as the brown and cracked rice, were captured using the frontlight system (Ring LED lights) with two different backlight systems: the dot patterned Flat dome LED light or a conventional white Flat LED light (example: CCS Co., Ltd., type: TH-63X60SW). During frontlight image capture the backlight system was turned off. The images captured by these two systems were compared for the different object types (paddy and brown rice) and background lighting systems to evaluate the performance and efficiency of the frontlight system.

3.2.2. Comparison of Frontlight and Backlight Image Performance

From the harvested paddy grain, we obtained a sub-sample of roughly 100 g. The sampling area in our compact prototype system is small; therefore, to avoid the problems of too much overlapping and grains touching, only a 1 g sub-sample was used to acquire the image for each representative sample type: paddy, paddy with a long rachis branch, grass, stems, and brown and cracked rice. The images taken with the front and back lighting were then evaluated in terms of target material detection.

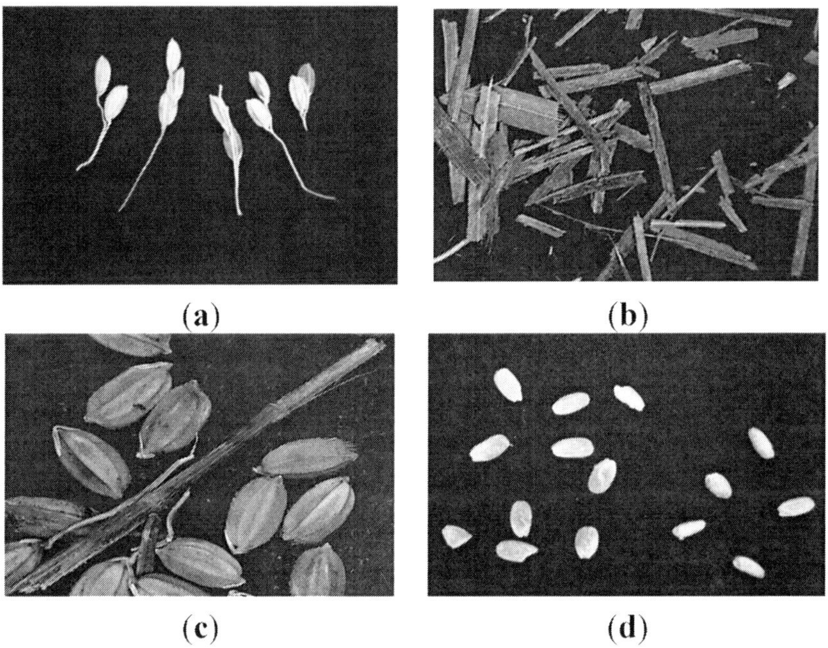

Figure 6. Object in harvested grains. (**a**) Long rachis branch; (**b**) grass; (**c**) stem; (**d**) brown and cracked rice

4. RESULTS AND DISCUSSION

4.1. Result of Frontlight System Performance

For images taken with our double lighting system, the background is blue-black as a result of the dot-patterned Flat dome light unit in the background, while the conventional system has a whitish background (Figure 7). As a consequence, the contrast between the target objects and background is lower in the conventional system. A clear boundary threshold between for the double lighting system using the saturation and hue distribution was found between the target objects (rough and brown rice) and the background compared to that for the conventional system (Figure 8). These results confirm that using the double lighting system makes it easier to distinguish between target objects and the background.

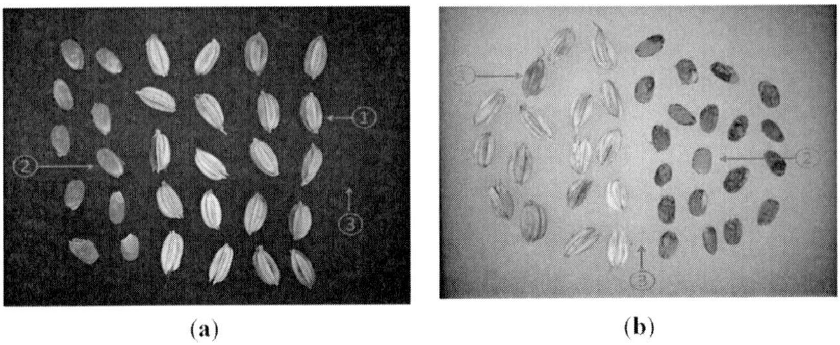

Figure 7. Frontlight images. (**a**) Double lighting system; (**b**) conventional system.

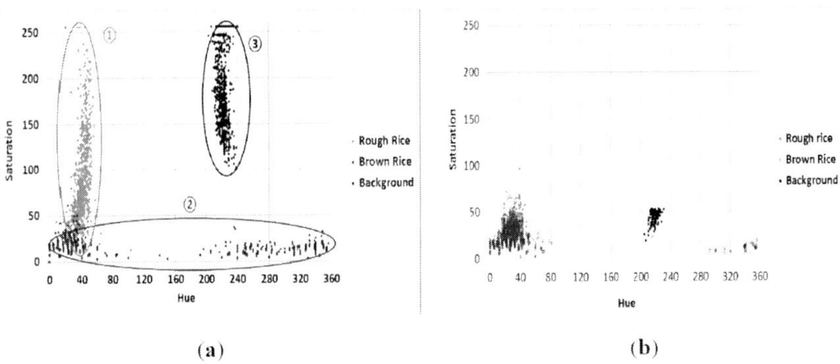

Figure 8. Saturation and hue distribution from frontlight images captured with the (**a**) double lighting system and (**b**) conventional system.

4.2. Comparison of Frontlight and Backlight Image Performance

While color information can be used to establish a threshold for paddy, grass and stem detection, the use of color information (HSI) to distinguish brown rice is difficult (Figure 8a). This is because some HSI color components of brown rice overlap with those of paddy HSI color components. Figure 9b below shows a frontlight brown rice image with 2 types of brown rice: (1) brown rice that is almost white in color and easy to detect; and (2) brown rice color that is almost brown in color, the same color as paddy. However, by using backlight images (Figure 10b), both types of brown rice are easily detected. Therefore, for husked rice detection, it is best to use the backlight images for detection purposes.

In many cases, agricultural product grading or detection needs only one lighting system, but in the case of harvested paddy, it has been shown that two lighting systems are necessary. For paddy, rachis branch, grass and stem detection, frontlight images are necessary, while backlight images and the dark silhouettes created on the bright background can detect paddy, long rachis branch, and brown rice.

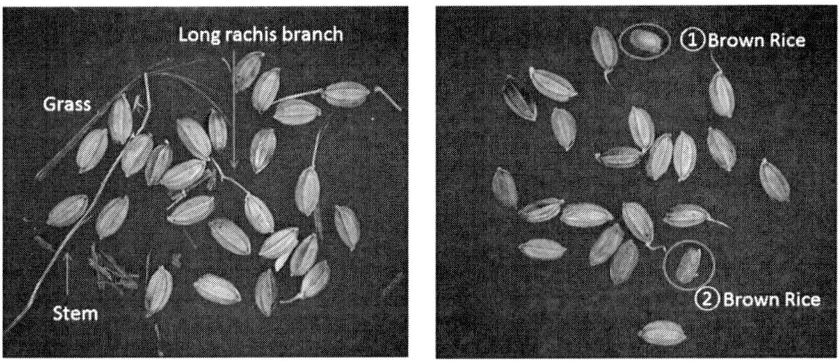

(a) (b)

Figure 9. Image using frontlight images. (**a**) Long rachis branch, grass and stem; (**b**) 2 types of brown rice.

(a) (b)

Figure 10. Image using backlight images. (**a**) Long rachis branch, grass and stem; (**b**) 2 types of brown rice.

5. CONCLUSIONS

In this study the principles of a new machine vision system approach to the monitoring of harvested paddy grain quality has been established. The system consists of a double lighting system (front and backlighting) set up along a coaxial axis. To maintain a compact and robust system capable of being mounted and operated in the harsh environment of a grain tank, Ring LED lights for frontlighting and Flat dome LED lights for backlighting were found to give the most uniform lighting distribution. Moreover, when using dot-patterned Flat dome LED lights for backlighting, having them turned off provides a higher-contrast background for frontlight images. Additionally, the rough surface of the botton of the object plate serves the same function as a diffuser panel, allowing for a more uniform light distribution, while the smooth front surface of the object plate aids in the cleaning process. Most importantly, this system can differentiate between four major target objects in freshly harvested paddy.

From a user and commercial perspective, the system is robust, compact, easy to assemble and low cost. As such, the proposed machine vision system shows great potential to be mounted in the grain tank of a head-feeding combine harvester for real-time monitoring of harvested paddy grain and decision support for combine harvester operations. Continued experimentation will be conducted to further enhance image processing and detection of target objects in the harvested paddy grain, such as damaged grain (husked and cracked rice) and unwanted materials (long rachis branch, grass and leaves, and stems).

ACKNOWLEDGMENTS

The authors would like to express their gratitude to Garry Piller for his editing and proofreading of this manuscript.

REFERENCES

1. Reyns, P.; Spaepen, P.; de Baerdemaeker, J. Site-specific Relationship between Grain Quality and Yield. *Precis. Agric.***2000**, *2*, 231–246.
2. Bramble, T.; Herrman, T.J.; Loughin, T.; Dowell, F. Single Kernel Protein Variance Structure in Commercial Wheat Fields in Western Kansas. *Crop Sci.* **2002**, *42*, 1488–1492.
3. Maertens, K.; Reyns, P.; de Baerdemaeker, J. On-line Measurement of Grain Quality with NIR Technology. *Trans. ASAE* **2004**, *47*, 1135–1140.

4. Paliwal, J.; Visen, N.S.; Jayas, D.S.; White, N.D.G. Cereal Grain and Dockage Identification Using Machine Vision. *Biosyst. Eng.* **2003**, *85*, 51–57.
5. Hobson, D.M.; Carter, R.M.; Yan, Y. Characterisation and Identification of Rice Grains through Digital Image Analysis. In Proceedings of the IEEE Instrumentation and Measurement Technology Conference, Warsaw, Poland, 1–3 May 2007; pp. 1–5.
6. Liu, W.; Tao, Y.; Siebenmorgen, T.J.; Chen, H. Digital Image Analysis Method for Rapid Measurement of Rice Degree of Milling. In *1997 ASAE Annual International Meeting Technical Papers*; Paper No. 973028; ASAE: St. Joseph, MI, USA, 1997.
7. Lloyd, B.J.; Cnossen, A.G.; Siebenmorgen, T.J. Evaluation of Two Methods for Separating Head Rice from Brokensfor Head Rice Yield Determination. In *2000 ASAE Annual International Meeting Technical Papers*; Paper No. 006074; ASAE: St. Joseph, MI, USA, 2000.
8. Matsui, M. Threshing Unit of Head Feeding Type Combine. *J. Jpn. Soc. Agric. Mach.* **1999**, *61*, 45–46.
9. Takahara, K.; Hayashi, S. Cleaner Control System for Head-feeding Combine Harvesters. *J. Jpn. Soc. Agric. Mach.* **1996**, *58*, 125–126.
10. Bebis, G.; Egbert, D.; Sham, M. Review of Computer Vision Education. *IEEE Trans. Educ.* **2003**, *46*, 2–21.
11. Brosnan, T.; Sun, D.W. Improving Quality Inspection of Food Products by Computer Vision—A Review. *J. Food Eng.* **2004**, *61*, 3–16.
12. Shantaiya, S.; Ansari, U. Identification of Food Grains and Its Quality Using Pattern Classification. In Proceedings of the 12th IEEE International Conference (ICCT-2010), Bhubaneswar, India, 3–5 December 2010; Volume 2.
13. Verma, B. Image Processing Techniques for Grading & Classification of Rice. In Proceedings of the 2010 International Conference on Computer & Communication Technology (ICCCT '10), Allahabad, India, 17–19 September 2010; pp. 220–223.
14. Yao, M.; Liu, M.; Zhen, H. Exterior Quality Inspection of Rice Based on Computer Vision. In Proceedings of the IEEE 2010 World Automation Congress (WAC), Kobe, Japan, 19–23 September 2010; pp. 369–374.
15. Novini, A. Fundamentals of Machine Vision Component Selection. In *Food Processing Automation II—Proceedings of the 1990 Conference Hyatt Regency*; ASAE: Lexington, KY, USA, 1990; p. 60.
16. Gunasekaran, S. Computer Vision Technology for Food Quality Assurance. *Trends Food Sci. Technol.* **1996**, *7*, 245–256.
17. CCS Inc. *Flat-Dome Lights LFXS2 Series Manual*; CCS Inc.: Kyoto, Japan, 2015, pp. 77–80.
18. CCS Inc. *Ring Lights LDR2 Series Manual*; CCS Inc.: Kyoto, Japan, 2015; pp. 9–12.
19. Miyamoto, M.; Kondo, N.; Ogawa, Y.; Yamamoto, K. Combine Harvester. Yanmar Co., Ltd; Kyoto University. Japan Patent No. 5,780,642, 24 July 2015.

CHAPTER 9

Runoff Harvesting and Storage for Rice Crop at Hamelmalo, Semiarid Region of Eritrea

Ramesh Prasad Tripathi*, Woldeselassie Ogbazghi, Semere Amlsom, Simon Measho

Department of Land Resources and Environment, Hamelmalo Agricultural College, Keren, Eritrea

ABSTRACT

Rice is staple food in Eritrea but it is not cultivated in the country due to semiarid conditions. However, possibilities exist for growing rice using runoff produced from nonagricultural hilly lands, which occupy >50% 80% area of all agricultural watersheds in Eritrea. Study was undertaken in 6 ha watershed at Hamelmalo to design and develop waterway for safe harvesting of runoff from 5.5 ha catchment into a pond for facilitating runoff farming of rice in 0.5 ha field at the outlet end receiving recurrent floods. Slope of the catchment ranged from 1% to 6%. The waterway was designed to intercept and carry runoff from two major drains in nonagricultural land together and delivering into a pond made adjacent to rice field. The waterway was about 323 m long with 3 m top width, 1.5 m bottom width and 0.3 - 0.8 m depth from surface. The pond was 60 m long, 9 m wide and 1 m deep with 1.5 m high earthen dam towards rice field using soil excavated from the pond. Embankment on the remaining sides was mango orchard land slopping towards pond. The dam base width was 4 m, top width was 2.5 m and height was 1.5 m from the ground surface. Two spill ways were provided in the dam at ends at the ground level to facilitate release of runoff from the pond for irrigation. Gross capacity of pond was

>1000 m^3. Combined effect of river water, percolation from pond and wetter rice field raised groundwater table from 3.25 m depth in June to 1.4 m by mid-crop season. This resulted in soil wetness exceeding field capacity below 0.7 m depth and greener crop. Rice yields exceeding 2000 $kg{\cdot}ha^{-1}$ were harvested under runoff farming conditions. Soil puddling was more conducive to rice crop than compaction under available soil and water resources.

Keywords: Compaction, Pond, Puddling, Rice, Runoff Farming, Waterway

1. INTRODUCTION

Rice is not a semiarid region crop because of its high water requirements. However, in Eritrea significant runoff produced from nonagricultural lands, which occupy >50% - 80% area of all agricultural watersheds, can be harvested and used as additional rainwater to facilitate rice farming [1] - [3]. Farmers cultivate crops on un-terraced sloppy lands that also contribute 60% - 70% of rainfall as runoff [2] [4] , which flows forming numerous channels and gullies damaging downstream field crops. In watersheds of size more than 5 ha, runoff frequently floods the downstream fields at the outlet rendering them unfit for raising a good crop during rainy season [3]. The same runoff could be boon to rice production if diverted through designed waterways into small ponds constructed on upstream side of such fields [3]. Crop yields in Eritrea are low (<0.2 - 0.6 $t{\cdot}ha^{-1}$) because >70% of the total rainfall is lost as runoff from traditional sloppy fields [1] [2] [5]. Storage of even part of the runoff in watersheds, which is also rich in nutrients and soil colloidal fractions, can serve as not only extra rainwater but also nutrients for crops [3] [6] - [8]. Rice cultivation through water harvesting in terraced fields was successfully demonstrated in Uganda [9]. Although rice likes wetter regime in the root zone, irrigations applied at 8 days interval were also optimum for rice growth in Sudan [10].

Rice needs water to meet its evapotranspiration (ET) requirements and to satisfy percolation from the soil moisture regime of field capacity to submergence necessary for optimum yields [11] [12]. High percolation from rice fields and ponds made for harvesting runoff to meet water requirements of rice will also recharge ground water table that may serve as source of sub-irrigation directly into the crop root zone [3] [13]. Tesfamichael [14] observed development of water table that fluctuated from 0.4 - 1.2 m depth from surface during October 2006 to March 2007 in about 6 ha valley farmland in Akriya, Asmara region of Eritrea, surrounded by 36.5 ha hilly terrain. Contribution of groundwater table to ET requirements of wheat was 90% from the water table fluctuating from 0.4 -

0.55 m depth from surface, which declined to 38% as water table dropped to 1.2 m depth by milk stage [14].

Runoff farming has been traditionally practiced as spate irrigation in Eritrea since more than 100 years ago [7]. Spate irrigation is a pre-planting system of irrigation by diverting seasonal rivers producing flash floods from highlands and mountainous areas to recharge soil profile in the lowlands. More than 50% of total irrigation in Eritrea is through spate irrigation practiced from eastern to western lowlands along the Red sea coast [15] [16]. Depending upon rainfall, catchment size and runoff from highlands, diversion dams are constructed in the rivers to channelize runoff into cultivated fields to recharge soil profile in several flash floods before planting a crop. Water use efficiency of spate irrigation systems in Eritrea has been low due to wild flooding of uncropped fields leading to over or inadequate watering [7] [16]. All spate irrigated fields in Eritrea could prove boon to rice production with high water use efficiency through improvements in existing diversion structures to regulate flood water supply and field level management. Rice could be grown during preplanting irrigation period from the third week of June to October with little intervention in the existing cropping system and flood water management. Objective of this study was to design and develop runoff harvesting system in agricultural watersheds associated with nonagricultural lands to facilitate runoff farming of rice in semiarid environments of Hamelmalo, Anseba region of Eritrea.

2. MATERIALS AND METHODS

Study was undertaken in 2013 and 2014 in a 6 ha partly cultivated watershed at Hamelmalo Agricultural College (15°52'20.6''N and 38°27'57.6''E at 1280 msl), located in the semiarid region of Eritrea. Annual rainfall in the past seven years ranged from 370 - 663.1 mm with a mean of 488 mm and average annual pan evaporation of 1931 mm. Highest mean monthly temperature occurred in May (35.7°C) and lowest in January (11.1°C). Total rainfall was 388 mm in 2013 and 429 mm in 2014 (Figure 1). Monsoon season in the two cropping years started from third week of June and ended by third week of September. Highest storm rainfall was 57.7 mm in July, 2013 and 48.5 mm in September, 2014.

2.1. Watershed Characteristics

The watershed land was surveyed using total station and GPS to determine slope, contour lines, ridge lines, boundary, area and location of natural drains. About 2.5 ha watershed land was cultivated and was on the downstream side and 3.5 ha uncultivated. The uncultivated land comprised of wild bushes and trees, dominated by

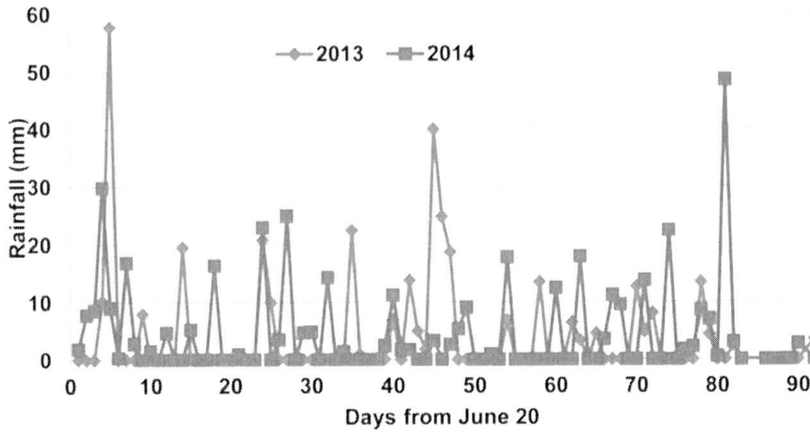

Figure 1. Rainfall during the crop season of 2013 and 2014.

Acacia and Zizyphus species and 4 buildings, each having cemented plan area of 800 m^2, together with an old mango orchard and a small building (40 m^2) adjacent to the selected rice field. Runoff from buildings was 100% of the rainfall. Soil texture was stony sandy loam in uncultivated portion, sandy loam in cultivated portion and loam in the rice field. Slope of the catchment contributing runoff for use by rice crop ranged from 1% - 6% and that of rice field was <1%. Average bulk density of soil was 1.69 $Mg{\cdot}m^{-3}$ in the uncultivated field and 1.4 Mg m^{-3} in the rice field. Presence of stones, beating action of raindrops and trampling by grazing animals over the years increased density of the uncultivated land. Two major drains were carrying runoff from uncultivated land on the upstream and leading to gully formation in downstream cropped fields before entering into Anseba river through the selected rice field. Anseba river is normally dry but flows whenever rainfall occurs in its catchment.

2.2. Design Considerations

Waterway was designed to intercept and carry runoff from the two drains of the uncultivated land together about 3 m upstream of the cultivated land into a pond constructed adjacent to the rice field (Figure 2). The two drains at the junction of cultivated and uncultivated lands were about 10 m wide and 40 m apart. Slope of the land, about 3 m upstream from the cultivated land, was <1% - 1.5%. The waterway was thus proposed from 3 m on upstream side of the cultivated land running about 50 m across the slope dividing the cultivated and uncultivated land before turning downslope towards the pond. Minor bending was allowed to accommodate local land surface conditions. To ensure non-erosive velocity of flow, channel bed slope was maintained zero by providing successive drops in bed level at the points of sudden change in slope. Equation (1) guided vertical interval between the successive channels [17].

$$VI = 0.12\ s + 0.3 \tag{1}$$

where VI is vertical interval (m) between channel beds, and s is per cent slope.

Flow in the waterway between successive drops was due to hydraulic gradient established by runoff. Length of each level section of the waterway and drop height was variable. Stones and gravels were placed at the drop section of the waterway to reduce scouring. Due to lack of data on rainfall intensity and duration, waterway and pond dimensions were decided from rainfall-runoff relationship developed from measurements made in another 4.29 ha model watershed at Hamelmalo farm under similar soil and vegetation conditions [2] and peak storm rainfall in the last 10 years. Peak storm rainfall in 2010 at Hamelmalo was 77.2 mm, therefore, waterway and pond were designed for the runoff that can be produced from a peak storm rainfall of 80 mm. Most storm rainfall amounts in the area are <30 mm. Following was the rainfall-runoff relationship on 3% slope used in the design.

$$\text{Runoff} = 0.52 \times \text{rainfall} - 2.0 \tag{2}$$

Both rainfall and runoff are in mm. The waterway length was 323 m, top width 3 m, bottom width 1.5 m and depth 0.3 - 0.8 m from surface. The pond length was 60 m, width 9 m and depth 1 m with 1.5 m high earthen dam on the downstream side adjacent to the rice field. Soil excavated from the pond was acceptable for construction of the dam. Embankment on the other three sides was not disturbed to allow natural local inflow of runoff from the adjoining mango orchard. Base width of the dam was 4 m, top width 2.5 m and height 1.5 m from the ground surface. Top width of 2.5 m was to facilitate thorough passage of farm machineries and ensure stability of the dam. Provision of two spill ways was made at ends of the dam at the ground level to allow irrigation and release of extra runoff from the pond. Gross capacity of the pond was >1000 m^3. Location of buildings, natural drains, Anseba River, rice field and designed waterway can be seen in Figure 2.

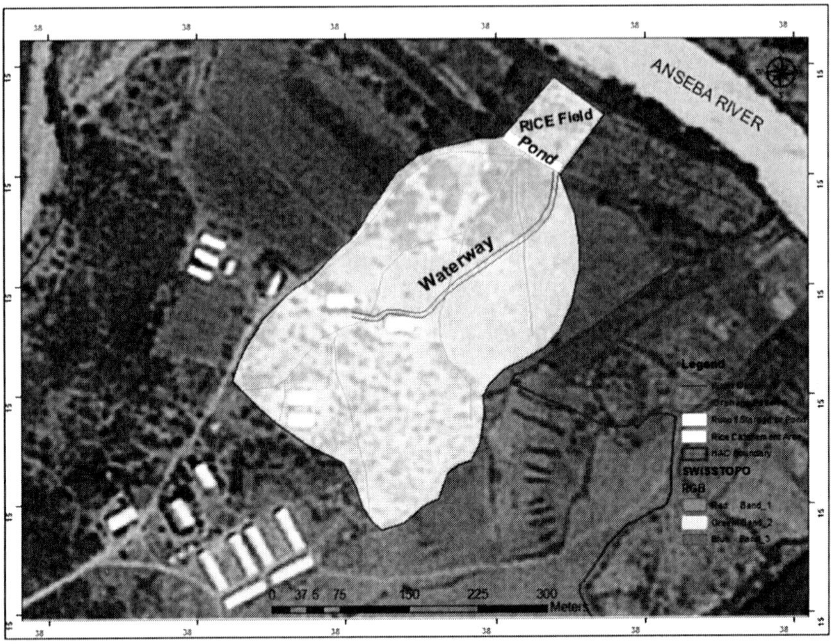

Figure 2. Rice field and catchment area contributing runoff.

2.3. Crop Response Evaluation

Effect of water harvesting on runoff farming of rice was evaluated through field experiments in 2013 and 2014. Since percolation is a major water management problem in rice farming, emphasis on treatment selections was to optimize level of soil puddling or compaction done to control percolation without affecting rice growth. Rice variety NERICA 4 was transplanted at 5 levels of puddling and 5 levels of compaction. Rice seedlings were raised 21 days before transplanting in the field. Puddling treatments were no puddling (T_0), puddling by one (T_1), two (T_2), three (T_3) and four (T_4) passes of puddler replicated 4 times [18]. To simulate puddling by local bullock drawn plough, wet tillage in the plots was done manually by spade followed by churning-cum-leveling by small wooden plank. Soil compaction treatments were no compaction (T_0), compaction by one (T_1), two (T_2), three (T_3) and four (T_4) passes of 600 kg tractor driven roller in 4 replications [19]. Both the experiments were separately laid out in complete randomized block design. Irrigations were given to maintain 20 - 50 mm submergence for 10 days for seedling establishment after transplanting in puddled and compacted fields. Fertilizer nutrients applied were 120 kg N and 46 kg P ha^{-1} through urea and DAP. Entire DAP was applied as basal dose during last puddling operation and remaining N was applied through urea in two equal splits at 20 days interval from transplanting. Rice is normally broadcasted under rainfed farming but transplanting was preferred to shorten crop duration in the field by raising nursery 21 days before commencement of the monsoon. After cessation of rainfall in September, water requirement of rice was much higher than 120 mm irrigation applied in 6 m × 1 m nursery plots. Submergence by rainfall or runoff was allowed to a maximum of 0.1 m. Results of 2013 experiment showed superiority of puddling on crop performance and, therefore, only puddling experiment was repeated in 2014.

3. RESULTS AND DISCUSSION

3.1. Construction of Waterway and Pond

As per design, waterway was constructed to intercept and carry runoff from 3 m upstream of the cultivated land, running 50 m across the slope with minor bending to accommodate local land surface situations before turning downslope towards the pond (Figure 3). The waterway downslope followed one of the drains almost centrally dividing the lower catchment. Soil excavated from the first 50 m waterway running across the slope was placed only on the downstream side of the embankment (Figure 3). Embankment on the upstream was natural ground level to facilitate runoff interception. However, embankments were formed on both sides after turning downslope until next 120 m. Thereafter slope of the cultivated land on the right side was perpendicular to the waterway and, therefore, embankment was again made only on the left bank to facilitate direct runoff interception from the right side. The waterway depth was varying with slope. Entire runoff from the cultivated land joined the waterway in its lower 1/4 section. Provision for inflow of runoff from left side of the waterway was also made at appropriate locations.

The pond of designed size (60 m × 9 m × 1 m) was constructed adjacent to rice field on the upstream side. Pond embankment (dam) adjacent to rice field was of base width 4 m, top width 2.5 m and height 1.5 m from ground surface. The remaining 3 sides were undisturbed mango orchard land sloping towards the pond, which facilitated not only direct inflow of local runoff into the pond but also increased pond capacity with rising water level above the land surface. Soil excavated from the pond was used for construction of the dam after removing gravels and stones. Two orifices of diameter 0.125 m were installed at ends of the dam for safe release of excess runoff from pond into 1 m wide (base width) field channel constructed on the two sides of the rice field. The dam was repeatedly compacted by moving a 600 kg tractor driven roller. Thorough passage of tractor and human traffic was allowed over the dam for its safety and proper use of the farmland. Entire work was done manually by students (Figure 4). In case of emergency, provision for natural spillway was also made on eastern side of the embankment from where overflowing runoff could be diverted into another safety channel of 1.5 m base width made along irrigation channel in the rice field. Natural inflow of runoff from sides was also intercepted and stored in this safety channel before allowing into irrigation channels or rice plots.

3.2. Performance of Waterway and Pond

The waterway has been performing well for the last 3 years. Soil texture of the rice field was slowly transforming from loam to sandy loam and loamy sand in lower depths below 0.8 - 1 m. Pond was thus draining quickly after every rainfall. In 2013, pond received runoff to almost its full capacity twice but in 2014, it received only once due to lower storm rainfall. However, waterway, pond and rice field channels assured additional water availability to the crop and zero runoff from the watershed. The incoming colloidal sediments with runoff in the pond are expected to clog soil pores with time allowing water to stay for longer time. Part of the sediments received in the pond could also be used as manure in the rice field or elsewhere as necessary.

Figure 3. Waterway to intercept runoff.

Figure 4. Students digging the pond.

3.3. Development of Water Table and Soil Wetness

Seepage from Anseba river, runoff stored in the pond and irrigation channels and wetness in the rice plots raised groundwater table in the rice field from 3.25 m depth in June end to 1.4 m by 4th week of August, which receded down to 1.7 m by crop maturity (Figure 5). Rate of rise of water table was faster than receding. The water table rose by 1.85 m in about 48 days but receded only 0.3 m in next 75 days. The water table was oscillating within 1.5 ± 0.1 m for about 2 months during grand growth to reproductive stages of rice crop.

The shallow groundwater table greatly affected wetness in the root zone. Soil water content in the third week of September showed upward gradient (Figure 6). Whereas surface soil was dry forming cracks, soil wetness below 0.7 m depth was near field capacity and was increasing with depth (Figure 6). Roots of NERICA rice were observed down to 0.8 m [3] and, therefore, parts of the rice roots were receiving water by capillarity from the fluctuating water table and the crop was greener until October (Figure 7). The rising groundwater table due to combined effect of water level in the river and percolation from the pond and rice field may serve as a natural source of sub-irrigation to the rice crop. The crop was harvested in November first week.

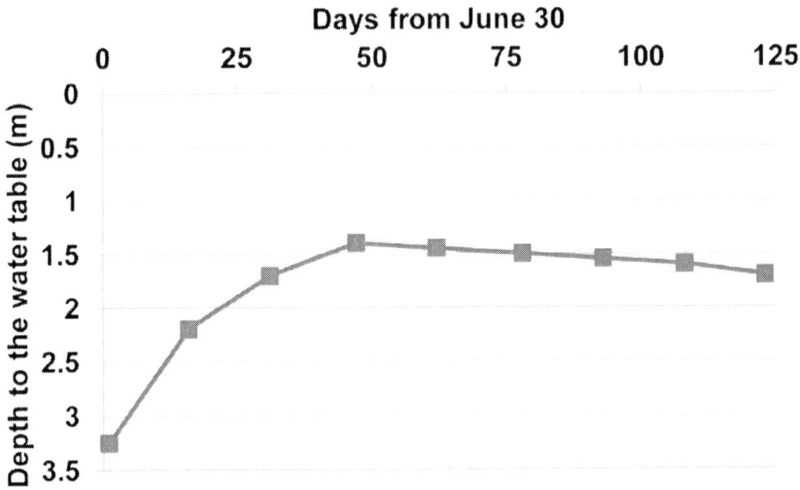

Figure 5. Two-year average water table fluctuation during the crop season.

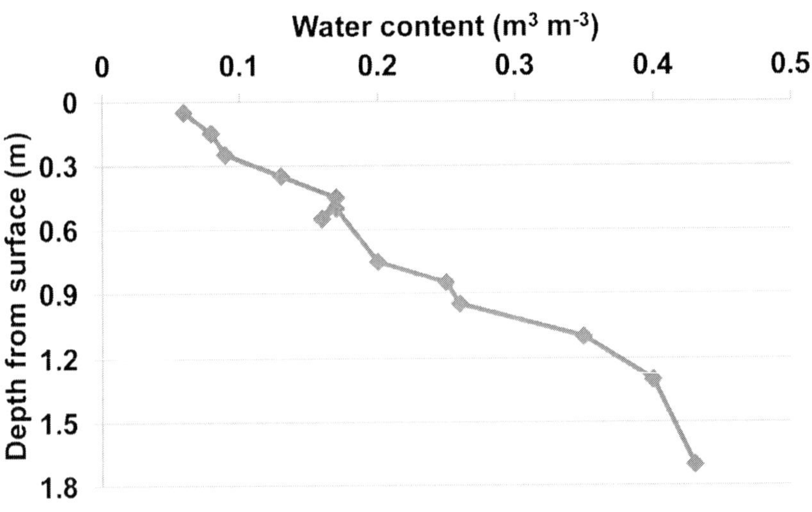

Figure 6. Soil moisture profile in the rice field.

Figure 7. Rice experimental crop on October 25, 2014.

3.4. Performance of Rainfed Rice

Rice yields in 2013 ranged from 782 kg·ha^{-1} under no puddling to 2307 kg·ha^{-1} under 4 puddlings (Table 1). Effect of 2 or 4 puddlings was statistically equal on crop performance. In 2014, yields increased from 850 kg·ha^{-1} under no puddling to 2438 kg·ha^{-1} under 2 puddlings. Yields in second year were also at par under 2 and 4 puddlings. Results thus show that 2 puddling operations by country plow would be sufficient for optimum rice yields in the loam soil of Hamelmalo under runoff farming. Similarly compaction, which had similar effect on percolation, improved rice yields from 690 kg·ha^{-1} under no compaction to 1980 kg·ha^{-1} under compaction by 4 passes of 600 kg roller. Yields were at par under 2 and 4 passes of roller. Results show that puddling was superior to compaction in terms of crop performance under runoff farming conditions of Hamelmalo. These yields are as against highly poor sorghum stands due to recurrent floods and consequent wetness and deposition of eroded materials in such fields. Since successful rice crop has been harvested for the first time in Eritrea under runoff farming conditions more multilocational trails are necessary to optimize soil, water and nutrient requirements in addition to development of less than 4 month duration NERICA varieties suitable for different rainwater management conditions.

Table 1. Effect of puddling and compaction on grain yield of NERICA 4 rice.

Puddling by spade	Grain yield in 2013 ($kg \cdot ha^{-1}$)	Grain yield in 2014 ($kg \cdot ha^{-1}$)	Compaction by 600 kg roller	Grain yield in 2013 ($kg \cdot ha^{-1}$)
T_0 (No puddling)	782	850	T_0 (no compaction)	690
T_1 (1 puddling)	1181	1488	T_1 (1 pass roller)	1234
T_2 (2 puddling)	1811	2438	T_2 (2 pass roller)	1430
T_3 (3 puddling)	1921	2313	T_3 (3 pass roller)	1550
T_4 (4 puddling)	2307	2063	T_4 (4 pass roller)	1980
Mean	1600	1830	Mean	1369
LSD at 5%	552	572	LSD at 5%	605

4. CONCLUSIONS

1) Water harvested from >50% - 80% nonagricultural lands of all agricultural watersheds in Eritrea can significantly meet water requirements of rice.
2) Earthen waterway and pond constructed to harvest runoff would be viable in arresting further land degradation.
3) Rice yields exceeding 2000 $kg \cdot ha^{-1}$ could be harvested under runoff farming conditions.
4) Soil puddling twice by local plow would be better option in loam soil but it should be verified for coarser texture soils.
5) River or any natural drain on the downstream rice field and upstream pond could raise groundwater table significantly to serve as source of sub-irrigation directly into root zone.
6) Efforts on rice cultivation could save hard currency to import rice for local consumption.

ACKNOWLEDGEMENTS

Authors are grateful to UNDP for providing funds from SGP/GEF grant for carrying out this research and to Mr. Tseneo Tsrusaki, Japanese International Cooperation Agency for supplying NERICA rice seed.

REFERENCES

1. MOA (Ministry of Agriculture) (2005) Area and Production by Zoba from 1992-2005. MOA, Asmara, Eritrea.
2. Tripathi, R.P. and Ogbazghi, W. (2010) Development and Management of a Hilly Watershed in Hamelmalo Agricultural College farm, as a Demonstration Site for Farmers and a Study Site for Students. Final Technical Report of the Project Financed by Eastern and Southern Africa Partnership Programme (ESAAP), Department of Land Resources and Environment, Hamelmalo Agricultural College, Keren, Eritrea, 59 p.
3. Tripathi, R.P., Ogbazghi, W., Amlesom, S. and Araia, W. (2014) Optimizing Tillage and Rain Water Conservation in the Soils of Hamelmalo Region of Eritrea for Arresting Soil Degradation and Achieving Sustainable High Crop Yields. Final Technical Report of the Project Financed by GEF/SGP, UNDP, Department of Land Resources and Environment, Hamelmalo Agricultural College, Keren, Eritrea, 112 p.
4. Temesgen, M., Rockstrom J., Savenije, H.H.C., Hoogmoed, W.B. and Alemu, D. (2008) Determination of Tillage Frequency among Smallholder Farmers in Two Semiarid Areas in Ethiopia. Physics and Chemistry of the Earth, 33, 183-191. http://dx.doi.org/10.1016/j.pce.2007.04.012
5. FAO (2005) Global Information and Early Warning System on Food and Agriculture World Food Programme. Special report FAO/WFP Crop and Food Supply Assessment Mission to Eritrea. FAO, Rome.
6. Elwell, H.A. and Stocking, M.A. (1988) Loss of Soil Nutrients by Sheet Erosion Is a Major Hidden Farming Cost. The Zimbabwe Science News, 22, 79-82.
7. Tesfai, M. (2001) Soil and Water Management in Spate Iirrigation Systems in Eritrea. Ph.D. Thesis, Department of Erosion and Soil and Water Conservation, Wageningen University and Research Center, The Netherlands, 211 p.
8. Ashouri, M. (2012) The Effect of Water Saving Irrigation and Nitrogen Fertilizer on Rice Production in Paddy Fields of Iran. International Journal of Bioscience, Biochemistry and Bioinformatics, 2, 56-59. http://dx.doi.org/10.7763/IJBBB.2012.V2.70
9. Goto, A., Nishimaki, R., Suzuki, S., Watanabe, F. and Takahashi, S. (2012) Terrace Development Applied as a Water Harvesting Technology for Stable NERICA Production in Uganda. Journal of Arid Land Studies, 22, 243-246.
10. Abu, S.T. and Malgwi, W.B. (2012) Effects of Irrigation Regime and Frequency on Soil Physical Quality, Water Use Efficiency, Water Productivity and Economic Returns of Paddy Rice. ARPN Journal of Agricultural and Biological Science, 7, 86-99.
11. Tripathi, R.P., Kushwaha, H.S. and Mishra, R.K. (1986) Irrigation Requirement of Rice under Shallow Water Table Conditions. Agricultural Water Management, 12, 127-136. http://dx.doi.org/10.1016/0378-3774(86)90011-9
12. Bajpai, R.K. and Tripathi, R.P. (2000) Evaluation of Non-Puddling under Shallow Water Tables and Alternative Tillage Methods on Soil and Crop Parameters in a Rice-Wheat System in Uttar Pradesh. Soil and Tillage Research, 55, 99-106. http://dx.doi.org/10.1016/S0167-1987(00)00111-2
13. Tripathi, R.P., Kushwaha, H.S. and Agrawal, A. (1987) A Simple Non-Weighing LysimeterInstallation with Rain Shelter. Agricultural and Forest Meteorology, 41, 275-288. http://dx.doi.org/10.1016/0168-1923(87)90084-0
14. Tesfamichael, F. (2007) Integrated Effect of Irrigation and Nitrogen on Wheat Productivity under Shallow Water Table Condition in Eritrea. M.Sc. Thesis, Department of Land Resources and Environment, Hamelmalo Agricultural College, Eritrea, 91 p.
15. Tesfai, M. and Stroosnijder, L. (2000) The Eritrean Spate Irrigation System. Agricultural Water Management, 48, 512-560.

16. Amlesom, S. (2005) Review of Irrigation Development in Eritrea. In: Mehari, T. and Ghebru, B., Eds., Irrigation Development in Eritrea: Potentials and Constraints, Proceedings of the Workshop of the Association of Eritreans in Agricultural Sciences (AEAS) and the Sustainable Land Management Programme (SLM), Eritrea, 14-15 August 2003, 5-11.
17. Abraham, K., Dawit, F., Kibrom, B., Natsnet, O. and Selemawit, W. (2014) Effect of Puddling on Soil Properties and Yield of Rice. Senior Research Project, Department of Land Resources and Environment, Hamelmalo Agricultural College, Eritrea, 40 p.
18. Asmelash, M., Filmon, D., Amanuel, G., Selam, E. and Natsnet, K. (2014) Effect of Compaction on Soil Properties and Yield of Rice. Senior Reseach Project, Department of Land Resources and Environment, Hamelmalo Agricultural College, Eritrea, 50 p.
19. ASAE (American Society of Agricultural Engineers) (1989) Design, Layout, Construction and Maintenance of Terrace Systems. ASAE Standard S268.3, St. Joseph, MI.

CHAPTER 10

Optimizing preplant irrigation for maize under limited water in the High Plains

I. Kisekka[a,*], A. Schlegel[b], L. Ma[c], P.H. Gowda[d], P.V.V. Prasad[e]

[a] *Kansa State University, Southwest Research-Extension Center, 4500 E. Mary St., Garden City, KS 67846, United States*

[b] *Kansas State University, Southwest Research-Extension Center, 1474 State Highway 96, Tribune, KS 67879, United States*

[c] *USDA ARS, Plains Area, Agricultural Systems Research Unit, 2150 Centre Ave., Fort Collins, CO 80526, United States*

[d] *USDA-ARS Grazing Lands Research Laboratory, El Reno, OK 73036, United StateseKansas State University, Department of Agronomy, Manhattan, KS 66506, United States*

ABSTRACT

Due to inadequate irrigation capacity, some farmers in the United States High Plains apply preplant irrigation to buffer the crop between irrigation events during the cropping season. The purpose of the study was to determine preplant irrigation amount and irrigation capacity combinations that optimize yield, water productivity, and precipitation use efficiency (PUE) and minimize soil water evaporation losses prior to planting. The CERES-Maize model embedded in the RZWQM2 model in combination with long-term climatic data from 1986 to 2014 for southwest Kansas were used for this research.

Experimental data from 2006 to 2009 was used to calibrate and validate the model. Model performance was satisfactory with high index of agreement (IA > 0.88). Relative root mean square error (RRMSE) ranged between 4.5% and 27%. Under very limited irrigation capacity (2.5 mm/day), applying 75–100 mm of preplant irrigation produced median yields that were 10–17% higher than not applying preplant irrigation. However, even at limited irrigation capacity the benefit of preplant irrigation were only realized if the seasonal yield potential was in the range of 6000 to 10,000 kg/ha corresponding to years with normal seasonal rainfall. Irrigation capacity had a stronger effect on maize grain yield compared to preplant irrigation amount. Preplant irrigation increased ET and transpiration under 2.5 mm/day irrigation capacity. Preplant irrigation amount did not have a substantial impact on water productivity at high and moderate irrigation capacity but had second order dominant effect under limited irrigation capacity. At low irrigation capacity (2.5 mm/day) increasing preplant irrigation increased median PUE up to 18% although the effect was second order dominant. Negligible water losses through deep percolation from 2.4 m soil profile were simulated. Increasing preplant irrigation resulted in significantly higher soil water evaporation losses prior to planting at all irrigation capacities. Overall preplant irrigation is beneficial under very limited irrigation capacity but is not necessary under sufficient irrigation capacity in most years. The decision to apply preplant irrigation should be evaluated and implemented carefully in combination with other agricultural water management technologies and strategies such as soil water monitoring, drip irrigation, and residue management.

Keywords: Preplant irrigation; Irrigation capacity; RZWQM; DSSAT-CSM; CERES-Maize; Maize yield; Evapotranspiration; Soil water evaporation

1. INTRODUCTION

In the U. S. south and central High Plains, groundwater levels in the Ogallala aquifer are declining due to water withdrawals for irrigation exceeding average annual recharge (McGuire, 2004). This results in diminished well capacities that eventually became incapable of meeting full crop water needs during the summer growing season. For this reason, some irrigators on fine and medium textured soils apply preplant irrigation from groundwater to ensure that the soil profile has adequate water before planting to help their low irrigation capacity systems to better keep up with the crop water needs during the growing season. Preplant irrigation (also known as preseason, dormant season, off-season, or winter irrigation) is a water management strategy in which water is applied prior to or several months before planting. This practice is common to the U.S. South and

Central High Plains (including areas in western Kansas, eastern Colorado and the panhandles of Texas and Oklahoma) as previously reported by Stone et al. (1994). Surveys conducted in western Kansas in the later part of the 20th century indicated over 60% of respondents used some form of preplant irrigation (Kromm and White, 1990). Recent droughts of 2011 and 2012 coupled with declining well capacities have stimulated interest again in preplant irrigation.

The major question for the producers is: when should preplant irrigation be used or when is it beneficial? Also, what is the best way to implement preplant irrigation in order to minimize nonproductive water losses that make this practice inefficient? Optimum preplant irrigation strategies target minimizing soil water evaporation, enhancing precipitation capture and storage and eliminating deep drainage. Preplant irrigation water is lost through two major processes; deep drainage and soil water evaporation. In wet years preplant irrigation might also use up storage that would have been used by spring rainfall which reduces rainfall storage efficiency. Stone et al. (2008) simulated efficiency of preplant irrigation and reported that available water in the soil at the time of the preplant irrigation application had the greatest effect on storage efficiency, which decreased dramatically when available water was greater than 60%. Stone et al. (2008) also reported that preplant irrigations made in the early spring were more efficient compared to irrigation made in the fall. Soil water during the off season is influenced by soil water at harvest which in turn is influenced by irrigation scheduling during the growing season. Given the significant effect of soil water on efficiency of preplant irrigation, a need exists for dynamic site specific decision support tools that can be used for predicting soil water over time and space and to assess need for and effect of preplant irrigation on soil water evaporation, transpiration, deep drainage, crop yield, and water productivity. Such a decision support tool can be developed from whole system agricultural models.

Whole system models integrate the physical, biological and chemical processes of an agricultural system and are very useful for extrapolating field research to different soils, climate and management, technology transfer and decision making (Ahuja et al., 2000). The need for such models and decision support tools will increase as farmers and other stakeholders demand quick transfer of research results in an integrated and usable form for site specific management. Calibrated and validated cropping systems models are an example of whole system agricultural decision support tools. Performance of these decision support tools can be enhanced by integrating them with field measurements such as soil water feedbacks. Examples of cropping systems models that can be used in assessing need and potential benefits of preplant irrigation include RZWQM2 (Ahuja et al., 2000), DSSAT-CSM (Hoogenboom et al., 2015), WOFOST (Van Diepen et al., 1989), APSIM

(Keating et al., 2003), and AquaCrop (Raes et al., 2009) Agricultural systems models have been successfully applied in assessing irrigation management in the U.S. Great Plains (Saseendran et al., 2008; DeJonge et al., 2012; Ma et al., 2012; Kisekka et al., 2016a ; Kisekka et al., 2016b).

In this paper, the RZWQM2 (Root Zone Water Quality Model) was selected for the following reasons: 1) process oriented, dynamic, and simulates the impact of agricultural management practices such as tillage, residue management, irrigation, and fertility on soil water, crop production, and water quality (Ahuja et al., 2000), 2) DSSAT v4.0 has been embedded into RZWQM2 which provides a suite of detailed biophysical crop models for simulating crop growth and development (Ma et al., 2006), 3) RZWQM2 has advanced features for assessing limited irrigation strategies such as water allocation limits by period, irrigation scheduling based on ET or soil water deficit, and 4) an automatic optimization algorithm called PEST (Doherty, 2009) has also been embedded into RZWQM2 to facilitate parameter estimation which allows for reproducible and objective model calibrations.

The purpose of the study was to determine preplant irrigation amount and irrigation capacity combinations that optimize yield, water productivity, and precipitation use efficiency (PUE) and minimize soil water evaporation losses prior to planting. The research involved combining short term experimental data with long-term historic climatic data (1986–2014), and crop simulation modeling to determine optimum preplant irrigation water management in the United States High Plains.

2. MATERIALS AND METHODS

2.1. Experimental data

A field experiment was conducted at the Kansas State University Southwest Research-Extension Center near Tribune Kansas (38.5°N, 101.7°W, and 1086 m above sea level) from 2006 to 2009. The soil at the study site is a deep well drained Ulysses silt loam (fine-silty, mixed, mesic Aridic Haplustolls). The climate of the study site is semi-arid with mean annual precipitation of 440 mm and mean annual evapotranspiration of 1943 mm (1986–2014). The growing season precipitation from 2006 to 2009 is given in Table 1. The experimental design was a factorial with preplant irrigation (0 and 76 mm), irrigation capacity (2.5, 3.8 and 5.0 mm/day) and seeding rate (80,000 plants/ha) replicated four times. Experimental plots were 36 m long by 18 m wide with irrigation as the main plot factor and seeding rate as subplots (3 × 36 m) with maize row spacing of 76.2 cm. Standard practices for maize production in the region were followed including preplant application of 269 kg N ha^{-1} as urea-ammonium nitrate (the model developed based on this experimental data was not calibrated for nitrogen). The full season maize hybrids used were 33B53 in 2006, Pioneer 33B54 in 2007, Pioneer 34B99 in 2008, and Pioneer 34B94 in 2009. Maize was planted on 01/May/2006, 10/May/2007, 26/Apr/2008, and 06/May/2009. Two middle rows of each plot were mechanically harvested. For model calibration and validation only experimental data corresponding to the 80,000 plants/ha seeding rates was used. Total irrigation applications are summarized in Table 1. Details and results from the field study are reported in Schlegel et al. (2012).

Table 1. Preplant and growing season irrigation and growing season precipitation for a field study from 2006 to 2009 at Kansas State University Southwest Research-Extension Center near Tribune, Kansas (Schlegel et al., 2012).

Year	Preplant irrigation (mm)	Growing season irrigation (mm)			Growing season precipitation (mm)
		2.5[a] (mm/day)	3.8 (mm/day)	5 (mm/day)	
2006	76	243	320	483	176
2007	76	183	257	397	205
2008	76	209	278	375	238
2009	76	225	299	453	364

[a] Irrigation capacity.

Irrigation was done using a linear move system modified to apply irrigation based on different treatment combinations. Preplant irrigation was applied in two events of 38 mm several days apart in early spring (April). In-season irrigations were initiated from late May to early September. In-season irrigation was scheduled as follows 35 mm weekly, 25 mm weekly and 35 mm bi-weekly for the 5.0, 3.8 and 2.5 mm/day irrigation capacities respectively. In case of sufficient rainfall, a scheduled irrigation was postponed. Soil water was measured using neutron attenuation with a field calibrated neutron probe. Seasonal crop water use (ETc) was estimated from a soil water balance as reported in Schlegel et al. (2012).

2.2. Model description

The RZWQM2 model (version 3.0) with embedded DSSAT-CSM CERES-Maize 4.0 crop growth module was used in this study (Ma et al., 2012). In RZWQM2, unsaturated soil water flow and redistribution was modeled using a one dimensional Richards's equation. The potential evaporation and transpiration demand of the atmosphere were computed using the Shuttleworth–Wallace ET model (Ahuja et al., 2000). The Shuttleworth–Wallace ET model in RZWQM2 is an extension of the Penman-Monteith ET model but the former takes into account incomplete canopy cover and plant height in potential evaporation and transpiration estimations. Actual root water uptake described by the sink term in Richards's equation was calculated numerically using the Nimah and Hanks (1973) procedure. Actual root water uptake, potential evaporation and potential transpiration were used in computing water stress factors used in CERES-Maize to modulate plant growth process like leaf growth as soil water gets depleted. The CERES-Maize model embedded in RZWQM2 is a radiation-based mechanistic crop model that predicted maize growth and development based on weather (precipitation, solar radiation, maximum and minimum temperature, to a lesser extent photoperiod); 6 cultivar/genetic coefficients (P1, P2, P5, G2, G3, and PHINT [as defined in Table 2]); and management practices (plant population, row spacing, planting date, and irrigation management) as described in Jones et al. (2003). More details about the RZWQM can be found in Ahuja et al. (2000).

Table 2. Soil and plant parameters for simulating soil water flow and maize growth and development in RZWQM-DSSAT at the Kansas State University Southwest Research-Extension Center near Tribune, Kansas.

Name	Description	Layer	Initial value	Lower bound	Upper bound	PEST[a]optimized value
		1	1.32	1	2	1.37
		2	1.32	1	2	1.32
		3	1.32	1	2	1.28
		4	1.32	1	2	1.28
	Bulk Density (g/cm3)	5	1.32	1	2	1.32
		6	1.32	1	2	1.32

		7	1.32	1	2	1.29
		8	1.32	1	2	1.32
Ks	Saturated Hydraulic Conductivity (cmh^{-1})	1	1.57	1	20	1.56
		2	3.24	1	20	3.23
		3	2.53	1	20	2.52
		4	1.00	1	20	1.00
		5	1.98	1	20	1.98
		6	3.51	1	20	3.49
		7	1.44	1	20	1.43
		8	1.82	1	20	1.82
$\theta_{1/3bar}$	Volumetric Soil Water Content at 1/3 bar ($cm^3\ cm^{-3}$)	1	0.33	0.28	0.38	0.34
		2	0.34	0.29	0.39	0.34
		3	0.30	0.26	0.35	0.29
		4	0.30	0.25	0.34	0.29
		5	0.33	0.28	0.38	0.32
		6	0.34	0.29	0.39	0.33
		7	0.24	0.20	0.28	0.24
		8	0.23	0.20	0.27	0.23
θ_{15bar}	Volumetric Soil Water Content at 1/3 bar ($cm^3\ cm^{-3}$)	1	0.16	0.13	0.18	0.16
		2	0.09	0.08	0.11	0.09
		3	0.15	0.13	0.18	0.16

		4	0.13	0.11	0.15	0.13
		5	0.14	0.12	0.16	0.14
		6	0.11	0.10	0.13	0.11
		7	0.14	0.12	0.16	0.14
		8	0.16	0.13	0.18	0.16
P1	Thermal time from seedling emergence to the end of Juvenile (degree days)		200.00	100.00	450.00	205.0
P2	Extent to which development is delayed due to increase in photoperiod		0.70	0.01	2.0	0.70
P5	Thermal time from silking to physiological maturity (Degree Days)		900.00	600	1000	867.0
G2	Maximum possible number of kernels per plant		516	440	1000	531.0
G3	Grain filling rate during the linear grain filling stage		12	5	16	12.5

	(mgday^{-1})					
PHINT	Phyllochron interval (degree days)		38.9	25	55	41.0

[a] Parameter Estimation algorithm.

2.3. Calibration and validation

Meteorological data was obtained from a weather station located near the study site and data accessed through the High Plains Regional Climate Center (http://www.hprcc.unl.edu/). The soil profile was set to 2.4 m comprising of eight layers 30 cm thick. Soil texture measured at the site was 12% sand, 63% silt, 25% clay within typical ranges of a Ulysses silt loam soil. Initial values and range of saturated hydraulic conductivity were obtained from the NRCS's Web Soil Survey for Ulysses silt loam soil. Initial values of cultivar coefficients were taken from one of the maize hybrids (7000 maize IB0031 PIO 3475) in the DSSAT-CSM database that produced close predictions to measured data during initial manual calibration. In this study, the treatment that received highest irrigation (75 mm preplant irrigation and 5 mm/day irrigation capacity) for each of the four years 2006–2009 was used for model calibration. The rest of the treatments (irrigation capacity of 3.8 and 2.5 mm/day with and without preplant irrigation) were used for model validation using data from 2006 to 2009 (Schlegel et al., 2012). One third of the data was used for calibration and the remaining two thirds were used for validation.

Following initial manual calibration of cultivar coefficients, all soil and plant parameters (P1, P2, P5, G2, G3, and PHINT), soil properties (bulk density, vertical *Ks*), and soil water capacity ($\theta_{1/3}$ and θ_{15}) were fitted using a gradient-based parameter optimization algorithm (PEST) embedded in the RZWQM2 model (Doherty, 2009). Default values for parameters related to carbon and nitrogen cycle processes were used. Parameter bounds in Table 2 were used to constrain automatic parameter estimation by PEST to ranges that were physically meaningfully for the study site based on expert knowledge. Parameters were log transformed to improve linearity to ensure the objective function was efficiently minimized. PEST Regularization was also used to minimize model overfitting. Following PEST optimization, the model was executed with uncertainty (with the best 26 parameter sets) to determine the parameter set that produced the least RMSE following similar procedures as reported in (Malone et al., 2010).

The model was executed continuously from January 1st, 2006 to December 31st, 2009 for each treatment. Observed grain yield, and seasonal crop water use (ET) were used to fit model parameters. Precise dates of critical growth stages were not recorded during the study but tasseling and physiological maturity are generally observed to occur in mid-July and late September respectively (Kisekka et al., 2016a ; Kisekka et al., 2016b). Initial soil water at planting was measured during the study. Model validation was done by comparing measured grain yield, seasonal ET and soil water to predicted values. The statistical goodness-of-fit indicators used in this study included Relative Root Mean Square Error (RRMSE), index of agreement (IA) and Relative Error.

2.4. Model application

The calibrated and validated model was applied to assess the effect of five preplant irrigation amounts (0, 25, 50, 75, 100%) applied in late spring between April 1st to 30th of each year on grain yield, ET, soil water evaporation, and precipitation use efficiency. This analysis was repeated for three irrigation capacities (2.5, 3.8, and 6.4 mm/day) over 29 years of historical weather data from January 1st, 1986 to December 31st, 2014. The model was continuously executed for the 29 years. Initial soil water on January 1st, 1986 was set to field capacity in the top 1.2 m of the soil profile and 60% of field capacity in the deeper profile up to 2.4 m. Planting was set to occur during the first week of May each year at a seeding rating of 80,000 seeds/ha. Nitrogen fertilizer was applied at a rate of 269 kg N ha^{-1} as urea-ammonium nitrate each year as pre-plant. Preplant and in-season irrigation were triggered whenever the soil water reached a defined threshold (60% of plant available water) and each irrigation application was limited to 25 mm. Frequency of irrigation was limited by irrigation capacity. Simulated maize yield was adjusted to 15.5% nominal moisture content to match nominal moisture content of measured yields.

2.5. Determining optimum preplant irrigation management strategies

The most efficient/optimum preplant irrigation management strategy was determined by evaluating two criteria including the mean-Gini dominance (MGD), and Stochastic Dominance Analysis (SD) (Tsuji et al., 1998; Kisekka et al., 2016a ; Kisekka et al., 2016b). The MGD decision criteria states that for example if two risky preplant irrigation management strategies A and B have the following characteristics, as shown in equation 1, then A is the more efficient of the two strategies:

$E(A) \geq E(B)$

and $E(A) - G(A) \geq E(B) - G(B)$ (1)

where E(.) is the expected value and G(.) is Gini-mean difference of distributions A and B.

The MGD decision criteria approach assumes that the decision maker is averse to risk and helps to avoid risky strategies. The Gini-mean difference in equation 1 is a function of Gini coefficient a measure of statistical dispersion. Gini coefficient is computed as half of the Gini-mean difference expressed as in equation 2 for unordered data following Dixon et al. (1987):

$$G = \frac{\sum_{i=1}^{n}\sum_{j=1}^{n}|x_i - x_j|}{2n^2\mu} \tag{2}$$

where |xi – xj| is the absolute difference of a randomly selected pair of values of a random variable, and μ is the arithmetic mean. For stochastic dominance analysis as an example, preplant irrigation management strategy A dominates strategy B if cumulative probability density function (CDF) of yield or water productivity gains from A lie to the right of the CDF of B over the entire probability interval 0–1.

3. RESULTS AND DISCUSSION

3.1. Model calibration and validation

Calibrated parameters for soil physical and hydraulic properties and maize cultivar coefficients are shown in Table 2. These values are within range of those reported by Stone et al. (2011) for Ulysses silt loam soils near Tribune Kansas. Bulk density values in Table 2 are within range of 1.24–1.46 g/cm^3 reported by USDA NRCS in western Kansas (http://websoilsurvey.sc.egov.usda.gov/App/WebSoilSurvey.aspx). Optimized saturated hydraulic conductivity values in Table 2 were also within range of 0.4–5 cm/h that was measured in the USDA NRCS soil surveys. Other studies that have used inverse modeling and measured soil water to characterize soil physical properties include (Šimůnek and Nimmo, 2005; Muñoz-Carpena et al., 2008 ; Kisekka et al., 2015).

Maize cultivar coefficients are within range of those reported by Kisekka et al., 2016a ; Kisekka et al., 2016b for maize hybrids grown in the US Central High Plains. Simulated grain yield from 2006 to 2009 for the treatment with preplant irrigation of 75 mm and irrigation capacity of 6.4 mm/day (well-watered treatment) were in agreement with measured yield (Fig. 1).

Although phenology data was not recorded, it was observed that anthesis occurred in mid-July (personal communication) which was in agreement with simulated phenology, which indicated that silking occurred on 07/10/2006, 07/19/2007, 07/12/2008 and 07/13/2009 respectively. Physiological maturity was generally observed to occur in late September although exact dates were not recorded, the corresponding model simulations were September 01, 2006, September 07, 2007, September 10, 2008 and September 12, 2009.

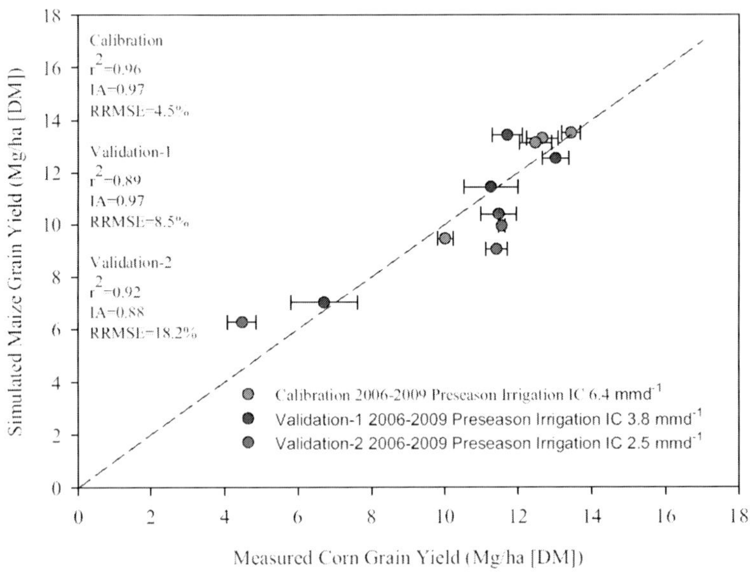

Figure. 1. Comparing simulated to measured maize grain yield in a preplant irrigation study that was conducted at the Kansas State University Southwest Research-Extension Center near Tribune, Kansas, IC refers to irrigation capacity, IA is index of agreement and RRMSE is the relative root mean square.

Seasonal ET was well simulated for all treatments with predicted values within 4–8% of the measured seasonal ET as shown by the goodness-fit-indicators in Fig. 2 and Table 3 with the exception of the 75 mm preplant irrigation and 6.4 irrigation capacity treatment in 2009. The ability of the model to predict grain yield and seasonal ET of maize with acceptable accuracy implies that this model could be applied for various assessments including generation of production functions, optimization of preplant irrigation management and strategic irrigation scheduling under deficit irrigation.

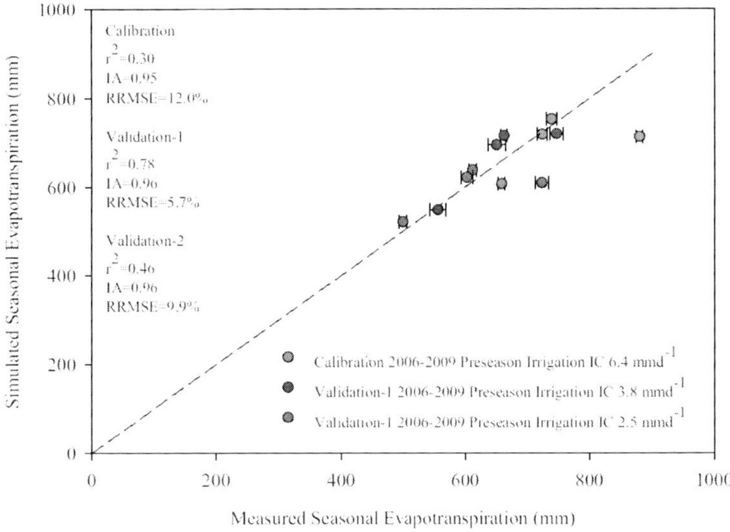

Figure. 2. Comparing simulated to measured seasonal evapotranspiration in a preplant irrigation study that was conducted at the Kansas State University Southwest Research-Extension Center near Tribune, Kansas, IC refers to irrigation capacity, IA is index of agreement and RRMSE is the relative root mean square.

Table 3. Maize grain yield and seasonal evapotranspiration validation goodness-of-fit indicators for treatments that did not receive 75 mm of preplant irrigation during a field study conducted at the Kansas State University Southwest Research-Extension Center near Tribune, Kansas and simulations were from CERES-maize in RZWQM2.

Irrigation capacity (mm/day) with no-preplant irrigation	Grain yield			Seasonal evapotranspiration		
	[a]r^2	[b]IA	[c]RRMSE (%)	r^2	IA	RRMSE (%)
6.4	0.64	0.96	9.0	0.32	0.95	10.4
3.8	0.67	0.84	19.9	0.36	0.93	13.0
2.5	0.44	0.60	32.0	0.48	0.94	12.0

[a] Coefficient of determination.

[b] Willmont Index of Agreement.

[c] Relative root mean square error.

Total soil water storage was also well simulated for both the calibration and validation treatments as shown in Fig. 3. However, the model had a tendency to under predict measured total profile soil water. Possible reasons for the discrepancy include uncertainty in conceptualization of unsaturated flow, measurement uncertainty due to heterogeneity in soil conditions e.g., the sensing volume of the neutron attenuation technique decreases as soil water content increases. Ma et al. (2016) also noted the challenge of spatial soil heterogeneity of soil properties in agricultural systems modeling and reported that best results were obtained when crop parameters and soil field capacity were fit simultaneously.

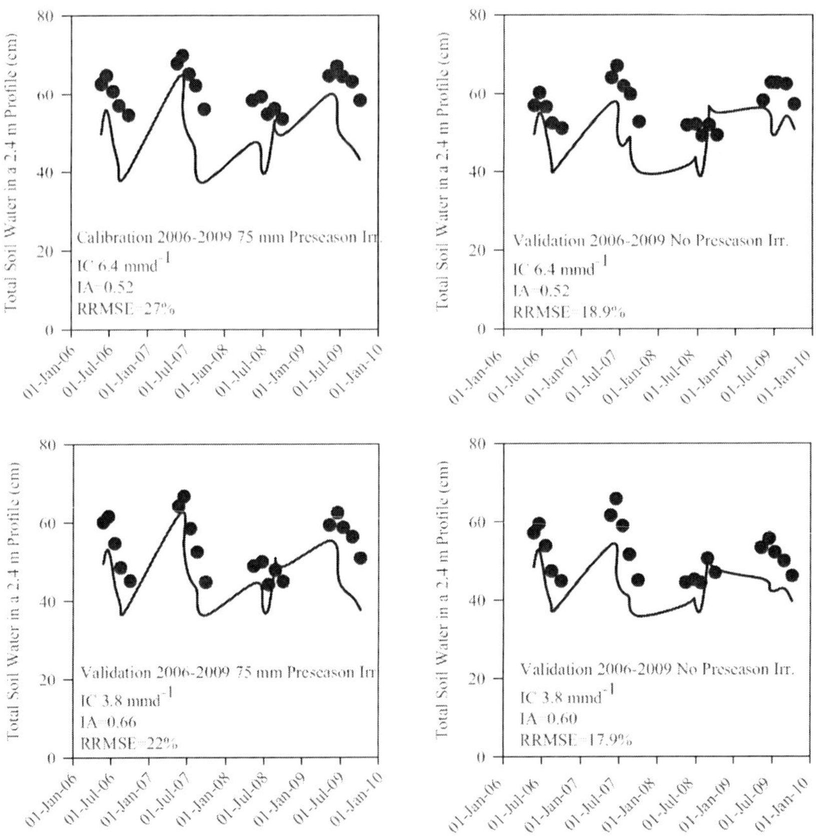

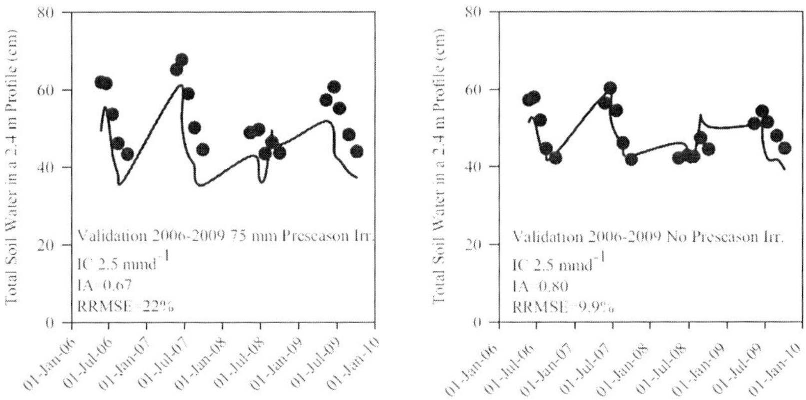

Figure. 3. Simulated and measured total soil profile soil water in a preplant irrigation study that was conducted at the Kansas State University Southwest Research-Extension Center near Tribune, Kansas, IC refers to irrigation capacity.

3.2. Optimizing preplant irrigation amount and irrigation capacity combinations

In order to determine preplant irrigation amount and irrigation capacity combinations that optimize maize grain yields, water productivity, and precipitation use efficiency (PUE) and minimize soil water evaporation losses the mean-Gini dominance (MGD), and Stochastic Dominance Analysis (SD) analysis techniques were used.

3.2.1. Maize grain yield

At very limited irrigation capacity of 2.5 mm/day, applying 50–100 mm of preplant irrigation produced median yields that were 10–17% higher than not applying preplant irrigation. Increases in yield were small under 3.8 and 6.4 irrigation capacity. It can be seen in Fig. 4 that the upper whiskers of the boxplots were larger under 2.5 mm/day compared to higher irrigation capacities, implying increased variability in high yields as irrigation capacity decreases. At very low irrigation capacity, the crop relies heavily on plant available water at planting that is made available to the crop to between rainfall and irrigation events during the growing season. Under limited irrigation capacity, irrigation might be triggered based on soil water status but the actual irrigation events may not occur. This is particularly common with center pivot systems in the High Plains that take several days to complete one revolution due to diminished well capacity from the Ogallala aquifer.

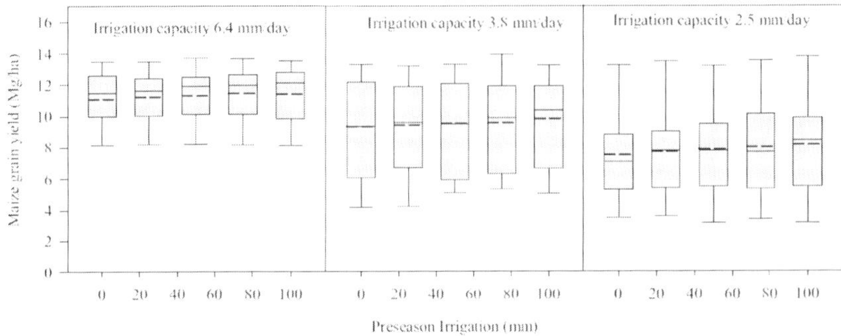

Figure. 4. Simulated maize grain yield under five preplant irrigation amounts and three irrigation capacities using historical weather data from 1986 to 2014 for Tribune, Kansas in a deep silt loam soil (solid line is median and dashed line is mean).

At low (2.5 mm/day) to moderate (3.8 mm/day) irrigation capacity, applying 100 mm of preplant irrigation was the optimum strategy (Table 5). At high irrigation capacity of 6.4 mm/day, applying 75 mm of preplant irrigation produced the highest expected mean grain yield and highest difference between expected grain yield and Gini mean difference as shown in Table 5.

Table 5. Maize grain yield (kg/ha) for different preplant irrigation amounts and irrigation capacities averaged over 29 years (1986–2014) of weather data at Kansas State University Southwest Research-Extension Center near Tribune, Kansas.

Preplant irrigation (mm)	Irrigation capacity					
	2.5 mm d^{-1}		3.8 mm d^{-1}		5.0 mm d^{-1}	
	$E(x)$	$E(x)$-$G(x)$	$E(x)$	$E(x)$-$G(x)$	$E(x)$	$E(x)$-$G(x)$
	Grain Yield (kg/ha)					
0	7464	3851	9306	5353	11083	8842
25	7675	4020	9407	5659	11225	8950
50	7814	4056	9491	5787	11321	9141

75	7948	4123	9526	5836	11450	9293
100	8098	4229	9798	6228	11424	9174

Stochastic dominance analysis results are shown in Fig. 5 and they confirm results from the MGD analysis. It is shown in Fig. 5 that applying 100 mm of preplant irrigation was dominant at the lowest irrigation capacity. However, the dominance of the 100 mm preplant irrigation scenario was second order since its cumulative density function (CDF) was not to the right of all the other CDFs over the entire range of possible yields. Under limited irrigation capacity (2.5 mm/day), preplant irrigation produced higher yields if the yield potential in a given cropping season was in the range of 6000 to 10,000 kg/ha. This corresponded to years with normal growing season (May to October) rainfall of 336 mm. In very wet years and very dry years there was no substantial differences in yields between different preplant irrigation amounts and irrigation capacity combinations. For irrigation capacity of 3.8 and 6.4 mm/day, the benefit from preplant irrigation were not consistent meaning preplant irrigation might not be necessary in most years. It can also be seen from Fig. 5 that there is a 90% chance that increasing irrigation capacity would increase yields since the CDFs of the highest irrigation capacity were to the right of those with lower irrigation capacity. If a producer has a constrained water supply, reducing the total number of irrigated land also termed as concentrating water (Howell et al., 2012) to increase irrigation capacity would increase median yield and also reduce year to year variability in grain yield (Klocke et al., 2015). However, as noted by Howell et al. (2012) concentrating water could reduce income potential in wet years.

3.2.2. Evapotranspiration and water productivity

Fig. 6 also shows that the median ET for all levels of preplant irrigation was higher than the 625 mm on average required to meet full maize water requirements in western Kansas under irrigation capacities of 6.4 and 3.8 mm/day (Stone and Schlegel, 2006). For limited irrigation capacity of 2.5 mm/day, at least 75 mm to 100 mm of preplant irrigation was required to attain median seasonal ET of 625 mm/day or higher. The goal in water limited cropping systems is to maximize ET since it is directly related to biomass production and grain yield (Stewart and Peterson, 2014). Under very limited water scenarios, late spring preplant irrigation can increase seasonal ET but the goal should be to partition more water to transpiration while reducing soil water evaporation to improve water productivity. Fig. 6 also shows that reducing irrigation capacity reduced seasonal ET due to reduced root uptake and transpiration.

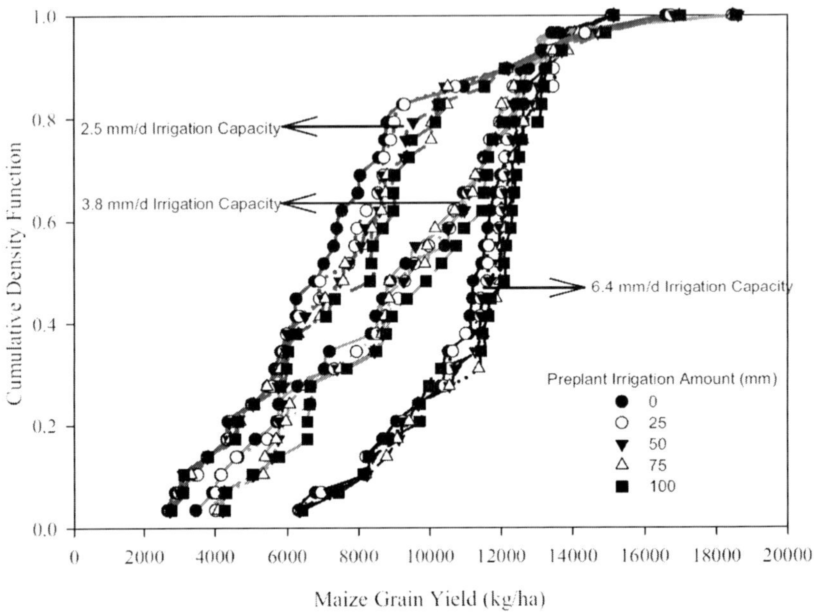

Figure. 5. Cumulative probability density function of maize grain yield for different amounts of preplant irrigation and irrigation capacity simulated using CERES-Maize in RZWQM for climatic data from 1986 to 2014 for Tribune, Kansas.

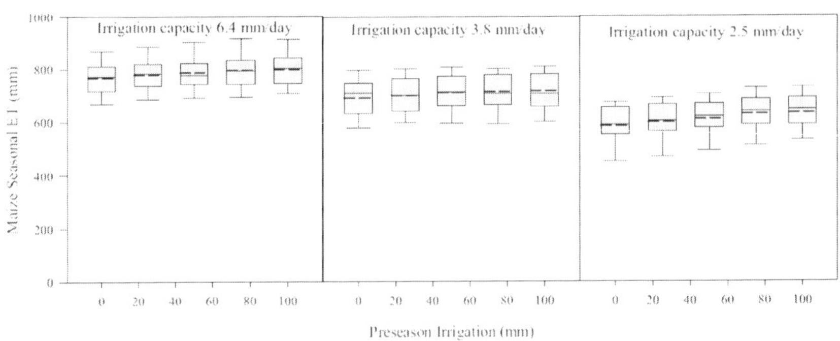

Figure. 6. Simulated seasonal maize evapotranspiration (solid line in boxplot is median and dashed line in mean) under five preplant irrigation amounts and three irrigation capacities using historical weather data from 1986 to 2014 for Tribune, Kansas in a deep silt loam soil.

Transpiration did not vary substantially at different amounts of preplant irrigation for irrigation capacities of 6.4 and 3.8 mm/day. This implies that extra water could have been lost to soil water evaporation or deep drainage. But results from this study indicated that deep drainage from the 2.4 m soil profile was negligible therefore most of the water was lost to soil water evaporation. However, for very limited irrigation capacity of 2.5 mm/day, median transpiration increased with increase in preplant irrigation amount (Table 4). The results suggest that preplant irrigation is beneficial under very limited irrigation capacities but is not necessary under sufficient irrigation capacity (≥3.8 mm/day).

Table 4. Mean seasonal transpiration and average number of seasonal irrigation events at different amounts of preplant irrigation simulated using 29 years of historical weather data (1986–2014) and the DSSAT-CSM embedded in the RZWQM2 at Kansas State University Southwest Research-Extension Center near Tribune.

Preplant irrigation amount (mm)	Seasonal Transpiration (mm)			Average number of seasonal irrigation events[1]		
	Irrigation capacity (mm/day)			Irrigation capacity (mm/day)		
	6.4	3.8	2.5	6.4	3.8	2.5
0	497 a	497 a	401 a	19	16	12
25	492 a	492 a	408 b	19	15	12
50	492 a	492 a	418 b	18	15	11
75	486 a	486 a	421 c	18	15	11
100	495 a	495 a	430 c	17	14	11

1 Each irrigation event was 25 mm.

Optimum preplant irrigation amounts were also analyzed based on water productivity as an indicator of how well water applied before planting was used for grain production. Table 6 shows that preplant irrigation amount did not have a substantial impact on water productivity at high and moderate irrigation capacity of 6.4 and 3.8 mm/day respectively. Stochastic dominance analysis revealed that similar to yield, there is a narrow range (11–15 kg/ha-mm) of water productivity where applying 100 mm preplant irrigation improved water productivity under the 2.5 mm/day irrigation capacity. Results in Fig. 7 also indicate that irrigation capacity has a more dominant effect on water productivity compared to preplant irrigation amounts.

Table 6. Maize water productivity (kg/ha-mm) for different preplant irrigation amounts and irrigation capacities averaged over 29 years (1986–2014) of weather data at Kansas State University Southwest Research-Extension Center near Tribune, Kansas.

Preplant irrigation (mm)	Irrigation capacity					
	2.5 mm d^{-1}		3.8 mm d^{-1}		5.0 mm d^{-1}	
	$E(x)$	$E(x)$-$G(x)$	$E(x)$	$E(x)$-$G(x)$	$E(x)$	$E(x)$-$G(x)$
	Grain Yield (kg/ha-mm)					
0	12.48	7.45	13.22	8.54	14.45	11.69
25	12.52	7.59	13.25	8.58	14.44	11.38
50	12.49	7.39	13.19	8.77	14.44	11.59
75	12.34	7.20	13.20	8.70	14.47	11.64
100	12.47	7.36	13.52	9.24	14.35	11.28

3.2.3. Precipitation use efficiency

Optimum combinations of preplant irrigation amounts and irrigation capacity were also evaluated based on their ability to enhance precipitation capture and storage expressed in form of precipitation use efficiency (PUE). PUE was calculated as a ratio of grain yield to precipitation. The precipitation was summed from harvest of previous year's crop to physical maturity of the crop in the current year (i.e., from October to September). Fig. 8 shows that preplant irrigation had strong effect on PUE at the lowest irrigation capacity. At low irrigation capacity (2.5 mm/day) increasing preplant irrigation increased median PUE up to 18%. This is due

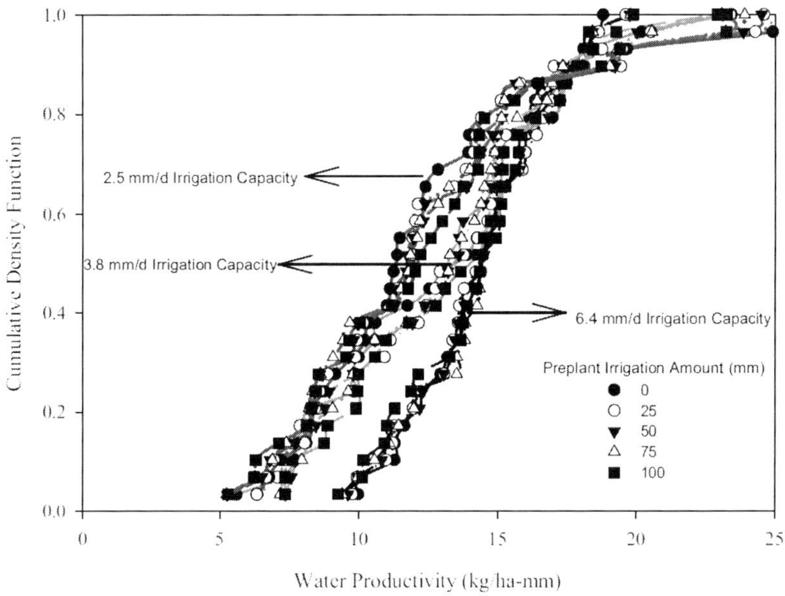

Figure. 7. Cumulative probability density function of maize water productivity for different amounts of preplant irrigation and irrigation capacity simulated using CERES-Maize in RZWQM for climatic data from 1986 to 2014 for Tribune, Kansas.

to the fact that applying 75–100 mm of preplant irrigation resulted in increased yield compared to the scenario with 0 mm of preplant irrigation at 2.5 mm/day irrigation capacity. A concern with preplant irrigation is to ensure that it does not reduce storage for precipitation, this study shows that if preplant irrigation is applied in late spring in combination with soil water monitoring (irrigation was triggered at 60% plant available in this study) it did not reduce PUE in fact it enhanced PUE at very limited irrigation capacity. It is worth noting from Fig. 8 that at moderate irrigation capacity of 3.8 mm/day preplant irrigation amounts of 100 mm improved PUE at lower levels. No effect of preplant irrigation on PUE at high irrigation capacity of 6.4 mm/day. It can also be seen in Fig. 8 that reducing irrigation capacity by irrigating larger land area compared to available water supply reduced PUE. We recommend that producers match size of irrigated land to available water resources in order to guarantee sufficient irrigation capacity that enhances PUE.

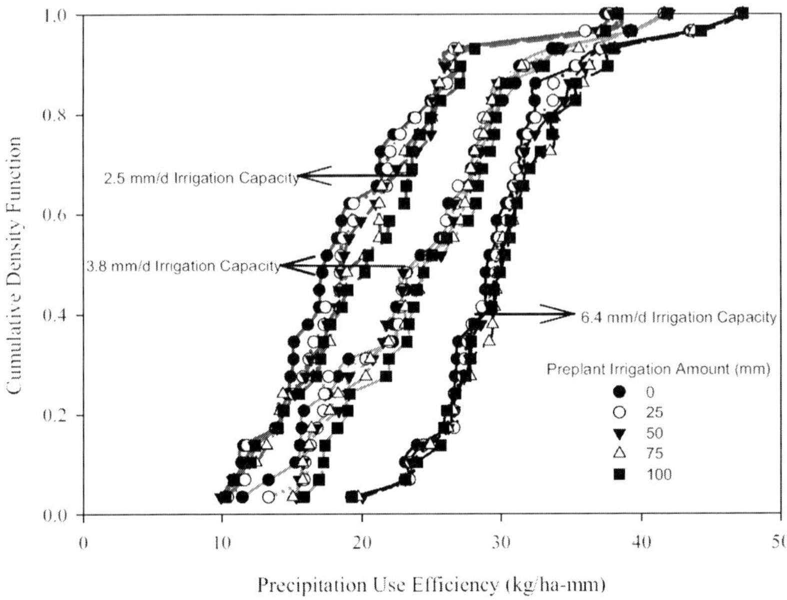

Figure. 8. Cumulative probability density function of precipitation use efficiency (PUE) for different amounts of preplant irrigation and irrigation capacity simulated using CERES-maize in RZWQM for climatic data from 1986 to 2014 for Tribune, Kansas.

3.2.4. Soil water evaporation losses prior to planting

Water losses due to soil water evaporation prior to planting increased with preplant irrigation amount at all irrigation capacities as shown in Fig. 9. Scenario with 0 mm preplant irrigation had the least soil water evaporation loss prior to planting. Increase in soil water evaporation with preplant irrigation is expected as there would be more water available for the evaporation process. Under sufficient irrigation capacity (≥3.8 mm/day), we recommend no preplant irrigation to minimize soil water evaporation losses. Under limited well capacity we recommend applying at least 75 mm of preplant irrigation in late spring (April) prior to planting. Also, this analysis indicated negligible losses through deep percolation from 2.4 m soil profile indicating the biggest potential water loss from preplant irrigation would be soil water evaporation.

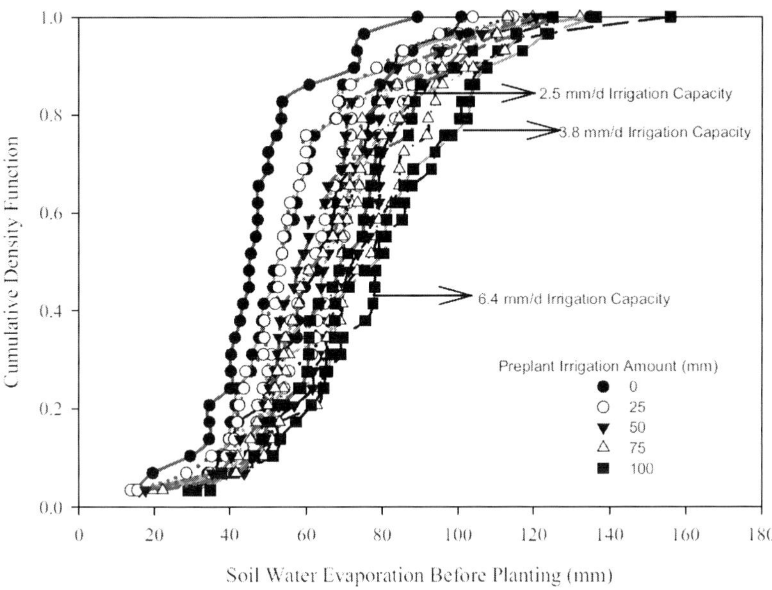

Figure. 9. Cumulative probability density function of soil water evaporation prior to planting in late spring for different amounts of preplant irrigation and irrigation capacity simulated using CERES-maize in RZWQM for climatic data from 1986 to 2014 for Tribune, Kansas.

Different irrigation application methods might influence storage efficiency of preplant irrigation. For example, subsurface drip irrigation (SDI) would reduce soil water evaporation losses compared to preplant irrigation done with a sprinkler irrigation system. Mobile Drip Irrigation (MDI) might also reduce soil water evaporation due to reduced surface wetting (Kisekka et al., 2016a ; Kisekka et al., 2016b). Lamm (2002) noted that SDI could be beneficial for applying preplant irrigation to increase seasonal capacity by allowing irrigation to be applied under near freezing temperatures which would not be possible with center pivot irrigation systems. Residue management might also improve the efficiency of preplant irrigation. Irrigators need to integrate technologies (e.g., soil water sensors) in implementation of preplant irrigation that help reduce water losses due to soil water evaporation and deep drainage but enhances precipitation capture and storage.

Although most of the earlier experimental research (Musick et al., 1971; Stone et al., 1983; Stone et al., 1987; Stone et al., 1994 ; Stone et al., 2008) showed that preplant irrigation was an inefficient use of water in most years, recent findings by Schlegel et al. (2012) have showed that under very limited well capacity (2.5 mm/day) preplant irrigation increased yield and was profitable. In addition, Ma et al. (2012) noted that preplant irrigation was necessary to ensure good germination in dry years. Musick and Lamm (1990) in their review of preplant irrigation in the South and Central High Plains listed conditions where preplant irrigation would have the best and least benefit, in general these conditions underscored the need to make preplant irrigation decisions based on knowledge of soil water and local rainfall patterns. The approach used in this study helps to build on earlier work by extrapolating short term experimental results to long term climatic conditions to understand effect of preplant irrigation on water limited maize production in the high plains under a wide range of growing conditions.

The decision to apply preplant irrigation should be evaluated carefully since it depends on irrigation capacity and cropping season yield potential. Optimum preplant irrigation for systems with 2.5 mm/day irrigation capacity is 100 mm. No preplant irrigation for systems with 3.8 and 6.4 irrigation capacity is required in most years. Overall, a producer would obtain higher yields, water productivity and PUE by increasing their irrigation capacity by reducing irrigated areas to match limited water supply.

4. CONCLUSIONS

A simulation study was conducted to assess the effect of preplant irrigation amounts and irrigation capacities on maize production in western Kansas region of the United States High Plains. The calibrated and validated CERES-maize model embedded in the RZWQM2 adequately predicted yield, ET, and soil water. The calibrated and validated model was applied to assess effect of five preplant irrigation amounts (0, 25, 50, 75 and 100 mm) and three irrigation capacities (2.5, 3.8 and 6.4 mm/day) on maize yield. Under very limited irrigation capacity of 2.5 mm/day, applying 75–100 mm of preplant irrigation produced median yields that were 10–17% higher than not applying preplant irrigation. However, even at limited irrigation capacity the benefit of preplant irrigation were only realized if the seasonal yield potential was in the range of 6000 to 10,000 kg/ha corresponding to years with normal seasonal rainfall. Irrigation capacity had a stronger effect on maize grain yield compared to preplant irrigation amount. Preplant irrigation increased ET and transpiration under 2.5 mm/day irrigation capacity. Preplant irrigation amount did not have a substantial impact on water productivity at high and moderate irrigation capacities but had second order dominant effect under limited irrigation capacity. At low irrigation capacity (2.5 mm/day) increasing preplant irrigation increased median PUE up to 18% although the effect was second order dominant. No effect on PUE at high irrigation capacity. Water losses due to soil water evaporation prior to planting increased with preplant irrigation amount at all irrigation capacities. Under very limited irrigation capacity (≤2.5 mm/day) we recommend applying 75–100 mm of preplant irrigation in late spring (April). No preplant irrigation is recommended at high irrigation capacity to minimize soil water evaporation losses. Deep percolation losses from 2.4 m soil profile were negligible. The decision to apply preplant irrigation should be evaluated and implemented carefully in combination with other agricultural water management technologies and strategies such as soil water monitoring, drip irrigation, and residue management to mitigate negative effects of soil water evaporation and deep drainage.

ACKNOWLEDGEMENTS

The authors would like to thank the following for providing funding for this research USDA Ogallala Aquifer Project, The Foundation for Food and Agricultural Research (Award # 430871), and USDA Project No. 2016-68007-25066, through the NIFA Water for Agriculture Challenge Area.

This is contribution number 17-181-J from the Kansas Agricultural Experiment Station. Authors thank the anonymous reviewers for their constructive comments.

REFERENCES

1. Ahuja, L.R., Rojas, K.W., Hanson, H.D., Shafer, M.J., Ma, L., 2000. Root zone waterquality model. In: Modeling Management Effects on Water Quality and CropProduction. Water resources publications, LLC, CO, USA.
2. DeJonge, K.C., Ascough, J.C., Andales, A.A., Hansen, N.C., Garcia, L.A., Arabi, M., 2012.Improving evapotranspiration simulations in the CERES-Maize model underlimited irrigation. Agric. Water Manage. 115, 92–103, http://dx.doi.org/10.1016/j.agwat.2012.08.013.
3. Dixon, P.M., Weiner, J., Mitchell-Olds, T., Woodley, R., 1987. Bootstrapping the Ginicoefficient of inequality. Ecology 68 (5), 1548–1551.
4. Doherty, J., 2009. PEST: Model Independent Parameter Estimation. WatermarkNumer. Comput., Corinda, Queensland, Australia, Available at http://www.pesthomepage.org.
5. Hoogenboom, G., Jones, J.W., Wilkens, P.W., Porter, C.H., Boote, K.J., Hunt, L.A.,Singh, U., Lizaso, J.I., White, J.W., Uryasev, O., Ogoshi, R., Koo, J., Shelia, V., Tsuji,G.Y., 2015. Decision Support System for Agrotechnology Transfer (DSSAT)Version 4.6. www.DSSAT.net.Howell, T.A., O'shaughnessy, S.A., Evett, S.R., 2012. Integrating multiple irrigationtechnologies for overall improvement in irrigation management.

In:Proceedings of the 2012 Central Plains Irrigation Conference, Central PlainsIrrigation conference proceedings, Colby, Kansas, pp. 170–186.
6. Jones, J.W., Hoogenboom, G., Porter, C.H., Boote, K.J., Batchelor, W.D., Hunt, L.A.,Wilkens, P.W., Singh, U., Gijsman, A.J., Ritchie, J.T., 2003. DSSAT croppingsystem model. Eur. J. Agron. 18, 235–265.
7. Keating, B.A., Carberry, P.S., Hammer, G.L., Probert, M.E., Robertson, M.J.,Holzworth, D., Huth, N.I., Hargreaves, J.N.G., Meinke, H., Hochman, Z., McLean,G., Verburg, K., Snow, V., Dimes, J.P., Silburn, M., Wang, E., Brown, S., Bristow,K.L., Asseng, S., Chapman, S., McCown, R.L., Freebairn, D.M., Smith, C.J., 2003.An overview of APSIM, a model designed for farming systems simulation. Eur.J. Agron. 18, 267–288.
8. Kisekka, I., Migliaccio, K.W., Mu˜noz-Carpena, R., Khare, Y., Li, Y.C., 2015. Modelingsoil water dynamics considering measurement uncertainty. Hydrol. Processes29, 692–711.
9. Kisekka, I., Nguyen, G., Aguilar, J., Rogers, D., 2016a. Mobile drip irrigationevaluation in maize. Kans. Agric. Exp. Stat. Res. Rep. 2 (7), http://dx.doi.org/10.4148/2378-5977.1253.
10. Kisekka, I., Aguilar, J., Rogers, D., Klocke, N., 2016b. Assessing deficit irrigationstrategies for maize using simulation. Trans. ASABE 59, 303–317.
11. Klocke, N., Currie, R., Kisekka, I., Stone, L., 2015. Maize and grain sorghum responseto limited irrigation, Drought, and Hail. Appl. Eng. Agric. 30, 915–924.
12. Kromm, D.E., White, S.E., 1990. Conserving Water in the High Plains. Kansas StateUniversity, Department of Geography, Manhattan, Kans.
13. Lamm, F.R., 2002. Advantages and disadvantages. In: International Meeting onAdvances in Drip/Micro Irrigation, Puerto de La Cruz, Tenerife, Canary Islands,December 2–5, p. 2002.
14. Ma, L., Hoogenboom, G., Ahuja, L.R., Ascough II, J.C., Saseendran, S.A., 2006.Evaluation of the RZWQM-CERES-Maize hybrid model for maize production.Agric. Syst. 87, 274–295.
15. Ma, L., Trout, T.J., Ahuja, L.R., Bausch, W.C., Saseendran, S.A., Malone, R.W., Nielsen,D.C., 2012. Calibrating RZWQM2 model for maize responses to deficitirrigation. Agric. Water Manage. 103, 140–149.
16. Ma, L., Ahuja, L.J., Trout, T., Nolan, B.T., Malone, R.W., 2016. Simulating maize yieldand bomass with spatial variability of soil field capacity. Agron. J., http://dx.doi.org /10.2134/agronj2015.0206.
17. Malone, R.W., Jaynes, D.B., Ma, L., Nola, B.T., Meek, D.W., Karle, D.L., 2010. Soil-testN recommendations augmented with PEST optimized RZWQM simulations. J.Environ. Qual. 39, 1711–1723.
18. McGuire, V.L., 2004. Water-level changes in the High Plains Aquifer,predevelopment to 2003 and 2002–2003. In: Fact Sheet FS-2004-3097. U.S.Geological Survey, Lincoln, Neb.
19. Mu˜noz-Carpena, R., Ritter, A., Bosch, D.D., Schaffer, B., Potter, T.L., 2008. Summercover crop impacts on soil percolation and nitrogen leaching from a wintermaize field. Agric. Water Manage. 95, 633–644.
20. Musick, J.T., Lamm, F.R., 1990. Preplant irrigation in the central and southern HighPlains: a review. Trans. ASAE 33, 1834–1842.
21. Musick, T.J., Sletten, W.H., Dusek, D.A., 1971. Preplant irrigation of grain sorghumin the southern high plains. Trans. ASAE 14, 93–97.
22. Nimah, M.N., Hanks, R.J., 1973. Model for estimating soil water, plant, andatmospheric interrelations: I. Description and sensitivity. Soil Sci. Soc. Am.Proc. 37, 522–527, http://dx.doi.org/10.2136/sssaj1973.03615995003700040018x.
23. Raes, D., Steduto, P., Hsiao, T.C., Fereres, E., 2009. AquaCrop- the FAO crop model tosimulate yield response to water: II. Main algorithms and soft waredescription. Agron. J. 101, 438–447.
24. Saseendran, S.A., Ahuja, L.R., Nielsen, D.C., Trout, T.J., Ma, L., 2008. Use of cropsimulation models to evaluate limited irrigation management options for cornin

a semiarid environment. Water Resour. Res. 44, W00E02, http://dx.doi.org/ 10.1029/2007WR006181.

25. Schlegel, A.J., Stone, L.R., Dumler, T.J., Lamm, F.R., 2012. Managing diminishedirrigation capacity with preseason irrigation and plant density for cropproduction. Trans. ASABE 55, 525–531.ˇ
26. Sim˚unek, J., Nimmo, J.R., 2005. Estimating soil hydraulic parameters from transientflow experiments in a centrifuge using parameter optimization technique.Water Resour. Res. 41, http://dx.doi.org/ 10.1029/2004wr003379W04015.
27. Stewart, B.A., Peterson, G.A., 2014. Managing green water in dryland agriculture.Agron. J. 106, 1–10, http://dx.doi.org/10.2134/agronj14.0038.
28. Stone, L.R., Schlegel, A.J., 2006. Crop water use in limited-irrigation environments.In: Proc. 2006 Central Plains Irrigation Conf., Colby, Kans.: Central PlainsIrrigation Association, pp. 173–184.
29. Stone, L.R., Carlson, C.G., Hanson, T.L., Gwin Jr., R.E., Gallagher, P., Horton, M.L.,1983. Amount of profile water in early spring resulting from increased profilewater in fall. SSSA J. 47, 305–309.
30. Stone, L.R., Gwin Jr., R.E., Gallagher, P.J., Hattendorf, M.J., 1987. Dormant-seasonirrigation Grain yield, water use, and water loss. Agron. J. 79, 632–636.
31. Stone, L.R., Schlegel, A.J., Lamm, F.R., Spurgeon, W.E., 1994. Storage efficiency ofpreplant irrigation. J. Soil Water Conserv. 49, 72–76.
32. Stone, L.R., Lamm, F.R., Schlegel, A.J., Klocke, N.L., 2008. Storage efficiency ofoff-season irrigation. Agron. J. 100, 1185–1192.
33. Stone, L.R., Klocke, N.L., Schlegel, A.J., Lamm, F.R., Tomsicek, D.J., 2011. Equations fordrainage component of the field water balance. Appl. Eng. Agric. 27, 345–350.
34. Tsuji, G.Y., Hoogenboom, G., Thornton, P.K. (Eds.), 1998. Understanding Options forAgricultural Production. Systems Approaches for Sustainable AgriculturalDevelopment, vol. 7. Kluwer Academic, Dordrecht, The Netherlands, http://dx.doi.org/10.1007/978-94-017-3624.
35. Van Diepen, C.A., Wolf, J., van Keulen, H., 1989. WOFOST: a simulation model ofcrop production. Soil Use Manage. 5 (1), 16–24, http://dx.doi.org/10.1111/j.1475-2743.1989.tb00755.x.

CHAPTER 11

Evaluation of new farming technologies in Ethiopia using the Integrated Decision Support System (IDSS)

Neville Clarke[a], Jean-Claude Bizimana[b], Yihun Dile[c], Abeyou Worqlul[d], Javier Osorio[d], Brian Herbst[b], James W. Richardson[b], Raghavan Srinivasan[c], Thomas J. Gerik[d], Jimmy Williams[d], Charles A. Jones[d], Jaehak Jeong[d*]

[a]*The Norman Borlaug Institute for International Agriculture, Texas A&M AgriLife Research, College Station, TX, United States*

[b]*Department of Agricultural Economics, Texas A&M University, College Station, TX, United States*

[c]*Spatial Sciences Laboratory, Texas A&M University, College Station, TX, United States*

[d]*Blackland Research Center, Texas A&M AgriLife Research, Temple, TX, United States*

ABSTRACT

This study investigates multi-dimensional impacts of adopting new technology in agriculture at the farm/village and watershed scale in sub-Saharan Africa using the Integrated Decision Support System (IDSS). Application of IDSS as an integrated modeling tool helps solve complex issues in agricultural systems by simultaneously assessing production, environmental, economic, and nutritional consequences of adopting agricultural technologies for sustainable increases in food production and use of scarce natural resources. The IDSS approach was applied to the Amhara region of Ethiopia, where the scarcity of resources and agro-environmental consequences are critical to agricultural productivity of small farm, to analyze the impacts of alternative agricultural technology interventions. Results show significant improvements in family income and nutrition, achieved through the adoption of irrigation technologies, proper use of fertilizer, and improved seed varieties while preserving environmental indicators in terms of soil erosion and sediment loadings. These pilot studies demonstrate the usefulness of the IDSS approach as a tool that can be used to predict and evaluate the economic and environmental consequences of adopting new agricultural technologies that aim to improve the livelihoods of subsistence farmers.

Keywords: IDSS; SWAT; APEX; FARMSIM; Small-scale water management interventions; Environmental sustainability

1. INTRODUCTION

The number of people who live below the poverty line in the world is estimated to have declined; however, there are still about 800 million people who are suffering from chronic hunger (FAO, 2013). Sub-Saharan Africa remains the region with the highest prevalence of undernourishment (FAO, 2013). The poor in sub-Saharan Africa usually live in rural areas (DFID, 2004 ; IFAD, 2011) and more than 90% earn less than $2/day (IFAD, 2011).

Agriculture is the leading economic sector in many developing countries despite production based on subsistence farming that suffers from low yields and high vulnerability due to climate change (Awokuse and Xie, 2015 ; Tibesigwa and Visser, 2015). In sub-Saharan Africa, 40–70% of rural households earn more than three-quarters of their income from on-farm sources (IFAD, 2011). For example, agriculture in Ethiopia accounts for 47% of GDP, 90% of exports and 85% of employment (IFAD, 2009). This suggests that investment in agriculture can contribute to food security and poverty reduction for the majority of the rural poor (DFID, 2004 ; WorldBank, 2008). Research across the world has proven that investment in agriculture can result in a sharp increase in economic development and poverty reduction (Gallup *et al.*, 1998; Pretty *et al.*, 2011; Thirtle *et al.*, 2001 ; WorldBank, 2008). Green Revolution in Asia during the 1970s is one well-known example, where the adoption of science-based technology pulled Asia from the age of famine to regional food surplus within 25 years (Djurfeldt *et al.*, 2005 ; Hazell, 2009).

Agriculture in sub-Saharan Africa is largely rainfed. Rainfed agriculture is the dominant source of staple food production (Cooper *et al.*, 2008; FAO, 2011 ; Rosegrant *et al.*, 2002) and covers 93% of the region's agricultural area (CA, 2007 ; FAO, 2002). There is a large yield gap (i.e. the difference between what is actually harvested from farmers' fields and what could potentially be achieved) in sub-Saharan agriculture. In tropical regions with reliable rainfall and sufficient nutrient application, average rainfed commercial agricultural yield exceeds 5–6 t/ha (CA, 2007 ; Rockström and Falkenmark, 2000) while the average rainfed yield in sub-Saharan Africa is less than 1.5 t/ha (Rosegrant et al., 2002).

Research has shown that the yield gaps in sub-Saharan Africa cannot be explained only by biophysical conditions (e.g. lower amounts of rainfall), but also due to sub-optimal management conditions (Licker *et al.*, 2010 ; Rockström *et al.*, 2002). Amede (2012) indicated that low yield in the rainfed agro-ecological landscapes of Ethiopia is typically not due to the lack of water alone rather a result of the inefficient management of water, soils, and crops. Supplementary irrigation combined with fertilizer application in a field experiment in the northern Burkina Faso increased the yield level three fold compared to the normal farmers' practices (Fox and Rockstrom, 2003). In semi-arid Kenya, Barron and Okwach, (2005) showed that supplementary irrigation with improved nutrient application can increase average yield by 70%. In-situ water harvesting practices contribute to dry spell mitigation and yield improvements. Tied-ridges, also known as furrow damming, are found to improve barley yields by 44% compared to traditional tillage methods in the semi-arid region of northern Ethiopia (Araya and Stroosnijder, 2010). Moreover, soil and water management interventions improve land productivity by restoring degraded land (Lal, 2001; Reij *et al.*, 2009 ; Vohland and Barry, 2009), maintain biodiversity (Norfolk *et al.*, 2012; Pandey, 2001 ; Vohland and Barry, 2009), and limit nutrient loss from fields by controlling soil erosion (Gebreegziabher *et al.*, 2009 ; McHugh *et al.*, 2007).

There is growing interests in developing policies by governments and international development agencies on the implementation of small-scale soil and water management interventions in most of the sub-Saharan African countries (Gebregziabher et al., 2013). However, the scale of adoption of these technologies is very limited although there is proven evidence that small-scale interventions can improve agricultural production. The adoption of small-scale soil and water management interventions is affected by various socio-economic factors. Gebregziabher et al. (2013) reported that there exist specific main determinants that affect the adoption of rainwater management technologies in the Upper Blue Nile basin in Ethiopia such as household demographic characteristics (such as age and gender), participation in off-farm activities, migration, ownership of livestock, ownership of land, and other factors. Boyd et al. (2000) show how improved access to markets and increased producer prices stimulate investment in in-situ water harvesting systems at the household level in Tanzania. Munamati and Nyagumbo (2010) showed that resource status and gender determined the success of in-situ water harvesting in the Gwanda district of Zimbabwe.

IDSS is comprised of a suite of previously validated, spatially explicit models (Soil and Water Assessment Tool (SWAT), Agricultural Policy/Environmental eXtender Model (APEX), and Farming Simulator (FARMSIM)), and databases that have been extensively applied in both U.S. and international settings. It provides an integrated approach linking production (APEX), economic (FARMSIM), and environmental (SWAT) consequences of the introduction of new technology, policy, and training for decision makers in agriculture at multiple temporal and spatial scales (Fig. 1). IDSS involves cropping systems analysis at the field scale, environmental risk assessment at the watershed scale, and nutritional and economic analyses of proposed interventions at the farm scale. The multi-scale and multi-aspects assessment is facilitated through passing information between SWAT, APEX, and FARMSIM such that proposed interventions are evaluated from different perspectives within a connected modeling framework. On the biophysical modeling side, SWAT and APEX share field/watershed information such as soil properties, historical weather data, or crop data. In addition, distributed parameters such as those related to water balance that are calibrated at the watershed scale are transferred to APEX for field scale analyses where observation data is unavailable. On the other hand, refined crop parameters at the field scale are transferred to SWAT to improve the calculation of evapotranspiration. Crop yield outputs on various intervention scenarios predicted by APEX are used in FARMSIM for feasibility assessment on the nutritional and economic point of view. Therefore, IDSS informs decision makers with unbiased assessment results.

SWAT was developed to predict the impact of land-management practices on water, sediment, and agricultural chemical yields in watersheds with varying soil, land use, and management conditions (Neitsch et al., 2012). SWAT can simulate hydrological cycles, vegetation growth, and nutrient cycling with a daily time-step by disaggregating a river basin into sub-basins and hydrologic response units (HRUs). HRUs are lumped land areas within the sub-basin comprised of unique land cover, soil and management combinations. SWAT has been applied in Ethiopian watersheds and demonstrated satisfactory performance (Dile *et al.*, 2013; Dile and

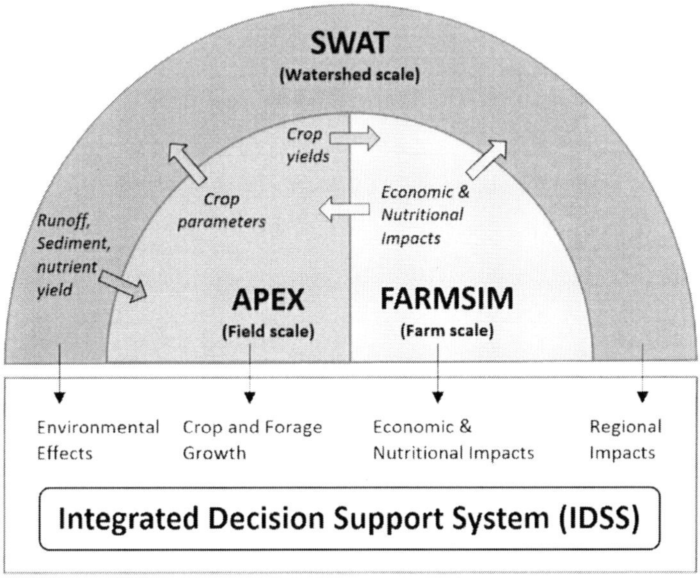

Figure. 1. Schematic of the IDSS framework.

Srinivasan, 2014; Betrie *et al.*, 2011; Setegn *et al.*, 2010a; Setegn *et al.*, 2010b ; White *et al.*, 2011). APEX is a biophysical simulation model that shares many of the attributes of SWAT. It is used to evaluate detailed crop management technologies and decisions that can affect agricultural production and environmental sustainability (soil, water, greenhouse gases, etc.) at the scales of individual fields, whole farms, or small watersheds (Ford *et al.*, 2015; Gassman *et al.*, 2010; Tuppad *et al.*, 2009 ; Wang *et al.*, 2012). FARMSIM is a farm level agro-economic model that simulates the nutritional uptake of farm family in a stochastic environment so the impacts of technology on the nutritional status of the farm family can be evaluated. FARMSIM is an extension of the Farm Level and Income Policy Simulation (FLIPSIM) model available in Microsoft Excel format which has been used extensively to simulate the impacts of alternative policies and farming systems on representative farms (e.g. Nyangito *et al.*, 1995; Nyangito *et al.*, 1996a; Nyangito *et al.*, 1996b; Richardson and Clair, 1981; Richardson *et al.*, 1983 ; Richardson *et al.*, 2000). FARMSIM is explicitly programmed to simulate small holder farms in developing countries. In addition, it simulates the family unit as the "first" consumer of farm production.

Currently there are many biophysical and socio-economic models that show the impacts of small-scale interventions (e.g. Abildtrup *et al.*,

2006; Chilonda and Van Huylenbroeck, 2001; Garg *et al.*, 2012 ; van de Giesen *et al.*, 2005). Moreover, there has been a number of models published for integrated hydrologic-agronomic-economic analysis (e.g. Cai *et al.*, 2003; Lefkoff and Gorelick, 1990 ; Volk *et al.*, 2008) but few model is capable of simulating hydrologic and agronomic components using comprehensive physically-based algorithms due to the complexity of the modeling system. The Spatial Decision Support System (Volk et al., 2008) is similar to IDSS in terms of implementing detail biophysical processes by integrating existing models and may be more advanced in assimilating data between process models. However, it is designed for ecologic-economic evaluation of land management or landuse changes. Little research has been done to show the impacts of these agricultural interventions using process-based simulation tools within an integrated decision framework on agricultural productivity and nutritional and economic impacts at the field scale and environmental consequences at the watershed scale. In this paper, the IDSS tool is applied to assess the effects of fertilizer, irrigation, genetically improved crop and crop rotation on production, environment, nutrition and economic wellbeing of small holder farmers in Fogera woreda of the Upper Blue Nile Basin, Ethiopia. This study demonstrates the use of IDSS for comprehensive assessment of complex agricultural technologies designed to increase food production, improve nutrition, enhance economic well-being, and minimize negative environmental consequences for small-holder farms.

2. MATERIALS AND METHODS

A pilot study was conducted for Ethiopia, in which the impacts of several levels of farming system intensification were evaluated at the kebele (i.e. the smallest administrative unit of Ethiopia) scale for productivity and nutritional/economic analyses and at the watershed scale for environmental impacts. The impacts of small-scale interventions were assessed in up to 13 scenarios including the current condition, fertilizer and irrigation management, cultivating generically improved varieties, and dry & wet season rotations combined with selected crops.

2.1. Description of study sites

Fogera woreda, located within the Gumera and Ribb river basins on the eastern side of Lake Tana in the South Gondar, Amhara region in Ethiopia (Fig. 2) was selected to demonstrate the capabilities of IDSS. A woreda is a second level administration unit in Ethiopia. The total land area of Fogerda woreda is 117,405 ha, with flat lands accounting for 76%, mountain and hills 11%, and valley bottoms 13%. (Gebey et al., 2012). Its total population is 233,529, with a rural population of 206,717. The male and female populations are similar in both rural and urban areas. The number of agricultural households is 42,746. The average land holding is about 1.4 ha, with minimum and maximum holdings of 0.5 and 3.0 ha respectively (ILRI, 2005; ILRI-IPMS Atlas, 2007). Two kebeles, Shena and Weg-Arba Amba, located within Fogera woreda were selected for case studies based on the following metrics.

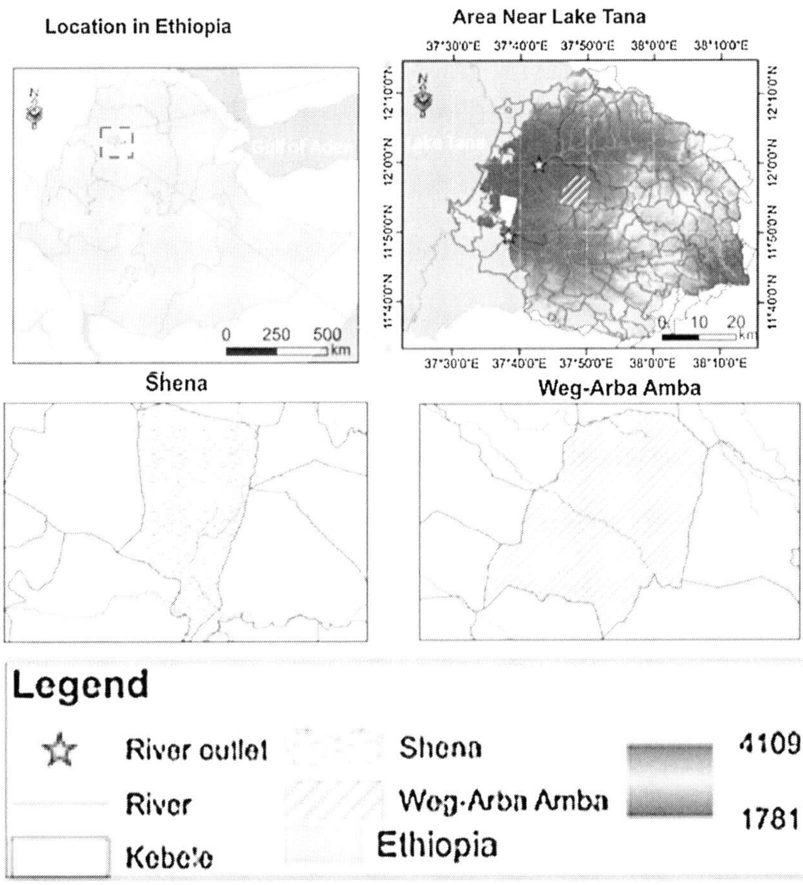

Figure. 2. Location of study sites on the eastern shore of Lake Tana in Ethiopia.

- Close proximity to an all-weather road for access to markets
- Close proximity to a river or streams for possible irrigation
- Diverse existing land uses, including crop land, grazing land, and woodland
- One peasant associations (or kebeles) in the near-level Lake Tana plain and another in hilly region of the woreda

The Shena kebele is located on nearly level land within the Ribb river basin near Lake Tana at an elevation of 1774 m to 1850 m. Land uses include cultivated cropland, dense shrub land, open forest, grassland, exposed sand/soil, and seasonal swamp. Cropping systems are dominated by rice (*Oryza sativa* L.), finger millet (*Panicum miliaceum* L.), horticultural crops, and noug (*Guizotia abyssinica (L.f) Cass*); however, in recent years noug and finger millet are being replaced by maize and rice. Livestock include cattle, sheep and chickens. Weg-Arba Amba kebele is located in the Gumera river basin on rolling land with elevations ranging from 1774 m to 2153 m. The kebele is predominantly grassland with significant amounts of cropland, dense shrub land, and open forest. Cropping and livestock systems below 2000 m are dominated by maize, teff, finger millet, noug, vegetables, cattle, and goats. Above 2000 m barley, horse beans, potatoes, sheep and cattle are produced. Most of the crop land is located at higher elevations. Croplands are developed on slopes ranging from 6 to 23%; the two dominant crops in the area are corn and teff. Most farmers currently grow only one crop per year during the rainy season. In this area, crop residues are immediately removed for animal feed after grain harvest, reflecting current and anticipated future needs for animal fodder.

2.2. Biophysical modeling

Cropping system models of IDSS were implemented to simulate current low-input and possible future higher input cropping systems. In particular, SWAT was used to simulate hydrology, soil erosion, and Baseline crop yields for the Ribb and Gumera river basins and APEX was used to simulate several current and potential cropping systems for the Shena and Weg-Arba Amba kebeles for 32 years of weather, thus incorporating weather risk for alternative cropping systems. Irrigation and fertilizer inputs, and the environmental consequences of both current and high-input systems were also investigated using IDSS. The topography and land use/land cover data were derived from the United States Geological Survey (USGS) sources. Daily weather data were obtained from the NOAA-USGS Climate Forecast System Reanalysis (CFSR) global daily weather database. This 32-year

(1979–2010) daily weather data contains precipitation, minimum and maximum temperatures, solar radiation, wind speed, and humidity (available at: globalweather.tamu.edu, last accessed on December 9, 2015). Soils data were obtained from the FAO 1:1 million world soils map and associated soil pedon data base (MoWR, 2009). Calibrated SWAT parameters were transferred to APEX to better simulate water balance at the field scale. Water quality data was not available and therefore the model was not refined for water quality related parameters. In turn, APEX crop parameters were transferred to SWAT to synchronize the two models on predicting water balance.

To demonstrate the role of the IDSS models on analyzing the impacts of crop management on crop growth and soil erosion in different landscapes, two levels of crop management were simulated for all the subbasins of the Rib and Gumera watersheds. The two input levels were: (1) the Baseline (low-input) maize, teff, and maize-rice-onion cropping system in use in the region, and (2) the same cropping systems with additional nitrogen fertilizer and irrigation in the dry season (with rates determined by the automatic fertilizer and irrigation routines in SWAT).

For simulating alternative cropping systems using APEX, a deep alluvial soil was used for the Shena kebele site and a shallow rocky soil was used for the Weg-Arba Amba kebele. The maize and teff fields on steep slopes were simulated with shallow clean tillage, no structural conservation practices, and complete residue removal (for animal feed) based on local practices. Rock barriers at the edges of many of the region's small fields produce field-edge sedimentation and reduce sediment delivery to streams in the region, which was represented in APEX by reduced slope lengths. To assess fertilizer and irrigation requirements and their effects in traditional grain cropping systems, continuous maize and continuous teff production were modeled as summarized in Table 1.

Table 1.Scenarios of agricultural interventions simulated with APEX.

Scenarios	Description	Sites
Baseline	current low-input, rainfed cropping system and management practices	Both
Baseline + N + I	a more intensive cropping system in which APEX was allowed to estimate and apply fertilizer and irrigation necessary to reduce nitrogen and water stress to near-zero	Both
Baseline + N + I + V	genetically improved varieties represented by crops with a 15%	Both

	increase in harvest index	
maize-chickpea	maize-chickpea relay cropping systems with current fertilizer inputs but supplemental irrigation of the chickpea during the dry season	Weg-Arba Amba
maize-squash	maize-squash relay cropping systems with current fertilizer inputs but supplemental irrigation of the squash during the dry season	Weg-Arba Amba
Maize-Rice-Onion-Soybean-Rice-Onion	2-year rotation of maize-rice-onion-soybean-rice-onion with current fertilizer inputs but supplemental irrigation of onion during the dry season	Shena

2.3. Economic and nutrition analysis

A farm scale Monte Carlo simulation model, FARMSIM, was used to perform economic and nutrition analysis for the two study regions. Two representative farms were simulated for the base and two alternative technology farming systems or scenarios. The FARMSIM model simulates a representative farm for five years (2013–2017) using stochastic market prices, crop yields, and livestock production values. The model is recursive in that the financial values at the end of 2013 are the beginning financial values for the start of 2014. Cattle, sheep, goats, and chickens are included in the model and their numbers are simulated over time based on parameters for fractions of animals consumed, died, culled, and stochastic birthrates.

Data to simulate farms in the two kebeles were provided by local experts in the Amhara Region and International Water Management Institute (IWMI) based in Addis Ababa, Ethiopia. The aggregate data for the kebele were used to describe and simulate each kebele as a representative farm due to problems associated with simulating individual small-holder farms such as heterogeneity and possibilities of outliers. The Shena farm represents the 2071 ha of cropland and 2151 families for the kebele. The Weg-Arba Amba farm represents 1514 families and 3640 ha of cropland (Table 2).

Table 2.Socio-economic characteristics for Shena and Weg-Arba Amba kebeles.

	Shena	Weg-Arba Amba
Number of families	2151	1514
Cropland (Ha)	2071[a]	3624
Crops:		
Maize (Ha)	1800	914
Rice (Ha)	2071	–
Onion (Ha)	1800	–
Teff (Ha)	–	2710
Livestock:		
Cows	4174	1893
Oxen	4611	1788
Hens	6000	3020
Ewes	700	150
Nannies	–	1853

[a]The total area of the cropland does not match the sum of the areas for individual crops because of continuous rotation of multiple crops.

The APEX model was simulated for 32 years using local weather data for 1979–2010 for each farming system to estimate annual crop yields under alternative weather conditions. The resulting 32 simulated grain and forage yields provided necessary data for FARMSIM to estimate probability distributions (PDFs) for yield since crop yield is a stochastic variable in FARMSIM. Stochastic annual yields for crops and forage in FARMSIM was simulated as multivariate empirical (MVE) PDFs maintaining the observed correlation for the stochastic yields generated by APEX (Richardson et al., 2000). Three MVE PDF's were developed for each farm as different yield distributions were used for their respective scenarios/farming system.

Results of the FARMSIM model are presented in different statistical variables, graphs and stoplight chart. The statistical variables include mean, standard deviation, coefficient of variation, minimum and maximum. The graphical presentations include probability density functions and cumulative probability density functions. The stoplight chart is simply way of presenting probabilities of getting different socio-economic outcomes using color coded charts. The color codes include green, yellow and red. Green represents the probability of getting the best outcome; yellow is the probability of getting moderate outcome and red is probability of getting the least socio-economic outcome.

2.4. Assumptions and limitation

Databases used in the analysis were taken from available sources and expert opinions. A variety of modeling related issues such as roads and markets and other infrastructure, training and education, and enabling policies that might be required to bring the proposed interventions to practice were not considered in the current study. The literature on technology indicates that the rate of adoption varies widely for alternative technologies. For the current analysis it was assumed that the farming systems interventions are adopted at the rate of 25% and this adoption rate is constant throughout the planning horizon. The model is capable of simulating any assumed adoption curve where the rate of adoption changes each year, such as, 5%, 10%, 15%, 20% and 25% over the 5 year planning horizon. To test the sensitivity of the adoption rate assumption the Baseline + N farming system was also simulated using a 50% adoption rate.

3. RESULTS AND DISCUSSION

3.1. Watershed scale simulations

For the Gumera and Rib river watersheds, mean stream flows totaled almost 60% of the mean total precipitation as the result of large amounts of runoff and subsurface flow during the wet season. These results are consistent with past studies in the Blue Nile basin (Bitew and Gebremichael, 2011; Setegn *et al.*, 2011 ; Tibebe and Bewket, 2011) and suggest that, depending on location within the watersheds, ample water may be available, either for capture in small reservoirs or in shallow aquifers, for use as supplemental irrigation during the dry season. In fact implementation of small reservoirs depends on the suitability of biophysical factors such as land use, soil, slope etc. (e.g. Dile et al., 2016). This study assesses the potential water resources to be captured by small reservoirs, but didn't identify locations that are suitable for implementing small reservoirs. Moreover, implementation of small reservoirs may also have impacts on the downstream social and ecological systems. Therefore, assessing suitable areas for implementation of small reservoirs and their impact on the downstream social and ecological system warrants further investigation.

The accuracy of multi-year estimation of monthly stream flows was evaluated by comparing SWAT outputs with measured data by the Ministry of Water, Irrigation & Electricity of Ethiopia at the Gumera river gauging station. The SWAT model parameters are calibrated using observed stream flow at the Gumera river gauging station, which is the closest gauging station to the study watershed. Daggupati et al. (2015) recommended calibrating ungauged watersheds using observed stream flow from a close-by gauged watershed. In many cases systematic errors were identified and addressed using the SUFI method, a sophisticated optimization approaches available in SWAT-CUP (Abbaspour, 2015). Fig. 3 shows measured and simulated monthly flows in Gumera River near its outlet to Lake Tana. The results ($R^2 = 0.76$ and NSE = 0.75) indicate that using globally available weather and landscape data, SWAT reasonably simulated monthly stream flows in this region of Ethiopia.

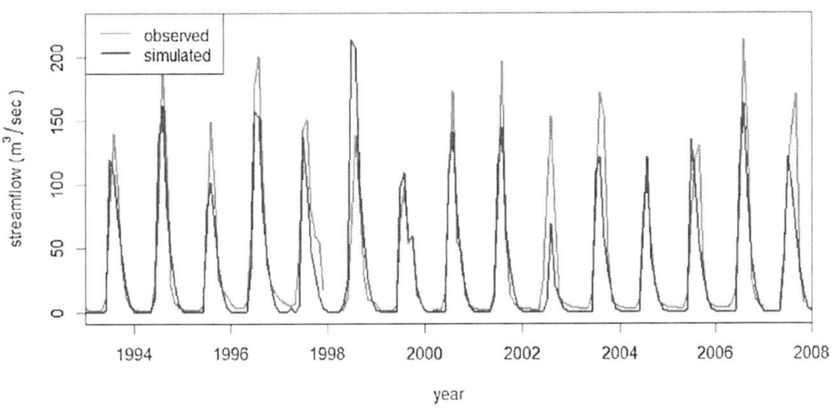

Figure. 3. Monthly measured and simulated stream flows in Gumera River observed near its outlet to Lake Tana.

SWAT results averaged over the entire Rib and Gumera river basins are summarized in Table 3. As reported by Ethiopian agricultural experts and other hydrologic studies of the region (Betrie *et al*., 2011; Setegn *et al*., 2009 ; Setegn *et al*., 2010a) significant erosion rates were estimated. The impacts of crop management scenarios on soil erosion are shown in Fig. 4; green indicates larger reductions in sediment yield while red indicates small or no reductions in sediment yield.

Table 3.Simulated cropping systems, vegetation types, and mean annual hydrologic and soil erosion components for the combined Gumera and Rib watersheds. Continuous maize and teff crops are grown in the main wet season without irrigation.

Land Use	Area	Precipitation	Irrigation	Runoff	Subsurface runoff	ET	Erosion
	(km^2)	(mm)	(mm)	(mm)	(mm)	(mm)	(t/ha)
Cropland							
Maize	1382	1303	0	402	351	525	85
Teff	944	1294	0	372	409	483	114
Maize-Rice-Onion	8	1313	639	386	237	1139	8
Grazing land	1049	1301	0	356	399	517	15
Forest land	5	1345	0	269	488	557	<1
Village	3	1278	0	472	279	503	<1

As shown in Fig. 4, a substantial increase in nitrogen fertilizer application results in a notable reduction in sediment delivery to streams in the sub-basins of the Rib and Gumera watersheds. Reduced erosion resulted from large increases in crop leaf area and biomass, which would reduce raindrop energy at the soil surface. Soil erosion happens because of the erosive forces of rain drops and surface flow of water (Arnold et al., 2012). Raindrop impact can detach soil particles on unprotected land surfaces between rills and initiate transport of these particles to the rills (Arnold et al., 2012). As the leaf area increases, the canopy cover increases and thereby reduces the erosive force of the raindrop. As such an increase in leaf area index and biomass, reduces soil erosion. These effects were greater on the sloping croplands between the Gumera and Rib watersheds (the green areas in the middle of the figure) than on the more level lands near the outlets of the watersheds or in the sub-basins with lesser amounts of cropland in the eastern and southern parts of watershed. In addition, the result indicates that sediment delivery to Lake Tana would be substantially reduced by implementation of more intensive cropping systems with appropriate conservation practices.

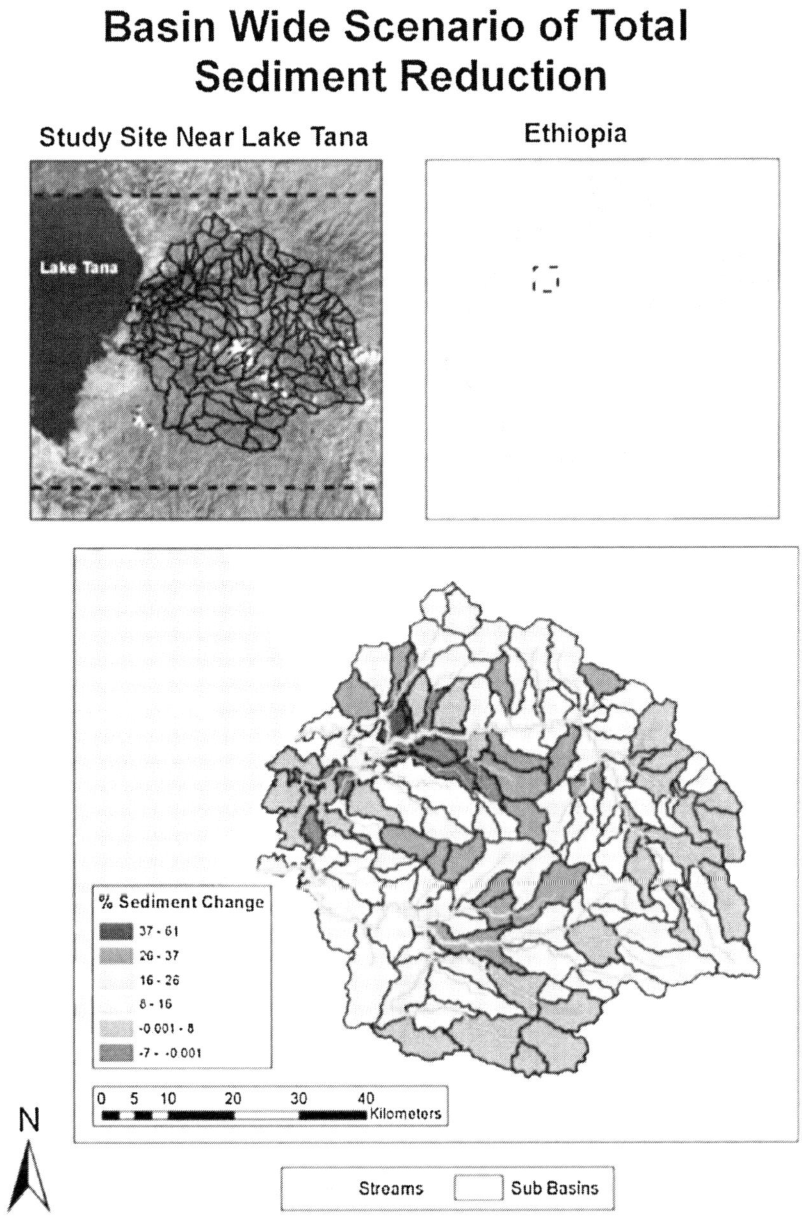

Figure. 4. Impacts of changing from low to high N fertilizer rates for crop rotations including maize, teff, rice, and onion for the same crop rotations on sediment delivery. Darker green colors indicate greater reductions in sediment yield. (For interpretation of the references to colour in this figure legend, the reader is referred to the web version of this article.)

3.2. Kebele scale simulations

3.2.1. Weg-Arba amba kebele

The Weg-Arba Amba kebele occupies an upland site with slopes ranging from 6 to 23%; the two dominant crops in the area are corn and teff. Most farmers currently grow only one crop per year during the rainy season. In this area, crop residues are immediately removed for animal feed after grain harvest, reflecting current and anticipated future needs for animal fodder. The results of these APEX simulations are summarized in Table 4. The Baseline scenario represented typical low-input continuous maize and continuous teff cropping systems for the Weg-Arba Amba kebele. As might be expected in a low-input farming system of this type, grain yields were quite low and strongly limited by nitrogen fertility. This is reflected by the large values of metrics for simulated nitrogen stress (NS) and water stress (WS).

Table 4.Weg-Arba Amba kebele simulated cropping systems.

Treatment[a]	Yield	Residue	FN	FP	Q	WYld	ET	IRGA	Y	WS	NS
	(t/ha)	(t/ha)	(kg/ha)	(kg/ha)	(mm)	(mm)	(mm)	(mm)	(t/ha)	(d)	(d)
Continuous Maize											
Baseline	2.4	2.3	56	40	406	681	562	0	30	14	65
Baseline + N + I	8.8	8.0	202	40	389	647	702	110	28	6	2
Baseline + N + I + V	10.2	6.8	202	40	389	646	703	110	28	6	2
Continuous Teff											
Baseline	0.6	1.1	24	17	372	648	596	0	33	23	31
Baseline + N + I	4.4	7.3	152	17	376	656	726	150	28	7	0
Baseline + N + I + V	5.0	6.8	152	17	376	656	726	150	28	7	0
Maize-Chickpea Relay Crop											
Maize	2.2	2.0	56	40	334	659	635	0	23	16	66

Chickpea	1.0	1.1	0	0	–	–	–	66	–	14	2
Maize-Squash Relay Crop											
Maize	2.1	2.0	56	40	331	666	739	0	24	16	67
Squash	0.3	0.6	40	0	–	–	–	184	–	3	46

[a]Grain/fruit yields (Yield), residue yields (Residue), fertilizer nitrogen (FN), fertilizer phosphorus (FP), runoff (Q), runoff + return flow to stream (WYld), evapotranspiration (ET), irrigation (IRGA), soil erosion (Y), water stress (WS), and nitrogen stress (NS).

IDSS/APEX crop model is sensitive to the abiotic stresses such as NS and WS that usually limit crop yields, and is able to dynamically estimate these stresses and simulate application of fertilizers and irrigation to minimize them. For example, for the Baseline + N + I scenario, IDSS applied 202 kg/ha of N fertilizer and 110 mm of supplemental irrigation for Maize (Table 4). This caused mean grain yields to increase from 2.4 to 8.8 tons/ha and greatly reduced the metrics for nitrogen (NS) and water stress (WS) substantially. Similar results were found for teff. Another effect of the N + I scenario was a small reduction in soil erosion due to the greater leaf area index of the well-fertilized crop. For the N + I + V scenario, the estimated crop yields increased with essentially no impact on stresses as the scenario represents genetically improved crops.

The Maize-Chickpea Relay Crop scenario caused a slight reduction in maize yield due to inter-crop competition. In addition, since all chickpea residue was removed after harvest for animal fodder, no effect of residual legume nitrogen was observed on maize yields. Despite the slight negative effect on maize yields, adding an irrigated, dry-season chickpea crop to the Baseline maize cropping system increased protein production with no additional fertilizer inputs. The Maize-Squash Relay Crop was extension to the Baseline unirrigated maize cropping system by adding squash to increase fresh vegetable production. Because the maize crop had taken up virtually all the available soil nitrogen by the time the squash was planted, and the rainy season normally ended before the squash matured, additional fertilizer nitrogen and irrigation were needed to obtain adequate squash yields.

As the shallow and rocky soils of the Weg-Arba Amba kebele suggest, severe soil erosion is occurring in the steep lands of the Ethiopian highlands due to little structural erosion control and over-grazing of crop residue (Betrie *et al.*, 2011; Setegn *et al.*, 2010a ; Setegn *et al.*, 2009). These simulations assumed clean tillage, no structural erosion controls, and removal of all crop residues for animal feed estimated annual erosion rates of over 20 tons per hectare. Even with adequate fertilizer, runoff (Q) plus subsurface return flow to streams are large, approaching total crop evapotranspiration (ET). This demonstrates that plenty of water is available for capture in small reservoirs for irrigation during the dry season. Simple low-pressure drip irrigation systems may be a useful technology to move farm families toward small-scale but intensive production of additional vegetable and/or grain legume crops.

3.2.2. Shena kebele

The Shena kebele site is located on the near-level floodplain near the shores of Lake Tana. In addition to simulating the low-input Baseline maize-rice-onion rotation, the effects of increasing levels of nitrogen fertilizer and growing improved crop varieties were assessed. A fourth two-year cropping system, consisting of a two-year maize-rice-onion-soybean-rice-onion rotation, was simulated to examine the effects of including a grain legume to increase protein production without increasing nitrogen fertilizer rates. Results of these simulations are summarized in Table 5.

Table 5.Shena kebele simulated irrigated maize, rice, Soybean, and onion rotation.

Treatment[a]	Yield	Residue	FN	FP	Q	WYld	ET	IRGA	Y	WS	NS	Protein
	(t/ha)	(t/ha)	(kg/ha)	(kg/ha)	(mm)	(mm)	(mm)	(mm)	(t/ha)	(d)	(d)	(kg/ha)
Baseline												
Maize	3.7	3.4	23	0	266	363	1591	591	0.7	6	47	300
Rice	4.2	4.3	55	23	-	-	-	0	-	1	45	352
Onion	1.8	0.1	23	0	-	-	-	111	-	1	21	-
Baseline + N + I												
Maize	9.1	8.6	178	0	272	370	1597	605	0.7	7	1	737
Rice	6	6.2	138	23	-	-	-	0	-	1	1	510
Onion	2.6	0.2	71	0	-	-	-	108	-	1	0	-
Baseline + N + I + V												

Maize	10.5	7.3	179	0	272	370	1607	604	0.7	7	1	850
Rice	6.9	5.5	150	23	–	–	–	0	–	2	1	586
Onion	3.7	0.1	73	0	–	–	–	117	–	2	0	–
Irrigated Maize-Rice-Onion-Soybean-Rice-Onion Rotation (2 Years)												
Maize	3.9	3.6	23	0	266	363	1631	578	0.6	3	47	316
Soybean	3.7	8	0	0	–	–	–	684	–	5	0	1628
Rice	4.7	4.9	55	23	–	–	–	0	–	3	34	400
Onion	1.9	0.1	23	0	–	–	–	111	–	1	19	–

[a]Simulated 2-year rotation of maize-rice-onion-soybean-rice-onion. Grain/bulb yields (Yield), residue yields (Residue), fertilizer nitrogen (FN), fertilizer phosphorus (FP), runoff (Q), runoff + return flow to stream (WYld), evapotranspiration (ET), irrigation (IRGA), soil erosion (Y), water stress (WS), nitrogen stress (NS), grain protein content (Protein).

The Maize-Rice-Onion Baseline scenario was a three-crop rotation of maize, rice, and onion. The rice was grown in paddies during the rainy season and received ample rainfall to maintain saturated or near-saturated soil conditions. Maize and onion were grown in the dry season and were irrigated at rates (IRGA) determined dynamically by APEX. Fertilizer rates for the Baseline cropping system were typical for the region, but APEX found that for this cropping system grain and onion yields are greatly limited by nitrogen stress (NS). In the Baseline + N scenario, fertilizer N requirements and crop response were investigated. In response to crop N stress, APEX increased average annual fertilizer N application from the 101 kg/ha Baseline rate to 387 kg/ha. The additional fertilizer reduced N stress (NS) to near zero and increased average yields of maize, rice, and onion by 246%, 143% and 144%, respectively, and almost doubled average annual protein production by the maize and rice crops. In addition, crop residue yields increased substantially, providing additional fodder for livestock.

The effects of growing crops with 15% greater harvest indexes were also simulated with increased fertilizer N (Baseline + N + V scenario) to examine the sensitivity of varietal improvement. As expected, APEX simulated greater yields, greater protein production, and slightly higher average nitrogen fertilizer requirements (402 kg N/ha) when varieties with increased yield potential were used (Table 5).

There was a substantial increase in protein production when maize was replaced with soybean every other year in the Baseline cropping system (without additional fertilizer nitrogen). Introduction of soybean slightly increased maize and rice yields compared to the Baseline maize-rice-onion rotation, but average annual protein production of the entire cropping system increased substantially (1628 kg/ha). This highlights the importance of including grain legumes in crop rotations to increase protein production of cropping systems where fertilizer nitrogen is limited.

3.3. Economic and nutritional impacts

3.3.1. Weg-Arba Amba kebele

The changes in economic benefits of the Weg-Arba Amba kebele for alternative farming systems are reported in Table 6. The Weg-Arba Amba kebele is non-irrigated in the Baseline and Alternative scenarios and has a 7650 Birr average annual net cash farm income (NCFI) per family which is equivalent to $352 USD based on the exchange rate in June 2016. For the Baseline + N +I scenario NCFI averages 103,080 Birr and for the Baseline + N + I + V scenario the value is 114,210 Birr. The increases in NCFI are due to higher grain and associated forage yields. Increased forage yields and reduced yield risk lead to increased livestock production and receipts. Average ending cash reserves in 2017 are projected to increase from 12,390 Birr in the Baseline to 552,260 Birr for the Baseline + N + I + V scenario.

Table 6.Economic and nutrition impacts of alternative farming systems for the Weg-Arba Amba kebele.

	Baseline	Base + N + I	Base + N + I + V
	(*1000 Birr)	(*1000 Birr)	(*1000 Birr)
Average values per family			
Net Present Value	100.7	408.74	447.44
Average Net Cash Farm Income	7.65	103.08	114.21
Average Ending Cash Reserves	12.39	492.71	552.26

Energy per Adult (cal/day)	1884	2091	2133
Protein per Adult (grams/day)	55.6	60.1	61.1
Fat per Adult (grams/day)	35.7	40	40.4
Calcium per Adult (grams/day)	0.25612	0.29144	0.29711
Iron per Adult (grams/day)	0.01718	0.01878	0.0192
Vitamin A per Adult (grams/day)	0.00118	0.0012	0.0012

Note: the values in table are multiples of 1000.

The nutritional benefits from improved yields and higher average annual NCFI per family are observed in terms of increased average daily consumption of energy, protein, fat, calcium, iron and vitamin A. The Baseline + N + I + V scenario leads to a 9.5% increase in average daily consumption of energy (calories) for a family in the Weg-Arba Amba kebele. FARMSIM uses the Organization for Economic Co-operation and Development (OECD) definition of an adult equivalent (AE) which represents a weighted aggregation of a family unit's consumption. An AE counts the first adult as 1 and each additional adult at 0.7 and counts children under 15 at 0.5.

The stochastic capabilities of FARMSIM provide probabilistic results for the output variables. The alternative farming systems shift the NCFI probability distribution to the right which reduces the chance of low average annual NCFI per family. The StopLight chart in Fig. 5 shows that there is an 83% chance annual NCFI is less than 10,000 Birr for the Baseline case. The StopLight Chart compares the target probabilities for one or more risky alternative (or scenarios) where the user must specify the two targets (Lower and Upper target). For the Baseline + N + I scenario the probability of annual NCFI being below 10,000 Birr is zero and there is a 57% chance that annual NCFI is greater than 100,000 Birr. Under a high technology adoption scenario (Baseline + N + I + V), the probability of annual NCFI exceeding 100,000 Birr is 75%.

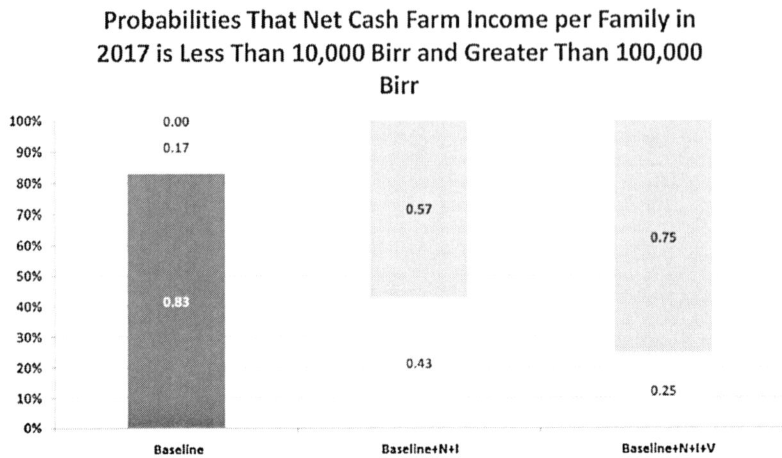

Figure. 5. StopLight Chart for per Family NCFI on a Weg-Arba Amba Farm for Alternative Farming Systems. The StopLight function presents the probability of: (a) exceeding the upper target (green); (b) being less than the lower target (red), and (c) observing the values between the targets (yellow). (For interpretation of the references to colour in this figure legend, the reader is referred to the web version of this article.)

The StopLight chart for daily energy consumption per AE is presented in Fig. 6. Introduction of irrigation and proper fertilization (Baseline + N + I) decreases the probability that daily energy consumption is less than 2,100 cal per AE from 100% to 60%. The high technology scenario (Baseline + N + I + V) further improves the probability of adequate energy consumption by increasing the probability of daily caloric intake per AE greater than 2,100 cal to 75 percent. Similarly, the daily consumption of protein, fat, calcium, iron and vitamin A increases as yields improve with the alternative farming systems.

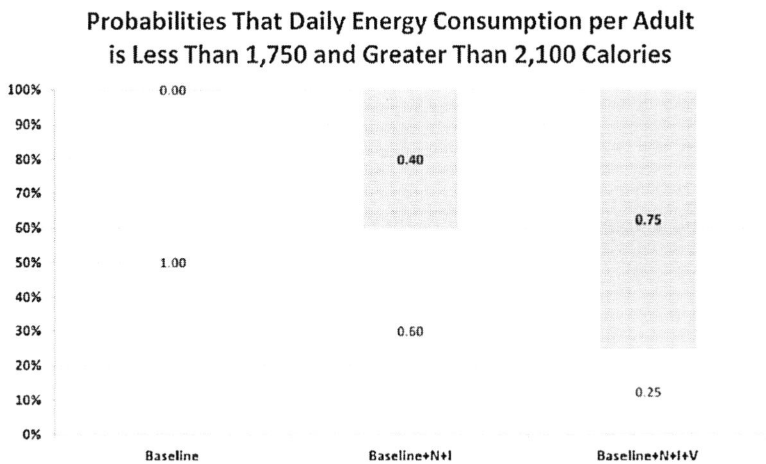

Figure. 6. StopLight Chart for Daily Energy Consumption per Adult Equivalent on a Weg-Arba Amba Farm for Alternative Farming Systems.

3.3.2. Shena kebele

The average annual NCFI per family under the Baseline is estimated at 19,010 Birr for the Shena kebele (Table 7). By optimizing crop production under the Baseline + N + I scenario, the average annual NCFI is 51,570 Birr and under the high technology adoption scenario (Baseline + N + I + V) the average annual NCFI is 58,960 Birr. The increases in NCFI are due to higher grain and onion yields and forage yields. Increased forage yields and reduced yield risk lead to increased livestock production and receipts. The increased average annual NCFI per family would lead to increases in ending cash reserves per family in 2017 (67,700 Birr for Baseline and 204,800 Birr for Baseline + N + I).

Table 7.Economic and nutrition impacts of alternative farming systems for the Shena kebele.

	Baseline	Baseline + N + I	Base + N + I + V
	(1000 Birr)	(1000 Birr)	(1000 Birr)
Average values per family			
Net Present Value	135.14	243.79	272.46

Average Net Cash Farm Income	19.01	51.57	58.96
Average Ending Cash Reserves	67.39	204.80	247.45
Energy per Adult (cal/day)	1913	1990	2071
Protein per Adult (grams/day)	58.5	60.9	62.4
Fat per Adult (grams/day)	38.20	41.20	42.00
Calcium per Adult (grams/day)	0.32600	0.38401	0.38560
Iron per Adult (grams/day)	0.01731	0.01757	0.01807
Vitamin A per Adult (grams/day)	0.00161	0.00163	0.00163

Note: the values in table are multiples of 1000.

The average daily consumption of calories for an AE is estimated at 1,913 cal for the Baseline, 1990 for the Baseline + N + I and 2071 for the Baseline + N + I + V scenario (The OECD defines an AE as a weight for aggregating a family unit. It counts the first adult as 1 and each additional adult at 0.7 and counts children under 15 t 0.5). Similarly, the daily consumption of protein, fat, calcium, iron and vitamin A increases as yields improve with the alternative farming systems. The probabilities of annual NCFI being less than 10,000 Birr and greater than 60,000 Birr are summarized in a StopLight chart in Fig. 7. Under the Baseline there is a 5% chance that annual NCFI for a farm family is less than 10,000 Birr and a zero chance it exceeds 60,000 Birr. For the Baseline + N + I scenario the probability that annual NCFI for a farm family exceeds 60,000 Birr is 26% and for the Baseline + N + I + V scenario the probability is 44%. Increased technology reduces the probability of low NCFI to zero and increases the probability of high NCFI.

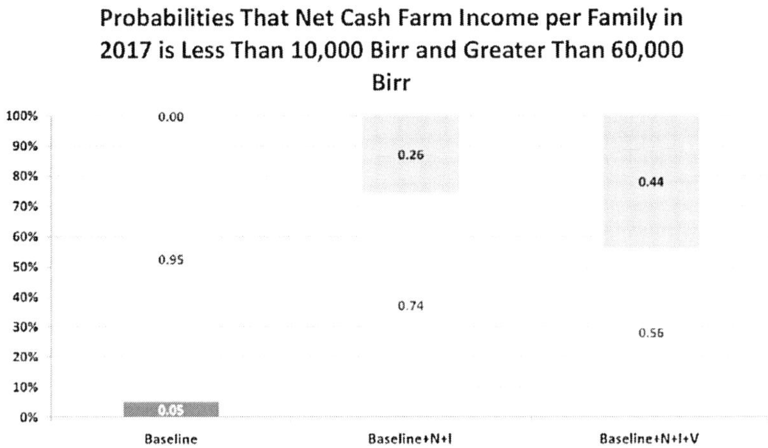

Figure. 7. StopLight Chart for per Family NCFI on a Shena Farm for Alternative Farming Systems.

The StopLight chart for daily energy consumption per AE is presented in Fig. 8. The introduction of improved production practices decreased the probability of energy deficiency in the diet. There is 85% chance that energy consumption per AE is between 1750 and 2000 cal per day for the Baseline, and a zero probability of energy consumption less than the high target value (2000 cal) for the Baseline + N + I + V scenario. Although stated as AE, the probable deficiency would likely have most serious consequences for the physical and cognitive development of children in the family. The Baseline + N + I scenario actually provides a 27% chance that average daily energy consumption per AE will exceed 2000 cal. The daily consumption of protein, fat, calcium, iron and vitamin A increases as yields improve with the alternative farming systems.

Introduction of proper plant fertilization and irrigation has the potential for greatly improving net income and nutritional intake for small-holder farmers in the Shena and Weg-Arba Amba kebele. Greater benefits will be observed for the Weg-Arba Amba farms because these farms are presently non-irrigated; average NCFI per family increases from 7650 to 103,080 Birr per year and net present value increases 400%. The Shena farms use irrigation there will still be a significant increase in average net income (27%) and net present value (180%) by optimizing their plant nutrition

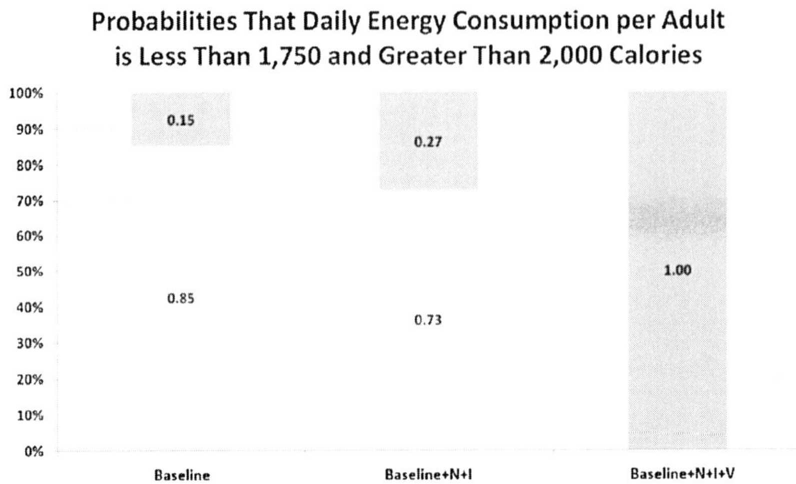

Figure. 8. StopLight Chart for Daily Energy Consumption per Adult Equivalent on a Shena Farm for Alternative Farming Systems.

program. Improvements in farming systems lead to an increase in daily intake of nutrients for an AE. The nutritional improvements are smaller (5%–10%) than the income increases because the kebele are producing adequate levels of grain and families are consuming adequate levels of nutrients in the Baseline case. Additionally, during years of low yields adequate cash should be available in the Baseline case to purchase food to meet current desired levels of consumption. Prolonged droughts would pose a significant problem because cash reserves would be reduced to zero after the first few years of drought.

3.3.3. Economic impacts for the Lake Tana Basin watershed

The economic impacts to the watershed analyzed with SWAT for the alternative farming systems are projected based on the per ha average annual net cash farm income projected for the representative farms and the number of ha of land with similar slope in the watershed. The watershed used for the analysis contains the Gumera and Ribb river basins. The watershed has about 8000 ha of cropland similar to less than 10% slope land in the Shena kebele. It is estimated that 750,000 ha of cropland that can be potentially irrigated in the watershed has from 10% to 20% slope like the maize/teff land in the Weg-Arba Amba kebele. Scaling up the average annual NCFI for the watershed, based on ha in the two kebeles, projected NCFI per hectare, and number of similar cropland ha is not ideal but it gives gross estimate of the economic impacts to the watershed.

The annual economic impacts, assuming widespread adoption of the Baseline + N + I farming system, is expected to be 11.2 million Birr and, for adoption of the Baseline + N + I + V farming system, is expected to be 13.3 million Birr (Table 8). The increases in average NCFI are largely due to the significant increase in net income for the non-irrigated maize/teff rotation and the very large number of ha planted to this rotation in the watershed. If full adoption in a short period was observed, there would most likely be a negative price impact which would decrease these estimated increases in NCFI. However, given a more normal adoption process the price effects will be much smaller as markets and population adjust to improved quantity and quality of food.

Table 8.Economic Benefits to the Watershed from Adoption of Alternative Farming Systems.

	Average annual Net Cash Farm Income			
Crop Rotation/Slope	Ha	Base	Base + N + I	Base + N + I + V
	(ha)	(1000 s Birr)		
Onion/Corn/Rice; fields < 10% slope	8000	76	207	237
Change from Base			131	160
Corn/Teff; fields 10–20% slope	750,000	890	11,995	13,290
Change from Base			11,105	13,214
Total Change in NCFI for Watershed			11,236	13,374

4. CONCLUSION

An integrated assessment was conducted using IDSS for multi-scale analyses of the impact of introducing new technology at the farm/village and watershed levels of scale. Regions that are agriculturally important in

Ethiopia were selected and analyzed from watershed down to the representative farm in two kebeles. SWAT, APEX, and FARMSIM were integrated by exchanging the output of one model to the other model as input for comprehensive assessment of agricultural technologies that are designed to increase food production, improve nutrition, enhance economic well-being, and minimize negative environmental consequences for small holder farms.

Even with conservative estimates of adoption, the results of the interventions were positive. The interventions implemented in this study are optimized to enhance the sustainable production of food under the conditions of the two locations and their related watersheds. The selection of the two sites was made to assure a reasonable access to markets and to compare different opportunities for the introduction of irrigation methods. Clearly, the successful introduction of the new farming practices and technologies would require substantial development of infrastructure such as roads and marketing support, supportive policy environments, and active programs such as those now being undertaken in Ethiopia to train and motivate changes in social mores for adoption of new technologies. This study includes the impact of traditionally low adoption rates, which might be enhanced by a more informed population.

The case studies demonstrate that the IDSS can be a useful tool to predict and evaluate the consequences of various interventions to improve the livelihoods of subsistence farmers while evaluating the environmental consequences at multiple levels of scale. The results demonstrate the ability to predict the outcomes of interventions using quantitative stochastic methods. IDSS provides an integrated systems approach to assessment of new investment options, including analysis of water use and protection to feed future populations. The ability to concurrently assess production, environmental, and economic and nutritional consequences of options for sustainable increases in production of food and use of scarce natural resources using an integrated modeling approach adds unique value to the current knowledge in agricultural research.

ACKNOWLEDGEMENT

The funding for this research was in part provided by The Bill and Melinda Gates Foundation and the Texas A&M AgriLife Research.

REFERENCES

1. Abbaspour, K.C., 2015. SWAT-CUP: SWAT Calibration and Uncertainty Programs –A User Manual. Swiss Federal Institute of Aquatic Science and Technology,Eawag, Dübendorf, Switzerland (Last accessed June 2016) http://swat.tamu.edu/media /114860/usermanual swatcup.pdf
2. Abildtrup, J., Audsley, E., Fekete-Farkas, M., Giupponi, C., Gylling, M., Rosato, P.,Rounsevell, M., 2006. Socio-economic scenario development for theassessment of climate change impacts on agricultural land use: a pairwisecomparison approach. Environ. Sci. Policy 9, 101–115
3. Amede, T., 2012. Rainwater Management for Sustainable AgriculturalIntensification in the Ethiopian Highlands. World Water Week 2012(Stockholm)
4. Araya, A., Stroosnijder, L., 2010. Effects of tied ridges and mulch on barley(Hordeum vulgare) rainwater use efficiency and production in NorthernEthiopia. Agric. Water Manag. 97, 841–847, Elsevier B.V
5. Arnold, J.G., Moriasi, D.N., Gassman, P.W., Abbaspour, K.C., White, M.J., Srinivasan,R., Santhi, C., Harmel, R.D., Van Griensven, A., Van Liew, M.W., Kannan, N., 2012.SWAT: Model use, calibration, and validation. Trans. ASABE. 55 (4), 1491–1508
6. Awokuse, T.O., Xie, R., 2015. Does agriculture really matter for economic growth indeveloping countries? Can. J. Agric. Econ. 63, 77–99
7. Barron, J., Okwach, G., 2005. Run-off water harvesting for dry spell mitigation inmaize (Zea mays L.): results from on-farm research in semi-arid Kenya. Agric.Water Manag. 74, 1–21
8. Betrie, G.D., Mohamed, Y.A., van Griensven, A., Srinivasan, R., 2011. Sedimentmanagement modelling in the Blue Nile Basin using SWAT model. Hydrol.Earth Syst. Sci. 15, 807–818
9. Bitew, M.M., Gebremichael, M., 2011. Assessment of satellite rainfall products forstreamflow simulation in medium watersheds of the Ethiopian highlands.Hydrol. Earth Syst. Sci. 15, 1147–1155
10. Boyd, C., Turton, C., Hatibu, N., Mahoo, H.F., Lazaro, E., Rwehumbiza, F.B., Okubal, P.,Makumb, M., 2000. The Contribution of Soil and Water Conservation toSustainable Livelihoods in Semi-Arid Areas of Sub-Saharan Africa. AgriculturalResearch & Extension Network, Network Paper No. 102
11. CA, 2007. In: Molden, D. (Ed.), Water for Food Water for Life: A ComprehensiveAssessment of Water Management in Agriculture. International WaterManagement Institute, Colombo
12. Cai, X., McKinney, D.C., Lasdon, L.S., 2003. Integratedhydrologic-agronomic-economic model for river basin management. J. WateWaterResour. Plann. Manage. 129 (1), 4–17
13. Chilonda, P., Van Huylenbroeck, G., 2001. A conceptual framework for theeconomic analysis of factors influencing decision-making of small-scalefarmers in animal health management. Rev. Sci. Tech. 20, 687–695
14. Cooper, P.J.M., Dimes, J., Rao, K.P.C., Shapiro, B., Shiferaw, B., Twomlow, S., 2008.Coping better with current climatic varability in the rain-fed farming systemsof the Sub-Saharan Africa: an essential first step in adapting to future climatechange? Agric. Ecosyst. Environ. 126, 24–35
15. Daggupati, P., Pai, N., Ale, S., Douglas-Mankin, K.R., Zeckoski, R.W., Jeong, J.,Parajuli, P.B., Saraswat, D., Youssef, M.A., 2015. A recommended calibrationand validation strategy for hydrologic and water quality models. Trans. ASABE58 (6), 1705–1719
16. DFID, 2004. Agriculture, Growth and Poverty Reduction. Oxford
17. Dile, Y.T., Srinivasan, R., 2014. Evaluation of CFSR climate data for hydrologicprediction in data-scarce watersheds: an application in the Blue Nile RiverBasin. J. Am. Water Resour. Assoc., http://dx.doi.org/10.1111/jawr.12182(in-press)

18. Dile, Y.T., Berndtsson, R., Setegn, S.G., 2013. Hydrological response to climatechange for gilgel abay river, in the lake tana basin-upper blue nile basin ofEthiopia A˜nel, J.A. (Ed.). PLoS One 8, e79296
19. Dile, Y.T., Karlberg, L., Daggupati, P., Srinivasan, R., Wiberg, D., Rockström, J., 2016.Assessing the implications of water harvesting intensification onupstream?downstream ecosystem services: a case study in the Lake Tanabasin. Sci. Total Environ. 542, 22–35
20. Djurfeldt, G., Holmen, H., Jirstrom, M., Larsson, R., 2005. The African Food Crisis:Lessons from the Asian Green Revolution (Cabi)
21. FAO, 2002. World Agriculture: Towards 2015/2030: Summary Report. Food andAgriculture Organization of the United Nations, Rome
22. FAO, 2011. The State of the World's Land and Water Resources for Food andAgriculture. Organization. Food and Agriculture Organization of the UnitedNations, Rome
23. FAO, 2013. The State of Food Insecurity in the World. The Multiple Dimensions ofFood Security (Rome)
24. Ford, W., King, K., Williams, M., Williams, J., Fausey, N., 2015. Sensitivity analysis ofthe Agricultural Policy/Environmental eXtender (APEX) for phosphorus loadsin tile-drained landscapes. J. Environ. Qual. 44, 1099–1110
25. Fox, P., Rockstrom, J., 2003. Supplemental irrigation for dry-spell mitigation ofrainfed agriculture in the Sahel. Agric. Water Manag. 61, 29–50
26. Gallup, J.L., Radelet, S., Warne, A., 1998. Economic Growth and the Income of thePoor (Cambridge, MA, USA)
27. Garg, K.K., Karlberg, L., Barron, J., Wani, S.P., Rockstrom, J.1, 2012. Assessingimpacts of agricultural water interventions in the Kothapally watershed,Southern India. Hydrol. Process. 26, 387–404
28. Gassman, P.W., Williams, J.R., Wang, X., Saleh, A., Osei, E., Hauck, L.M., Izaurralde,R.C., Flowers, J.D., 2010. The Agricultural Policy/Environmental (APEX) modelan emerging tool for landscape and watershed environmental analyses. Trans.Asabe 53, 711–740
29. Gebey, T., Lemma, T., Hoekstra, D., Tegegne, A., Alemu, B., 2012. ParticipatorySmallholder Dairy Value Chain Development in Fogera Woreda, Ethiopia:Experiences from IPMS Project Interventions. IPMS Case Study. ILRI, Nairobi,Kenya
30. Gebreegziabher, T., Nyssen, J., Govaerts, B., Getnet, F., Behailu, M., Haile, M.,Deckers, J., 2009. Contour furrows for in situ soil and water conservationTigray, Northern Ethiopia. Soil Tillage Res. 103, 257–264
31. Gebregziabher, G., Rebelo, L.-M., Notenbaert, A., Ergano, K., Abebe, Y., 2013.Gebrehaweria Gebregziabher, Lisa-Maria Rebelo, An Notenbaert, KebebeErgano and Yenenesh Abebe (Colombo, Sri Lanka)
32. Hazell, P.B., 2009. The Asian Green Revolution. Int. Food Policy Res. Inst
33. IFAD, 2009. Federal Democratic Republic of Ethiopia: Country Program Evaluation
34. IFAD, 2011. Rural Poverty Report: New Realities New Challenges: NewOpportunities for Tomorrow's Generation (Rome)
35. ILRI (International Livestock Research Institute), 2005. Improving Productivity andMarket Success (IPMS)-Pilot Learning Sites (PLS), Diagnostic and ProgramDesign, Fogera Woreda. Ethiopia, Addis Ababa
36. ILRI (International Livestock Research Institute)-IPMS Atlas 2007, Fogera Woreda,Amhara
37. Lal, R., 2001. Potential of desertification control to sequester carbon and mitigatethe greenhouse effect. Clim. Change 51, 35–72
38. Lefkoff, L.J., Gorelick, S.M., 1990. Simulating physical processes and economicbehaviour in saline, irrigated agriculture: model development. Water Resour.Res. 26 (7), 1359–1369
39. Licker, R., Johnston, M., a, Foley J., Barford, C., Kucharik, C.J., Monfreda, C.,Ramankutty, N., 2010. Mind the gap: how do climate and agriculturalmanagement

explain the yield gap of croplands around the world? Global Ecol.Biogeogr. 19, 769–782

40. McHugh, O.V., Steenhuis, T.S., Fernandes, E.C.M., 2007. Performance of in siturainwater conservation tillage techniques on dry spell mitigation and erosioncontrol in the drought-prone North Wello zone of the Ethiopian highlands. SoilTillage Res. 97, 19–36
41. MoWR, 2009. Land Use and Soil Data: Ethiopian Minstry of Water Resources.Ethiopia, Addis Ababa
42. Munamati, M., Nyagumbo, I., 2010. In situ rainwater harvesting using dead levelcontours in semi-arid southern Zimbabwe: insights on the role ofsocio-economic factors on performance and effectiveness in Gwanda District.Phys. Chem. Earth Parts A/B/C 35, 699–705 (Elsevier Ltd.)
43. Neitsch, S.L., Arnold, J.G., Kiniry, J.R., Williams, J.R., 2012. Soil and WaterAssessment Tool Theoretical DocumentatiDocumentation: Version 2005. Grassland Soil andWater Research Laboratory, Agricultural Research Service, Blackland ResearchCenter, Texas Agricultural Experiment Station, Temple, Texas
44. Norfolk, O., Abdel-Dayem, M., Gilbert, F., 2012. Rainwater harvesting andarthropod biodiversity within an arid agro-ecosystem. Agric. Ecosyst. Environ.162, 8–14, Elsevier B.V
45. Nyangito, H.O., Richardson, J.W., Mukhebi, A.W., Mundy, D.S., Zimmel, P., Namken,J., 1995. Whole farm economic evaluation of East Coast fever immunizationstrategies on farms in the Uasin Gishu District of Kenya. Comput. Electron.Agric. 12, 19–33
46. Nyangito, H.O., Richardson, J.W., Mukhebi, A.W., Zimmel, P., Namken, J., Berry, B.P.,1996a. Whole farm simulation analysis of economic impacts of East CoastFever immunication strategies on mixed crop-livestock farms in Kenya. Agric.Syst. 51, 1–27
47. Nyangito, H.O., Richardson, J.W., Mundy, D.S., Mukhebi, A.W., Zimmel, P., Namken,J., 1996b. Economic impacts of East Coast Fever immunization on smallholderfarms, Kenya: a simulation analysis. Agric. Econ. 13, 163–177
48. Pandey, D.N., 2001. A bountiful harvest of rainwater. Science 293 (80-), 1763
49. Pretty, J., Toulmin, C., Williams, S., 2011. Sustainable intensification in Africanagriculture. Int. J Agric. Sustainability 9 (1), 5–24
50. Reij, C., Tappan, G., Smale, M., 2009. Agroenvironmental Transformation in theSahel: Another Kind of Green Revolution IFPRI Discussion Paper 00914. FoodPolicy, Washington, DC, USA
51. Richardson, J.W., Clair, J.N., 1981 The Farm Level Income and Policy SimulationModel: FLIPSIM. Departmental Technical Report, DTR No. 81 2, College Station
52. Richardson, J.W., Lemieux, C.M., Nixon, C.J., 1983. Entry into farming: the effects ofleasing and leverage on firm survival. 15. South. J. Agric. Econ. 15, 139–145
53. Richardson, J.W., Klose, S.L., Gray, A.W., 2000. An applied procedure for estimatingand simulating multivariate empirical (MVE) probability distributions infarm-level risk assessment and policy analysis. J. Agric. Appl. Econ. 2, 299–315
54. Rockström, J., Falkenmark, M., 2000. Semiarid crop production from a hydrologicalperspective – gap between potential and actual yields. CRC Crit. Rev. Plant Sci.19, 319–346
55. Rockström, J., Barron, J., Fox, P., 2002. Rainwater management for increasedproductivity among small-holder farmers in drought prone environments.Phys. Chem. Earth 27, 949–959
56. Rosegrant, M., Cai, X., Cline, S., Nakagawa, N., 2002. The Role of Rainfed Agriculturein the Future of Global Food Production. Food Policy, Washington DC, USA
57. Setegn, S.G., Srinivasan, R., Dargahi, B., Melesse, A.M., 2009. Spatial delineation ofsoil erosion vulnerability in the Lake Tana Basin, Ethiopia. Hydrol. Process. 23,3738–3750

58. Setegn, S.G., Dargahi, B., Srinivasan, R., Melesse, A.M., 2010a. Modeling of sedimentyield from anjeni-gauged watershed, ethiopia using swat model. J. Am. WaterResour. Assoc. 46, 514–526
59. Setegn, S.G., Srinivasan, R., Melesse, A.M., Dargahi, B., 2010b. SWAT modelapplication and prediction uncertainty analysis in the Lake Tana Basin,Ethiopia. Hydrol. Process. 367, 357–367
60. Setegn, S.G., Rayner, D., Melesse, A.M., Dargahi, B., Srinivasan, R., 2011. Impact ofclimate change on the hydroclimatology of Lake Tana Basin, Ethiopia. WaterResour. Res. 47, W04511
61. Thirtle, C., Irz, X., Lin, L., Mckenzie-hill, V., Wiggins, S., 2001. Relationship betweenchanges in agricultural productivity and the incidence of poverty in developingcountries. Growth (Lakeland)
62. Tibebe, D., Bewket, W., 2011. Surface runoff and soil erosion estimation using theSWAT model in the Keleta Watershed, Ethiopia. Land Degrad. Dev. 22, 551–564
63. Tibesigwa, B., Visser, M., 2015. Small-scale Subsistence Farming, Food Security,Climate Change and Adaptation in South Africa: Male-Female HeadedHouseholds and Urban-Rural Nexus
64. Tuppad, P., Winchell, M., Wang, X., Srinivasan, R., Williams, J., 2009. ArcAPEX:ArcGIS interface for Agricultural Policy Environmental eXtender (APEX)hydrology/water quality model. Int. Agric. Eng. J. 18, 59
65. Vohland, K., Barry, B., 2009. A review of in situ rainwater harvesting (RWH)practices modifying landscape functions in African drylands. Agric. Ecosyst.Environ. 131, 119–127
66. Wang, X., Williams, J., Gassman, P., Baffaut, C., Izaurralde, C., Jeong, J., Kiniry, J.,2012. The EPIC and APEX models: use, calibration and validation. Transactions55, 1447–1462
67. van de Giesen, N., Stomph, T.J., de Ridder, N., 2005. Surface runoff scale effects inWest African watersheds: modeling and management options. Agric. WaterManag. 72, 109–130
68. Volk, M., Hirschfeld, J., Dehnhardt, A., Schmidt, G., Bohn, C., Liersch, S., Gassman,P.W., 2008. Integrated ecological-economic modelling of water pollutionabatement management options in the Upper Ems River Basin. Ecol. Econ. 66(1), 66–76
69. White, E.D., Easton, Z.M., Fuka, D.R., Collick, A.S., Adgo, E., McCartney, M.,Awulachew, S.B., Selassie, Y.G., Steenhuis, T.S., 2011. Development andapplication of a physically based landscape water balance in the SWAT model.Hydrol. Process. 25, 915–925
70. WorldBank, 2008. Agriculture for development. In: Agriculture. World BankWashington, DC, USA

CHAPTER 12

Unlocking the Energy Potential of Manure—An Assessment of the Biogas Production Potential at the Farm Level in Germany

Mattes Scheftelowitz [1,*,†] and Daniela Thrän [1,2,†]

[1]*Deutsches Biomasseforschungszentrum gGmbH, Leipzig 04347, Germany*

[2]*Department Bioenergy, Helmholtz Centre for Environmental Research GmbH, Leipzig 04347, Germany*

ABSTRACT

Residues from animal husbandry are one of the major greenhouse gas (GHG) emission sources in agriculture. The production of biogas from agricultural residues can reduce GHG emissions through an improved handling of the material streams such as manure storage. Additionally, biogas can substitute fossil energy carriers in the provision of heat, power, and transport fuels. The aim of this work is to estimate the manure potential for biogas production in Germany under the consideration of the farm size of livestock production. In Germany, cattle and pig farming is of major relevance with more than 130,000 farms throughout the country. To unlock the biogas potential of manure, the low energy density of manure, depending on the dry matter content, needs to be considered, meaning that biogas installations need to be built close to the manure production on the farm site. This not only results in a high number of biogas plants, but also due to the wide range of farm sizes in Germany, a huge number of very small biogas plants. Small biogas installations have higher specific investment costs. Together with the relatively low methane yields from manure, costs for power generation would be very high. Co-substrates with higher methane yield can lower the costs for biogas. Thus, the use of a co-substrate could help to use small manure potentials. Biogas plants with the necessary minimum size of 50 kW_{el} installed power could be established at farms representing 12% of all cattle and 16.5% of all pigs respectively in Germany. Using excrement from pigs, farms representing 16.5% of the total amount of pigs could establish a biogas plant. The use of manure in combination with energy crops can increase the size of biogas plants on a farm site significantly. At cattle farms, the share would increase to 31.1% with 40% co-substrate and to 40.8% with 60% co-substrate. At pig farms, the share would increase to 36% if co-substrates were used.

Keywords: cattle manure; pig manure; co-substrate; biogas potential; biogas plant size

1. INTRODUCTION

The reduction of greenhouse gas (GHG) emissions from anthropogenic activities is one of the major challenges for society in terms of sustainable development on a global level [1]. In 2012, the consumption of fossil fuels for energy provision was responsible for three quarters of worldwide GHG emissions, and 83% of GHG emissions in Germany [2,3]. Agriculture and animal husbandry is another main emitter [4,5]. The use of excrements from animal husbandry for biogas production can reduce GHG emissions [6]. To achieve the envisaged reduction target of GHG emissions of 80% by 2050, a transition of the energy system towards a greater use of renewable energy is one central element [7]. Activities to introduce energy production from renewables have been initiated in many countries. In Germany, a target has been set to cover 40%–45% of the overall power demand with renewables by 2025 [8].

However, GHG reductions in Germany are not only an issue for the energy sector but also one for the agricultural sector. In 2011, agriculture accounted for 7.7% of the total greenhouse gas emissions in Germany [3]. With nearly 9 million tons of CO_2-equivalents per year, more than 10% of the GHG emissions from German agriculture were caused by manure storage. Liquid manure and droppings contain a high percentage of biomass, which is usually re-used as a fertilizer. The storage is necessary because of the continuous production of excrement from livestock farming and its discontinuous re-use as a farm fertilizer, which is restricted to certain times of the year. Manure storage facilities are often realized in open silos and the stored manure off-gases equate to amounts of methane, which causes 1% of the overall German GHG emissions [9]. By using manure in biogas plants, these emissions can be significantly reduced. The conversion of manure to biogas mainly includes the degradation of organic carbon to methane. There is no competition with regard to the potential use as a fertilizer, since the digestate can also be spread over the field, and the nutrients are even more available to plants.

Biogas potential from liquid manure and/or litter for energetic utilization in Germany is 139 million t per year corresponding to a total biogas yield of 4 billion m^3/yr [10]. Assumptions about the overall biogas potential in Germany take into account that not all excrement can be used for biogas generation, for example, in the case of pasture management or small quantities of livestock. The production of biogas from those agricultural residues reduces GHG emissions from manure storage. Biogas can substitute fossil energy carrier and has a technical fuel potential of approximately 90 PJ/yr. The largest share of this potential is from cattle droppings with around 60%, followed by pigs accounting for around 30 PJ/yr. Liquid manure from chicken contributes to this potential with 3 PJ/yr [10]. The economical potential of the listed manure is not considered due to its many influencing factors.

Supported by political instruments over the last decade, 7800 biogas plants have been built, which are currently producing 18% of the renewable power in Germany from different feedstock. The most widely used substrates are maize silage, whole crop silage, and manure and dung. [11]. Even though, the use of manure has been put forward for biogas production by the Renewable Energy Sources Act [12], less than half of its potential is used [13]. To unlock the remaining biogas potential of manure, additional obstacles need to be considered. Due to the high water content of manure, the methane yield is comparably low. The range of dry matter content varies from 0.9% to 23%, depending on livestock and husbandry [14,15]. Therefore, biogas plants using manure need to be operated close to the farm site. This means that the size of the livestock production at each individual farm will determine the size of the biogas plant. Considering the higher specific investment efforts for smaller installations (economies of scale), the size of the livestock production significantly influences investment and is an important factor for unlocking the energy potential of manure.

The aim of this work is to estimate the distribution of the overall potential for the biogas provision of manure. For this calculation, the size of livestock production on farm sites in Germany is considered. Considering the wide range of farm sizes, the demand of the biogas plants will also differ in size. Assuming a “minimum biogas plant size” for an economically feasible biogas production, how the biogas potential related to manure only, and the potential that can be tapped into from the co-fermentation of manure and energy crops can be used, is discussed.

Three scenarios were analyzed: the potential biogas plant size (1) without using co-substrates, (2) with 40% of maize, and (3) 60% of maize as co-substrate. The three scenarios were chosen by political circumstances. Biogas plants that were constructed after 2012 were allowed to use a maximum of 60% maize by the amendment of the Renewable Energy Sources Act of 2012 [12]. Since the amendment in 2014, no extra remuneration was provided for the use of renewable raw material such as maize [16] (scenario 0% co-substrate). For the amendment of the Renewable Energy Sources Act in 2016, it is expected that a maximum of 50% maize as co-substrate is allowed [17] (scenario 40% of maize).

2. METHODOLOGY AND DATA

To calculate the biogas production potential of manure at the farm level, we (1) identified and classified the most relevant animal husbandry systems for Germany in terms of their manure/dung quality, (2) calculated the farm-specific manure production for the classified systems with its related biogas potential, (3) assumed the investment costs for the different plant capacities for typical biogas plant concepts for manure, and (4) conducted different feedstock scenarios taking into account a threshold value of the minimum capacity of the biogas plants. Since the remuneration is determined by the Renewable Energy Sources Act, biogas plants cannot produce any expensive electricity. The threshold value is therefore defined as the minimum size of biogas plants economical feasible with the remuneration by the Renewable Energy Sources Act [18].

2.1. Identifying Relevant Livestock Systems

To evaluate the biogas potential of excrement from animal husbandry for Germany, the livestock species cattle, and pigs were considered.

Sheep, goats, and horses are mostly kept free-range, meaning that excrement cannot be collected. Since 2009, the cage production of poultry is no longer allowed in Germany, except for those farms with cage production, which were built before 2009 [19]. For the above types of animal husbandry, it is difficult to collect the excrement in a usable form for the purposes of energy production. Therefore, the livestock species of sheep, goats, horses, and poultry are not considered in this potential analysis.

Based on the survey results among biogas plant operators, conducted by the Germany Biomass Research Center, the most important feedstock could be determined. With 62% cattle, slurry is the most widely used form of manure for energy production, followed by pig slurry with a share of 14%. All other types of manure have minor shares and are therefore less important for biogas production [20].

The annual amount of manure or dung depends on the type of animal husbandry. The main types can be divided into straw-bedded animal husbandry and animal husbandry on slatted floors. The straw-bedded animal husbandry in loose housing stables mainly produces dung, based on straw, manure, and urine. The liquid phase (slurry) has a very low dry matter content [14]. Therefore it has also a low energy density [12,21]. Meanwhile, the range of liquid manure from cattle varies from 4% to 13% and, from pigs, from 1.5%–12% dry matter content. Slurry has just 2% maximum dry matter content (see Table 1). Therefore, it is irrelevant for biogas production and not considered in the following calculations.

Table 1. Range of dry matter content in different types of excrements [15,22,23,24].

Type of Excrement	Dry Matter Content in%
Dung cattle	25
Manure cattle	4–13
Slurry cattle	2
Dung pigs	23–25
Manure pigs	1.5–12
Slurry pigs	0.9–2

In barns with slatted floors, manure, and urine drop down into a pit below the floor where the liquid manure is collected.

2.2. Data on Livestock

Data on livestock breeding is published by the federal statistical office for Germany. The data comes from the agricultural census in 2010. The statistics provide livestock species, farm size categories in terms of the number of animals per farm, and types of husbandry (*i.e.*, manure or dung producing). Seven farm size classes are considered for the different livestock breeding systems [25]. For our calculations, we use the average farm size and neglect the variation within the size classes. For the different breeding systems, we also consider: data for cattle, which is differentiated into dairy cattle and all other cattle, and data for pigs, with piglets and all other pigs.

For cattle classed under other types of husbandry other than tethering of cattle and loose housing stables, it is assumed that these cattle are kept free-range, and they are not considered for the potential calculation.

For pigs classed under other types of husbandry, it is assumed that the main type of excrement is dung, as free-range husbandry for pigs is uncommon.

2.3. Data on Manure and Dung

Data on the annual quantity of manure and dung is provided by the Association for Technology and Structures in Agriculture (KTBL). KTBL provides the parameters of costs, time, and amounts for all the relevant parts of agriculture including different types of livestock and forms of livestock farming.

The annual quantity is given in m^3/yr per animal (see Table 2). By providing data per animal, different production cycles for different livestock species are taken into account. The amount of excrement is differentiated into manure and dung, depending on the form of livestock farming on slatted floors or in loose housing stables.

Table 2. Potential of dung and manure depending on livestock species and the type of livestock breeding [26,27].

Livestock and Form of Livestock Farming	Dung/Manure Per Animal in m^3/yr
Cattle, slatted floors	11.76/manure
Cattle, loose housing stable	8.3/dung
Dairy cattle, slatted floors	18/manure
Dairy cattle, loose housing stable	14.4/dung
Pig (with piglets)	2.15/dung
Pig (with piglets)	5/manure
Pig (pup)	0.3/dung
Pig (pup)	0.6/manure
Pig (fattened pig)	1.4/dung
Pig (fattened pig)	1.5/manure

As the statistics are not differentiated into younger pigs and fattened pigs, an average of 1.1 m^3 of liquid manure per year and animal on slatted floors and 0.85 m^3 of dung on solid floors is assumed [26,27].

For cattle, it is assumed that the breeding types "tethering of cattle on slatted floors" and "loose housing stables with slatted floors" produce the same amount of manure.

2.4. Data for Methane Yields

Potential methane yields from manure compared to usual biogas substrates are shown in Table 3 in m^3 of methane per ton of fresh matter of substrate. The values given are an average. In fact, in reality the values have a broader bandwidth for the methane yield, depending on the dry matter content, the water content, the composition of substrates, and other factors. Data on methane yields for the different types of biogas substrates are given in the annex of the Ordinance on the Generation of Electricity from Biomass [28]. Due to the lack of opportunity to determine the methane yield for each substrate, biogas plant operators in Germany are obliged to use the values provided to receive the remuneration for electricity production.

Table 3. Methane yields from different substrates [28].

Substrate	Methane Yield in m^3/t_{FM}
Cattle dung	53
Cattle manure	17
Pig dung	45
Pig manure	12
Whole crop silage (rye)	103
Maize silage	106

2.5. Biogas Potential Calculation

The amount of installable electrical power from a biogas (combined heat and power) plant (chp) was calculated for each type of livestock, farm class, and husbandry system.

In addition, two different scenarios were calculated with 40% and 60% of maize silage in relation to the amount of produced dung or manure.

The possible installable rated power in kW_{el} per livestock is calculated based on P_n, the net calorific value of methane, an average electrical efficiency of 37%, and 8760 full load hours. The conversion factor for MJ to kWh is 0.277 (see all parameters in Table 4).

Table 4. Parameters for calculating the potential electrical capacity from biogas.

Parameter	Description	Parameterization and Data Source
kW_{POT}	Potential of electrical capacity (kW)	Potential of maximum installable electrical power
C_f	Farm class	Section 2.1.
HUS	Husbandry system	Section 2.1.
n_{AP}	Number of animal places	Section 2.1.
$\dot{m}_{EAP}$	Excrement per animal place (in m^3/place per year)	Manure or dung per animal place depending on livestock and type of animal husbandry [27]
$\dot{m}_{my\ pi}$	Methane yield (in m^3/t_{FM} per year)	Methane yield per t of fresh mass depending on livestock and type of manure, Section 2.4
NCV	Net calorific value of methane	36 MJ/m^3
fc	Conversion factor	Conversion factor for MJ to kWh (0.2777777)
η_{el}	Electrical efficiency	For chp up to 500 kW of installed power, an average electrical efficiency of 37% is assumed [29]
FLH	Full load hours	8760 h/a, max. for y
ρi	Dung/manure density	Dung density depending on livestock: cattle 0.8 t/m^3 [22], pigs 0.91 t/m^3, manure density 1 t/m^3 [23]
COSUB	Co-substrate	Maize silage
p_{COSUB}	Share of Co-substrate (in %)	Depending on the scenario, 40% or 60% of Co-substrate maize silage, depending on the mass of manure or dung

First, the average biogas potential per farm site was calculated by dividing the number of animals per farm class and husbandry system. Then, it is multiplied with the manure or dung per animal place and with the density of the excrements. This is then multiplied with the specific methane yield of the substrate, which is multiplied with net calorific value, a conversion factor, and the average of electrical efficiency of small combined heat and power plants. (1) In the next step, the amount of manure or dung and its methane potential was calculated divided by the maximum full load hours per year and divided by the number of animal placesFor the scenarios with co-substrate, the calculation was complemented with calculations for the co-substrate depending on the mass of excrement produced per livestock, the number of animals, the farm class, and the husbandry system (2).

The potential for the installable electrical power of manure or dung for biogas per farm class is obtained as:

$$kW_{POT}(Cf,\ HUS) = \frac{\frac{\sum_{i=1}^{n} n_{AP}(Cf,HUS)}{\sum_{i=1}^{n} n_{FARM}(Cf,\ HUS)} * \frac{m_{EAP} * \rho i * m_{MY} * NCV * fc * \eta_{el}}{FLH} * 100}{\sum_{i=1}^{n} n_{AP}} \quad (1)$$

The potential for the installable electrical power of manure or dung and the co-substrate for biogas per farm class are obtained as:

$$kW_{POT}(Cf, HUS) = \frac{\frac{\sum_{i=1}^{n} n_{AP}(Cf, HUS)}{\sum_{i=1}^{n} n_{FARM}(Cf, HUS)} * \frac{m_{EAP} * pi * m_{MY} * NCV * fc * \eta_{el} * p_{COSUB}(m_{EAP}) * m_{MY} * NCV * fc * \eta_{el}}{FLH} * 100}{\sum_{i=1}^{n} n_{AP}} \quad (2)$$

2.6. Assumed Investment Costs for Biogas Concepts

Biogas plants operated by agricultural feedstock, including manure, other residues, and energy crops typically consist of a fermentation reactor, a gas storage unit, and the energy provision step and peripheral equipment. The main process technology applied is the continuous stirred tank reactor system. The substrate is usually fed into the digesters by means of pumps (for liquid substrates), feeding systems for solid matter (e.g., energy crops), or mixing tanks. The insulated digesters that are mainly operated at mesophilic temperatures have rubber domes for gas collection. Due to German regulations, the retention times are more than 100 days in most cases, with the resulting overall organic loading rates being respectively low. Gas cleaning (*i.e.*, desulphurization) and safety installations (*i.e.*, excess gas burners) are required in addition. The gas produced is collected in rubber domes (one and two layer systems) on top of the digesters and converted on site at the biogas plant into power, fed into the grid, the produced heat is used to supply some heat demand close to the plant [30]. The capacity of a biogas plant is typically given as "installed capacity of the power provision unit (kW_{el})." In sum, there is a reasonable technical effort behind the conversion of manure into energy so that economies of scale are a relevant issue for the feasibility of biogas concepts for manure processing. Past experiences from Germany have placed average investment costs of 3500 EUR/kW for a 1000 kW_{el} biogas plant and up to 9000 EUR/kW_{el} for a 50–75 kW_{el} biogas plant [10,31]. The degression of costs as a function of the installed electrical capacity is shown in Figure 1. This cost function led to the assumption that biogas plants at farm scale should not fall below a minimum capacity of 50 kW_{el}.

2.7. Considering Different Scenarios

The relatively small potential of manure can be supplemented with other renewable raw materials such as maize silage. By using additional substrates for the fermentation process, the required capacity for biogas plants at a farm site can change dramatically. We calculated mixed biogas systems using manure and so-called co-substrates in three scenarios: without co-substrate, with 40%, and with 60% maize silage as a co-substrate.

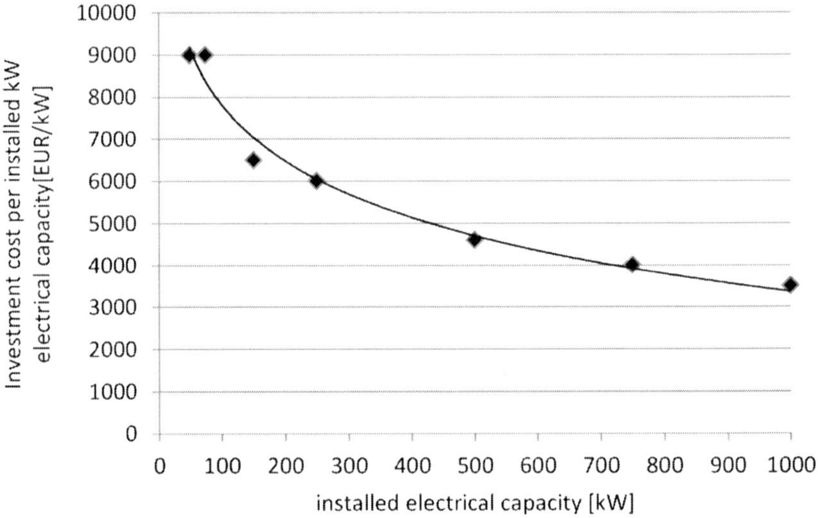

Figure 1. Degression of costs as a function of the installed electrical capacity, based on FNR e.V. and Dederer [18,31].

3. RESULTS AND DISCUSSION

The biogas potential is distributed over more than 100,000 farms with very different livestock sizes for cattle and pigs:

- For the evaluation of biogas potential in farms with cattle, 95.8% of the livestock was considered. The remaining 4.2% do not have any biogas potential due to their type of livestock breeding without collectable excrement. Around 130,000 farms in total breed cattle in Germany with a total number of 12.37 million cattle. The number of cattle per farm ranges from 1 to 9 individuals to more than 500 cows per farm. The largest share of cattle is bred on medium- to small-sized farms with 200 to 499 cows per farm. Large farms with more than 500 cows, representing 2% of all farms with cattle have a share of 17% of the total number of cattle [25].
- For the evaluation of the manure and dung potential from pigs, 100% of the livestock were considered. Pigs are bred on 68,000 farms in Germany, totaling 28.7 million pigs. The range of farm sizes varies from small farms with up to 49 pigs to farms with more than 5,000 pigs. The majority of pigs are reared on farms with 1000 to 1999 pigs. Similar to the cattle farms, the largest pig farms have, in relation to the number of

pigs that they breed, a relatively large share of the pig population, accounting for 16% [25].

The calculation of the related biogas plant capacity for different farm classes is given in Table 5. It indicates the average biogas capacity for the class while discarding the farm size distribution within the classes. Nevertheless, the information provides robust results about the required biogas size categories. The results range from 1 to 400 kW_{el}. The distribution is given in Figure 2 for cattle and in Figure 3 for pigs. The figures show the total amount of cattle and pigs as percentage and the installable electrical capacity.

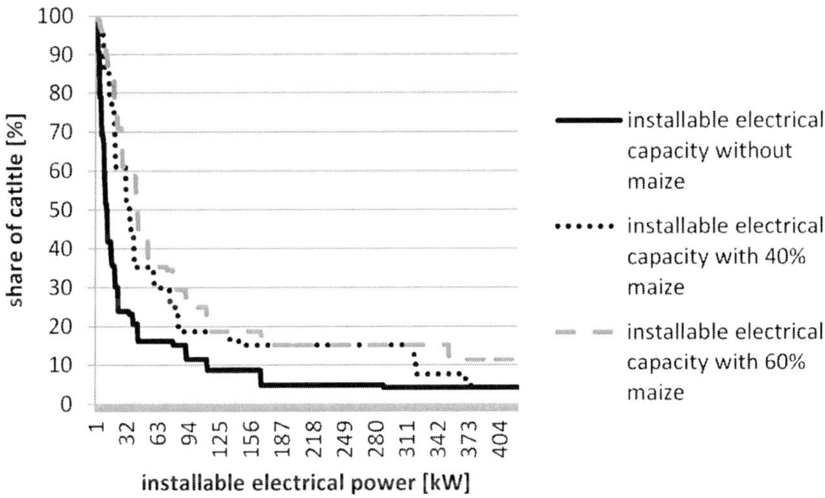

Figure 2. Installable electrical power from biogas at cattle farms depending on the number of cattle and the percentage of co-substrate.

Based on cattle manure, 7.2% of the livestock could provide feedstock for more than 100 kW_{el} for biogas production alone, whereas additional farms with 4.8% of the livestock could install between 75 and 100 kW_{el}. A biogas power plant with at least 50 kW_{el} could be installed at farms with 12% of all cattle with manure and dung only. With the use of 40% maize silage, the share increases to 31.1%. If 60% maize silage is used, the share of livestock increases to 40.8% (see Figure 2).

For 3.1% of the livestock of pigs, a biogas power plant with at least 100 kW_{el} could be installed, by using only manure or dung as a substrate.

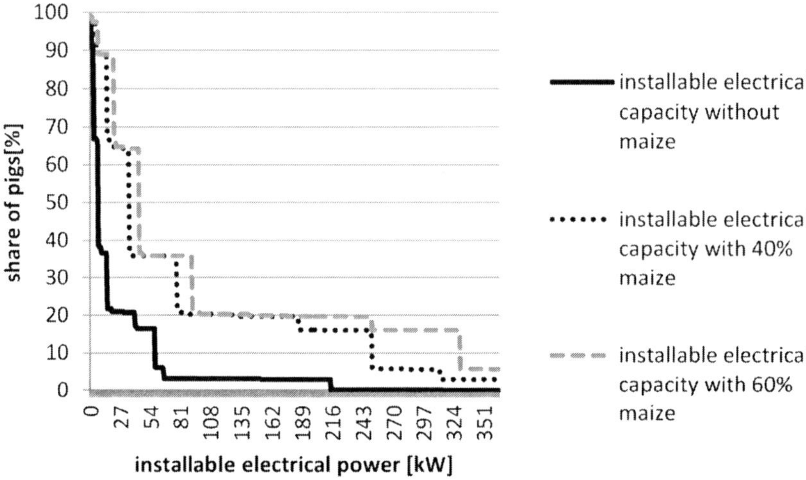

Figure 3. Installable electrical power from biogas at pig farms depending on the number of pigs and the percentage of co-substrate.

Table 5. Overview of farm types with the total number animal places, the percentage of livestock, and the farm-specific biogas potential.

Farm Type	Main Form of Excrement Produced	Farm Size (Total Number of Animal Places)	Percentage of Total Livestock (Cattle or Pigs) (%) for Manure Production	Farm-Specific Manure Production (m^3/yr)	Farm-Specific Biogas Production (m^3/a Methane)	Farm-Specific Biogas Plant Capacity (kW_{el}) (Manure and Dung Only)
Pigs with piglets	manure	1–49	0.03%	156	1873	0.8
Pigs with piglets	manure	50–99	0.08%	580	6963	2.9
Pigs with piglets	manure	100–399	0.49%	1107	13,280	5.6
Pigs with piglets	manure	400–999	0.38%	5538	66,450	28.1
Pigs with piglets	manure	1000–1999	3.52%	7863	94,354	39.9
Pigs with piglets	manure	2000–4999	2.66%	12,897	154,760	65.4
Pigs with piglets	manure	more than 5000	2.81%	41,965	503,580	212.7
Pigs	manure	1–49	0.28%	12.9	156	0.1
Pigs	manure	50–99	0.73%	73.6	883	0.4
Pigs	manure	100–399	6.67%	224.3	2691	1.1
Pigs	manure	400–999	22.22%	683.0	8195	3.5
Pigs	manure	1000–1999	27.61%	1408.1	16,898	7.1
Pigs	manure	2000–4999	14.86%	2886.3	34,637	14.6
Pigs	manure	more than 5000	10.31%	10,359.7	124,316	52.5
Pigs with piglets	dung	1–49	0.04%	61	2751	1.2
Pigs with piglets	dung	50–99	0.04%	227	10,213	4.3
Pigs with piglets	dung	100–399	0.45%	433	19,487	8.2
Pigs with piglets	dung	400–999	0.48%	2167	97,507	41.2
Pigs with piglets	dung	1000–1999	0.38%	3077	138,453	58.5
Pigs with piglets	dung	2000–4999	0.67%	5046	227,091	95.9
Pigs with piglets	dung	more than 5000	0.09%	16,421	738,941	312.1
Pigs	dung	1–49	0.82%	9	413	0.2
Pigs	dung	50–99	0.65%	54	2440	1.0
Pigs	dung	100–399	1.81%	165	7435	3.1
Pigs	dung	400–999	1.58%	503	22,640	9.6
Pigs	dung	1000–1999	0.72%	1037	46,680	19.7
Pigs	dung	2000–4999	0.25%	2126	95,686	40.4
Pigs	dung	more than 5000	0.12%	7632	343,422	145.1
Dairy cattle	manure	1 to 9	0.1%	98	1613	0.7
Dairy cattle	manure	10 to 19	0.8%	208	3463	1.5
Dairy cattle	manure	20 to 49	6.5%	568	9648	4.1
Dairy cattle	manure	50 to 99	9.6%	1236.9	20,517	8.7
Dairy cattle	manure	100 to 199	5.1%	2313.0	39,321	16.6
Dairy cattle	manure	200 to 499	2.5%	5267.6	89,550	37.8
Dairy cattle	manure	500 and more	2.8%	14,376.0	244,392	103.2
Cattle	manure	1 to 9	0.1%	64.0	1088	0.5
Cattle	manure	10 to 19	0.6%	162.3	2759	1.2
Cattle	manure	20 to 49	4.7%	374.3	6362	2.7
Cattle	manure	50 to 99	9.8%	803.7	13,663	5.8
Cattle	manure	100 to 199	9.7%	1561.7	26,549	11.2
Cattle	manure	200 to 499	6.3%	3292.4	55,971	23.6
Cattle	manure	500 and more	3.8%	12,688.4	215,690	91.2
Dairy cattle	dung	1 to 9	0.3%	60.7	3218	1.4
Dairy cattle	dung	10 to 19	0.9%	164.4	8712	3.7
Dairy cattle	dung	20 to 49	1.9%	363.2	19,251	8.1
Dairy cattle	dung	50 to 99	1.1%	772.4	40,937	17.3
Dairy cattle	dung	100 to 199	0.7%	1480.3	78,457	33.1
Dairy cattle	dung	200 to 499	1.1%	3371.3	178,679	75.5
Dairy cattle	dung	500 and more	0.6%	9200.6	487,634	206.0
Cattle	dung	1 to 9	0.5%	36.1	1915	0.8
Cattle	dung	10 to 19	1.6%	91.6	4856	2.1
Cattle	dung	20 to 49	5.0%	211.3	11,200	4.7
Cattle	dung	50 to 99	6.0%	453.8	24,051	10.2
Cattle	dung	100 to 199	5.4%	881.8	46,735	19.7
Cattle	dung	200 to 499	4.9%	1859.0	98,525	41.6
Cattle	dung	500 and more	1.0%	7168.4	379,927	160.5

Assuming an installed electrical capacity with a minimum of 50 kW_{el}, farms with 16.5% of the pig livestock could install a biogas plant. With an increasing share of maize silage as a co-substrate, the share of the livestock would increase to 35.9% (40%–60% maize silage).

4. CONCLUSIONS

The biogas production potential of manure is significant with 90 PJ/yr but spread over a large number of farms. By contrast, the individual potential at the farm site is comparably low in many cases. Our calculations clearly indicate that the farm size is the limiting factor for manure or dung processing to biogas. Due to the fact that manure has a very high percentage of water and therefore cannot be transported, biogas facilities for manure processing are always comparably small units that are individually located on farm sites. They are characterized by comparably high investment costs, which bring, however, a double-saving GHG effect as a result of (1) avoided emissions from manure storage and (2) avoided emissions through bioenergy provision as a substitute for fossil fuels. However, if we assume small and cost-intensive plants based on the available technology down to an installed capacity of 50 kW_{el}, we see a clear limitation due to farm structure, meaning that only a small proportion of the potential can be tapped into, equating to around 12% of the manure potential for cattle and 16.5% of the manure potential for pigs. To unlock the remaining potential, different strategies have to be taken into consideration.

The co-fermentation of manure and dung with energy crops, for example, can reduce this limitation and release more manure for biogas production. This option was investigated here and showed a considerable effect: With co-fermentation (40%–60% maize silage), the share of excrement, which can be converted into biogas in plants with a capacity of more than 50 kW increases to 31% or to 41% of cattle, respectively. For pigs, it also more than doubles with 36% of pigs. Even if this increase is rather impressive, there is still a huge amount of biomass that cannot be tapped into by this kind of co-fermentation.

Additional strategies could include management or technical adaptations, *i.e.*, a co-operation of farms could also increase the manure availability at a single plant (the potential of this measure strongly depends on local conditions; therefore, an assessment of the overall effect of this option cannot be carried out with the approach that we developed). In the long term, technology adaptations could be an option, for example by additionally pretreating the manure to convert it into an intermediate with a higher energy density, making transportation more feasible.

Our calculations clearly show a need for action to reduce the GHG emissions from livestock farming in Germany and the need to think about different strategies to unlock the potential of efficiently producing biogas of manure.

REFERENCES

1. IPCC. *Climate Change 2014 Synthesis Report Summary for Policymakers*; IPCC: Geneva, Switzerland, 2014.
2. WRI. *Climate Analysis Indicators Tool (CAIT) (version 9.0)*; World Resources Institute: Washington, DC, USA, 2012.
3. UBA. *Berichterstattung Unter Der Klimarahmenkonvention Der Vereinten Nationen Und Dem Kyoto-Protokoll 2013 Nationaler Inventarbericht Zum Deutschen Treibhausgasinventar 1990–2011*; Umweltbundesamt: Dessau-Roßlau, Germany, 2013.
4. Freibauer, A. *Biogenic Greenhouse Gas Emissions from Agriculture in Europe-Quantificatio and Mitigation*; Universität Hohenheim: Hohenheim, Germany, 2002.
5. Monteny, G.J.; Bannink, A.; Chadwick, D. Greenhouse Gas Abatement Strategies for Animal Husbandry. *Mitig. Greenh. Gas Emiss. Livest. Product.* 2016, *112*, 163–170.
6. Dämmgen, U.; Webb, J. The development of the EMEP/CORINAIR guidebook with respect to the emissions of different nitrogen and carbon species from animal production. *Mitig. Greenh. Gas Emiss. Livest. Product.* 2006, *112*, 241–248.
7. Grübler, A.; Ishitani, H.; Johansson, T.; Marland, G.; Moreira, J.R.; Rogner, H.-H. B: Energy Primer. IPCC—Intergovernmental Panel on Climate Change. Available online: http://www.ipcc-wg2.gov/publications/SAR/SAR_Chapter% 20B.pdf (accessed on 11 July 2014).
8. Bundesregierung. Bundesregierung Erneuerbare Energien-Ein Neues Zeitalter Hat Begonnen. Available online: http://www.bundesregierung.de/Webs/Breg/DE/Themen/Energiewende/ EnergieErzeugen/ErneuerbareEnergien-Zeitalter/_node.html;jsessionid=BEFF78D3A8ED2D8ECC35E6B4A09DE74E.s4t1#doc516106bodyText2 (accessed on 22 July 2015).
9. Haenel, H.D.; Rösemann, C.; Dämmgen, U.; Poddey, E.; Freibauer, A.; Wulf, S.; Eurich-Menden, B.; Döhler, H.; Schreiner, C. *Calculations of Gaseous and Particulate Emissions from German Agriculture 1990–2012*; Johann Heinrich von Thünen-Institut: Braunschweig, Germany, 2014.
10. Seyfert, U.; Bunzel, K.; Thrän, D.; Mauky, E.; Fritsche, B.; Schreiber, A.; Liebetrau, J.; Schmidt, T.; Ulbricht, T.; Lenz, V. *DBFZ Report Nr. 8*; Deutsches Biomasseforschungszentrum gemeinnützige GmbH: Leipzig, Germany, 2011.
11. Scheftelowitz, M.; Thrän, D.; Hennig, C.; Krautz, A.; Lenz, V.; Liebetrau, J.; Daniel-Gromke, J.; Denysenko, V.; Hilebrand, K.; Naumann, K.; *et al. Entwicklung Der Förderung Der Stromerzeugung Aus Biomasse Im Rahmen Des EEG*; DBFZ Report Nr. 21; Nelles, M., Ed.; DBFZ Deutsches Biomasseforschungszentrum gemeinnützige GmbH: Leipzig, Germany, 2014.
12. Federal Government of Germany 2012a. Act on Granting Priority to Renewable Energy Sources (Renewable Energy Sources Act-EEG). Federal Law Gazette on 4 August 2011 (BGBl. I, P. 1634 et Seq.). Vol. Federal Law Gazette on 4 August 2011 (BGBl. I, p. 1634 et seq.). Available online: http://www.bmu.de/english/renewable_ energy/doc/47883.php (accessed on 23 July 2015).
13. Thrän, D.; Krautz, A.; Scheftelowitz, M.; Lenz, V.; Liebetrau, J.; Daniel-Gromke, J.; Zeymer, M.; Nelles, M. Auswirkungen Der Gegenwärtig Diskutierten Novellierungs-Vorschläge Für Das EEG-2014. In *Biogas 2014: 7. Innovationskongress; Tagungsband; 22–23 Mai 2014 Im Zentrum Für Umweltkommunikation Der Deutschen*

Bundesstiftung Umwelt (DBU) Osnabrück; 11–22; Profair Consult: Hildesheim, Germany, 2014.

14. Kaltschmitt, M.; Hartmann, H.; Hofbauer, H. *Energie Aus Biomasse: Grundlagen, Techniken Und Verfahren. 2*; Corrected Aufl. 2009. 2., korr. Nachdruck 2009; Springer: Heidelberg, Germany, 2009.
15. LKVBB. Untersuchung von Gülle-Notwendig Für Den Fachgerechten Einsatz Als Dünger Oder Als Substrat in Biogasanlagen. Landeskontrollverband Brandenburg e.V. Available online: http://www.lkvbb.de/hauptinf-Dateien/Guelle%20mit%20Tabellen.pdf (accessed on 13 January 2016).
16. BMWi. *Gesetz Für Den Ausbau Erneuerbarer Energien (Erneuerbare-Energien-Gesetz-EEG 2014)*; Bundesministerium für Wirtschaft und Energie: Berlin, Germany, 2014.
17. BMWi. *Referentenentwurf Des BMWi (IIIB2) Entwurf Eines Gesetzes Zur Einführung von Ausschreibungen Für Strom Aus Erneuerbaren Energien Und Zu Weiteren Änderungen Des Rechts Der Erneuerbaren Energien*; Bundesministerium für Wirtschaft und Energie: Berlin, Germany, 2016.
18. Dederer, M. *Betriebswirtschaftliche Bewertung Kleiner Biogasanlagen*; ALB Tagung: Hohenheim, Germany, 2012.
19. TierSchNutztV. *Verordnung Zum Schutz Landwirtschaftlicher Nutztiere Und Anderer Zur Erzeugung Tierischer Produkte Gehaltener Tiere Bei Ihrer Haltung (Tierschutz-Nutztierhaltungsverordnung—TierSchNutztV)*; Bundesministerium für Ernährung und Landwirtschaft: Bonn, Germany, 2014.
20. Scheftelowitz, M.; Rensberg, N.; Denysenko, V.; Daniel-Gromke, J.; Stinner, W.; Hillebrand, K.; Naumann, K.; Peetz, D.; Henning, C. *Stromerzeugung Aus Biomasse (Vorhaben IIa) Zwischenbericht Mai 2015*; Deutsches Biomasseforschungszentrum gemeinnützige GmbH: Leipzig, Germany, 2015.
21. Shen, X.L.; Huang, G.Q.; Yang, Z.L.; Han, L.J. Compositional characteristics and energy potential of Chinese animal manure by type and as a whole. *Appl. Energy* 2015, *160*, 108–119.
22. Schultheiss, U.; Döhler, H.; Bach, M. *Festmistaußenlagerung 1. überarbeitete Auflage*; Kuratorium für Technik und Bauwesen in der Landwirtschaft KTBL: Darmstadt, Germany, 2011.
23. Reinhold, G. *Masse-Und Trockensubstanzbilanz in Landwirtschaftlichen Biogasanlagen-Langfassung Der Veröffentlichung Reinhold, G.: Genau Bilanzieren in Neue Landwirtschaft Heft 12/2005, S. 68 Bis 72*; Thüringer Landesanstalt für Landwirtschaft TLL: Jena, Germany, 2015.
24. Albert, E.; Förster, F.; Ernst, H. *Wirksamkeit Organischer Dünger—Berechnung Der Pflanzenbaulichen Wirksamkeit Der Nährstoffe Organischer Dünger*; Freistaat Sachsen, Sächsische Landesanstalt für Landwirtschaft: Dresden, Germany, 2007.
25. BMEL. *Statistisches Jahrbuch über Ernährung Landwirtschaft Und Forsten 2014*; Bundesministerium für Ernährung und Landwirtschaft: Bonn, Germany, 2015; Volume 58.
26. Achilles, W.; Brenndörfer, M.; Döhler, H.; Eckel, H.; Eurich-Menden, B.; Frisch, J.; Fritzsche, S.; Froeba, N.; Fuebbeker, A.; Funk, M.; *et al. Betriebsplanung Landwirtschaft 2004/2005. KTBL-Datensammlung: Daten für die Betriebsplanung in der Landwirtschaft*; Landwirtschaftsvlg Münster: Münster, Germany, 2004.
27. Frisch, J.; Fritzsche, S.; Fröba, N.; Funk, M.; Gaio, Ch.; Grimm, E.; Grube, J.; Hartmann, S.; Hartmann, W.; Klocpfer, F.; *et al. Betriebsplanung Landwirtschaft 2012/2013 KTBL Datensammlung*; Kuratorium für Technik und Bauwesen in der Landwirtschaft KTBL: Darmstadt, Germany, 2012; Volume 23.
28. BMU. *Ordinance on the Generation of Electricity from Biomasse (Biomasse Ordinance—BiomasseV) Consolidated Version of the Ordinance in the Version Applicable as of 1 January 2012*; BMU: Berlin, Germany, 2012.
29. ASUE. *BHKW-Kenndaten 2011*; ASUE Arbeitsgemeinschaft für Sparsamen und Umweltfreundlichen Energieverbrauch e.V.: Berlin, Germany, 2011.

30. Thrän, D.; Liebetrau, J.; Daniel-Gromke, J.; Jacobi, F. *Smart Bioenergy. Technologies and Concepts for a More Flexible Bioenergy Provision in Future Energy Systems*; Thrän, D., Ed.; Springer-Verlag: Heidelberg, Germany, 2015.
31. FNR e.V. Faustzahlen Biogas. Available online: http://biogas.fnr.de/daten-und-fakten/faustzahlen/ (accessed on 16 September 2015).

CHAPTER 13

Aerobic and Anaerobic Transformations in Estrogens and Nutrients in Swine Manure: Environmental Consequences

Ashok K. Singh

Department of Veterinary Population Medicine, College of Veterinary Medicine, University of Minnesota, St. Paul, MN 55108, USA;

ABSTRACT

Pig manure is an excellent fertilizer and rich source of organic carbon and nitrogen compounds such organic nitrogen (O-N) (95% of total nitrogen) that is plant-unavailable-nitrogen (PUN) and mineralized nitrogen (about 1% of total nitrogen) such as ammonium (NH_4^+) and nitrate (NO_3) that are plant-available-nitrogen (PAN). In addition, manure also contains two forms of estrogens: (i) poorly estrogenic thus essentially nontoxic conjugated estrogens (cEs) such as estrone (cE1), estradiol (cE2) and estriol (cE3); and (ii) highly estrogenic and toxic free estrogens (fEs) such as fE2, fE1 and fE3. This study showed that aerobic processing reduced concentrations of total carbon (TC), O-N, PAN and NH_4^+/NH_3 ratio, transiently hydrolyzed cEs (cE2 > cE1 > cE3) into corresponding fEs, transiently increased estrogenic activity and potential toxicity, and rapidly degraded fEs (fE2, fE1 > fE3), thus reducing the estrogenic activity in manure. Unlike aerobic processing, anaerobic processing stabilized and increased PAN and NH_4^+/NH_3ratio, thus increasing the manure's fertilizer value. However, anaerobic processing, relative to aerobic processing, poorly hydrolyzed cEs (reducing transient toxicity and increasing reserve toxicity potential) and poorly degraded fEs (increasing toxicity) in manure. Thus, aerobic and anaerobic environments have distinct effects on manures' PAN and estrogenic activity, presenting an interesting dilemma: anaerobic incubation that increases manures' PAN does not effectively degrade estrogens, while aerobic incubation that effectively degrades estrogens (after transiently increasing their estrogenic activity) also decreases PAN, thus making manure less profitable. New techniques are need to fully use manure as organic fertilizer.

Keywords: manure; aerobic; anaerobic; estrogens; plant available nitrogen; ammonium; ammonia; organic nitrogen

1. INTRODUCTION

The United States has approximately 1.3 million farms holding livestock and, about 100,000 of which, are considered swine Concentrated Animal Feeding Operations (CAFOs) where large number of pigs are held and raised in a small area [1]. On average, a pig produces about 3.9 kg of manure (a slurry of urine and feces)/day [2], thus about 2.3×10^8 kg of manure is generated each day by the pig industry. This is equivalent to an influx into the environment of about 3.6×10^9 kg nitrogen and approximately 3.5×10^{11} kg organic carbon each day. Compared to the smaller operations, CAFOs have limited land available for effective use/storage of the generated waste as fertilizer [3]. In addition, CAFOs do not require treatment of manure as long as it is not disposed directly into water bodies [2]. As a result, the waste load arising from CAFO may exceed the absorption capacity of the local environment, necessitating storage of large quantities of manure at the site of production [4]. The negative environmental implications of CAFO are related in significant part to the large volume of manure contaminants coming from a small area.

A serious concern regarding pig CAFO is that manure, despite being rich in nutrients, also contains high concentrations of estrogens that, even at low doses, have been proven to have long-term adverse health effects in humans and aquatic/wild animals [5,6]. The ecological hazard posed by steroidal hormones resulting from swine production is not clearly known. Mammals excrete two main forms of steroid hormones: (i) the biologically active free-estrogens (fEs) are excreted through feces (accounting for 10% of total steroid excretion); and (ii) the biologically inactive conjugated steroids (cEs) are excreted through urine [7]. Since cEs are the predominant form of estrogens found in manure [7,8], there is a presumption that manure may have poor estrogenic activity. However, earlier studies have shown that storage may increase manure's estrogenic potency due to the hydrolysis of conjugated hormones by glucuronidase and sulfatase enzymes present in manure microorganisms [9,10]. Swine manure contains aerobic and anaerobic bacteria that, under aerobic and anaerobic incubations, respectively, metabolize estrogens and convert organic nitrogen (O-N) into mineralized nitrogen (nitrate (NO_3), nitrite (NO_2), ammonium (NH_4^+), ammonia (NH_3), *etc.*) (Figure 1) [11].

However, possible effects of incubation conditions on relative alterations in estrogen concentrations, estrogenic activity and nutrient concentrations in manure are not fully understood. The aim of the present study was to establish the time-course of change in estrogen concentrations, estrogenic activity and nutrient (organic nitrogen and ammoniated nitrogen) concentrations in swine manure stored in aerobic and anaerobic conditions. Possible environmental consequences of these transformations has also been discussed.

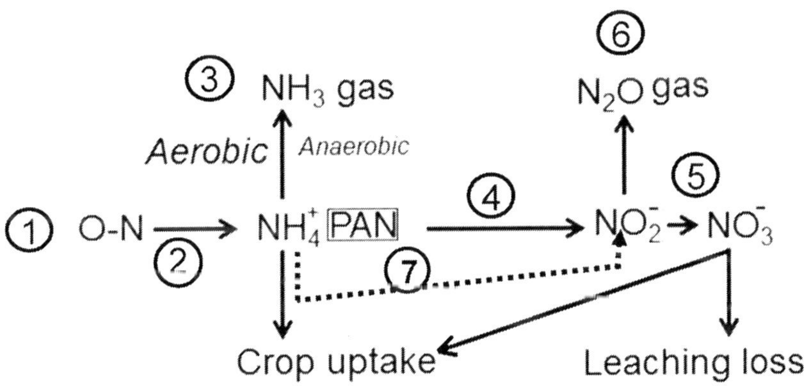

Figure 1. Biotransformation of organic nitrogen (O-N) in manure. (1) Decay of dead plants and animals release amino acids and proteins; (2) Ammonification of O-N by Saprotrophs, resulting in formation of NH_4^+ (in acidic pH); (3) In aerobic environment the pH is basic, thus NH_4^+ is converted into NH_3 gas that enters the atmosphere. In anaerobic process, the pH is acidic that favors NH_4^+; (4) Oxidative nitrification by nitrifying bacteria (*Nitrosomonas)* at acidic pH; (5) Nitrification by *Nitrobacter*; (6) Bacteria such as *Pseudomolas denitrificans convert* NO_3^- into N_2O gas; (7) Anoxic oxidation of NH_4^+. Abbreviations: NH_4^+: ammonium, PAN: plant available nitrogen, NO_3^-: nitrate, NO_2^-: nitrite, N_2O: nitrous oxide gas, NH_3: ammonia gas.

2. MATERIALS AND METHODS

Collection and processing of manure: Fresh manure samples (n = 27) were collected from swine facilities housing finishers. The samples containing 40% to 80% solid were mixed thoroughly and divided into three aliquots. Aliquot-1 was analyzed for estrogenic activity, estrogen concentrations and nutrients; and aliquot-2 and -3 were subjected to aerobic and anaerobic incubations, respectively, as described in Supporting Information.

Anaerobic incubation of manure*:* 20 mL of thoroughly mixed slurry of each sample was transferred to amber conical flask fitted with gas intake tube, a gas collection port and chemically inert air-tight cap. Multiple flasks could be connected to a single gas source. The flasks' headspaces were immediately evacuated and filled with nitrogen gas. For manure sterilization, a second set of flasks were mixed with sodium azide (50 g/L) to inhibit biodegradation. Samples were analyzed at different time intervals from start of incubation. At regular intervals, the outgoing air from collecting port was passed through 20 mL 0.1 N H_2SO_4 in order to collect the volatilized NH_3. The experiment was run in a constant-temperature room at 25 °C. NH_3 volatilization was determined using a colorimetric method based on the Berthelot reaction [12].

Aerobic Incubation of manure: Aerobic incubation were prepared in a similar manner to the anaerobic experiments, except the manure liquid or slurry was constantly oxygenated using a 16-gauge Teflon connected to compressed air (breathing-air grade) and bubbled at a continuous rate. Samples were incubated at room temperature. Samples were analyzed at different time intervals from start of incubation. The procedure described above were followed.

Sample processing and analysis: Each slurry sample was mixed thoroughly and split into two equal parts (Figure S1). Part-1 was analyzed for estrogenic activity (in terms of estradiol equivalent—E2Eq), organic carbon (OC), NH_4^+-N, NH_3, NO_2, NO_3 and O-N. E2Eq was measured using the Yeast Estrogen Screen (YES) assay that was validated for pig manure (Figures S2 and S3). NH_4^+-N, NH_3, NO_2, NO_3 and O-N were analyzed using standard procedures described in supplementary section. Part-2 manure was mixed with internal standards and poured into a fritted glass column. Conjugated estrogens were eluted with 100 mL of pure water and free estrogens were eluted with 100 mL methanol [8]. The two effluxes were concentrated to 5 mL under nitrogen and reduced pressure. Individual cEs (E2 glucuronate (E2G), E2 sulfate (E2S), E1G, E1S, E3G and E3S) were analyzed using an Applied BioSystems binary High Performance Liquid Chromatograph (HPLC) equipped with a triple-stage quadruple mass spectrometer (Figure S4A). Individual fEs (estradiol (fE2), estrone (fE1) and estriol (fE3)) were derivatized to the silyl ethers in using pyridine:bis(trimethylsilyl)-trifluoroacetamide before analysis by GC-MS (Agilent 7000 series gas chromatograph Mass Spectrometer, Agilent, Palo Alto, CA, USA) (Figure S4B) [8]. The detailed analytical procedures are described in the Supplementary section (Tables S1 and S2). Analysis of pre-incubation samples established baseline for estrogens and estrogenic activity. Changes in free estrogen concentration were assessed by subtracting basal values from the corresponding test-values (a negative value indicated decrease in concentration). Changes in cEs concentrations were determined by subtracting test-values from basal values. Data were recorded as mean ± SD using Microsoft Excel. Statistical analysis was performed using ANOVA followed by Tukey's Multiple Comparison Test or *t*-test at <0.05 significance level.

3. RESULTS

3.1. Fresh Manure Characteristics

Manure slurry used in this study contained 40% to 80% solid contents. Total O-N, NH_4^+ and NH_3, NO_2^-, NO_3^-, total carbon (TC) and Kjeldahl nitrogen (KN) concentrations, and NH_4^+/NH_3 and C/N ratio values were 53 ± 43 (% DM), 0.5 ± 0.3 ($gN\cdot kg^{-1}$), 0.1 ± 0.08 ($gN\cdot kg^{-1}$), <MDL, 23 ± 21 ($gN\cdot kg^{-1}$), 51 ± 20 ($gC\cdot kg^{-1}$) 37 ± 29 ($gN\cdot kg^{-1}$), 5.1 ± 1.8, 17 ± 7, respectively. Free and conjugated estrogens distributed between the liquid and solid fractions of manure (Table 1). Total fE concentrations in liquid and solid fractions of manure were 33 ± 10 $\mu g\cdot L^{-1}$ and 646 ± 230 $\mu g\cdot L^{-1}$, respectively. Total cEs concentrations in liquid and solid fractions of manure were 255 ± 103 $\mu g\cdot L^{-1}$ and 355 ± 170 $\mu g\cdot L^{-1}$, respectively. This suggests that cEs were greater than fEs in liquid, while fEs were greater than cEs in solid fraction of manure analyzed in this study. Contrary to the perception, manure samples exhibit estrogenic activity.

3.2. Effects of Aerobic Digestion

Manure Carbon and Nutrients: Aerobic digestion transiently increased manure pH that peaked at day-10 and then returned to the basal level at day-30. (Figure 2). O-N concentrations decreased in biphasic manner: a rapid decrease at a rate of 0.9 ± 0.5 $gN\cdot kg^{-1}\cdot day^{-1}$ for up to day-5 was followed by a slow decrease at a rate 0.2 ± 0.1 $gN\cdot kg^{-1}\cdot day^{-1}$. Ammonium ($NH_4^+$) concentrations decreased (rate of 0.03 ± 0.02 $gN\cdot kg^{-1}\cdot day^{-1}$)

gradually with minimal values of 0.1 gN·kg^{-1} (Figure 3iiib). Ammonia (NH_3) concentrations increased in two stages (Figure 3iiia): a rapid increase from 0.05 to 1 gN·kg^{-1} occurred at day-1 and the values remained unchanged for up to day-15, then the values increased further to 2 gN·kg^{-1} and remained elevated throughout the experiment. This resulted in a significant decrease in the NH_4^+/NH_3 ratio (Figure 4i). Nitrate (NO_3^-) levels increased rapidly and peaked at day-10, followed by a gradual decrease reaching to 75% of basal values at day-30 (Figure 4i). Nitrite (NO_2^-) was not detected in this study. A delayed decrease in TC contents and C/N ratio in manure samples reaching to 50% of basal values occurred at day-30 (Figure 4i).

Table 1. Hormone levels in manure liquid and solid samples.

	Estradiol (E2)		Estrone (E1)		Estriol (E3)		Total
	Conjugated μg L	Free μg L	Conjugated μg L	Free μg L	Conjugated μg L	Free μg L	μg L
			Manure Liquid				
n = 23	43 ± 17	13 ± 11	165 ± 4 *	3 ± 7	15 ± 13 #	17 ± 12	310 ± 117
	G 78%		G 20%		G 90%		
E2Eq Ph							13 ± 12
E2Eq Ah							71 ± 22 †
			Manure Solid				
n = 23	143 ± 63	210 ± 111	229 ± 138	413 ± 211	45 ± 8 #	18 ± 13 *	973 ± 471
	G 80%		G 30%		G 90%		
E2Eq Ph							53 ± 11
E2Eq Ah							97 ± 32 †
K_D	3.3 ± 1.7	16 ± 5 *	1.4 ± 0.9	105 ± 21 *	3 ± 2	1 ± 0.8	

Values are mean ± SD; * $p < 0.05$ when compared from control values; # $p < 0.05$ when compared with E2 values; † $p < 0.05$ when Ph (prior to hydrolysis) values were compared with Ah (after hydrolysis) values; Ph: prior to hydrolysis; Ah: after hydrolysis.

Manure Estrogens: Aerobic incubation gradually decreased conjugated estrogen concentrations (E1 = E2 > E3) with $t_{1/2}$values ranging from 8 to 10 days (Figure 5I: cE1—plot 5, cE2—plot 6 and cE3—plot 7). Approximately 15%, 40% and 50% cE3 (Figure 5I plot 7), cE1 (Figure 5I plot 6) and cE2 (Figure 5I plot 5), respectively, remained in manure at day-30 of aerobic incubation. A transient increase (37% ± 10%, Figure 5I plot 1) in E2Eq levels and fE concentrations (fE2: 36% ± 10%, Figure 5I—plot 2; fE1: 31% ± 8%, Figure 5I—plot 3; and fE3: 11% ± 5% increase, Figure 5I plot 4) occurred in manure. The fE2 and E2Eq values peaked at day-5 and decreased gradually reaching to the control values at day-15 to day-20. As shown in Figure 5III, gEs and corresponding sEs exhibit comparable sensitivity to aerobic (a) incubation (for 30 days) in manure. A comparable change in E2Eq and fE2 values indicate that fE2 may solely account for the increase in estrogenic activity in manure samples subjected to aerobic digestion. In sterilized manure, only 10%

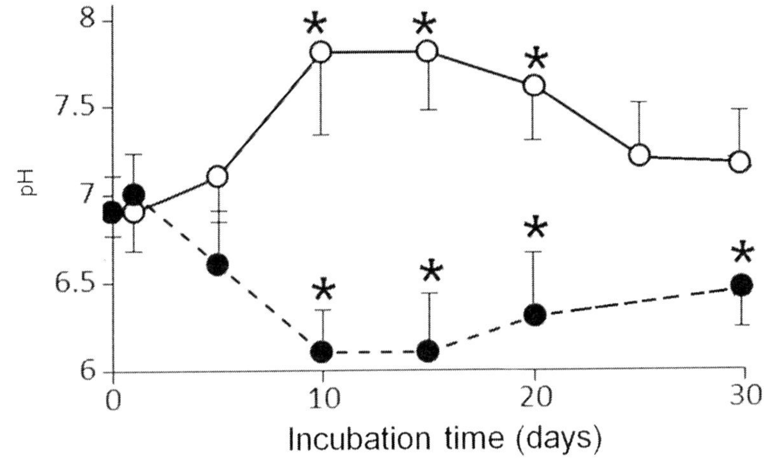

Figure 2. Effects or aerobic (solid line) and anaerobic (broken line) incubation on manure pH (n = 27). Values are mean ± SD. * $p < 0.05$, significant when compare with control.

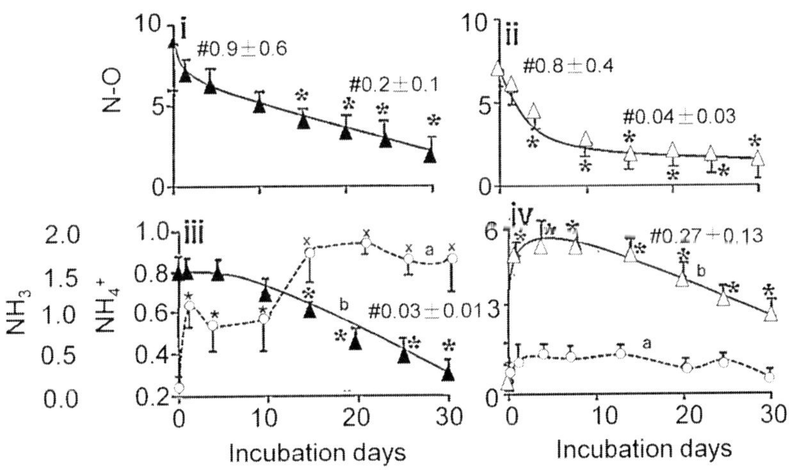

Figure 3. Effects of aerobic and anaerobic incubations on time-course of change in pH, organic nitrogen (O-N), ammonium (NH_4^+) and ammonia (NH_3) in manure samples. i-*aerobic* and ii-*anaerobic*: Time-course of change in O-N in manure samples subjected to aerobic and anaerobic incubations; iii-*aerobic* and iv-*anaerobic*: Time-course of change in NH_4^+ (plot b) and NH_3 (plot a) concentrations in manure samples subjected to aerobic or anaerobic incubations; values are mean ± SD, n = 27, * significant ($p < 0.05$) compared with control.

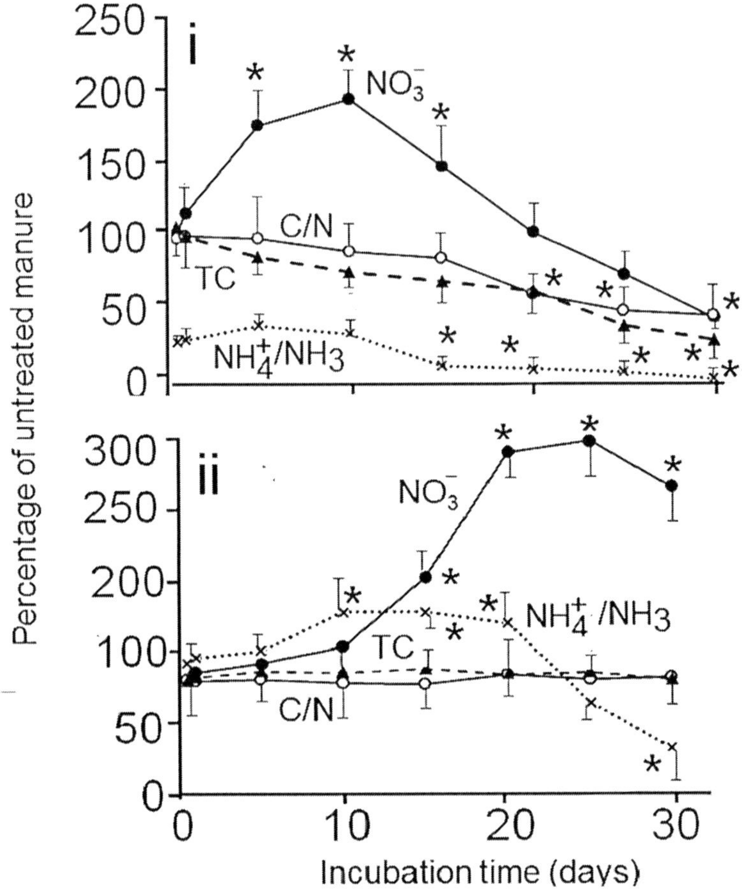

Figure 4. Effects of aerobic (i) and anaerobic (ii) incubations on time-course of change in NO_3^-, NH_4^+/NH_3 ratio, total carbon (TC) and carbon/nitrogen (C/N) ratio in manure samples. Values are mean ± SD, $n = 27$, * significant ($p < 0.05$) compared with control.

decrease (Figure 5IV and V) in estrogen concentrations were observed. This represents chemical degradation of estrogens.

3.3. Effects of Anaerobic Digestion

Manure Carbon and Nutrients: Anaerobic incubation significantly decreased manure pH at day-10 that remained suppressed throughout the experiment (Figure 2). O-N exhibited a rapid decrease occurring for up to day-5 (rate 0.8 ± 0.4 gN·kg^{-1}·day^{-1}), followed by a slower decrease (rate 0.04 ± 0.03 gN·kg^{-1}·day^{-1}) (Figure 3ii). A rapid increase in NH_4^+concentrations (peak value at day-5) (Figure 3ivb), a gradual increase in NH_4^+/NH_3 ratio (peak value at day-10) (Figure 4ii) no change in NH_3 concentrations (Figure 3iva) was observed. NO_3^{-1} concentrations remained unchanged for up to 10 days and then increased gradually that peaked at day-20 and remained elevated thereafter. The TC levels and C/N ratio remained unchanged throughout the experiment.

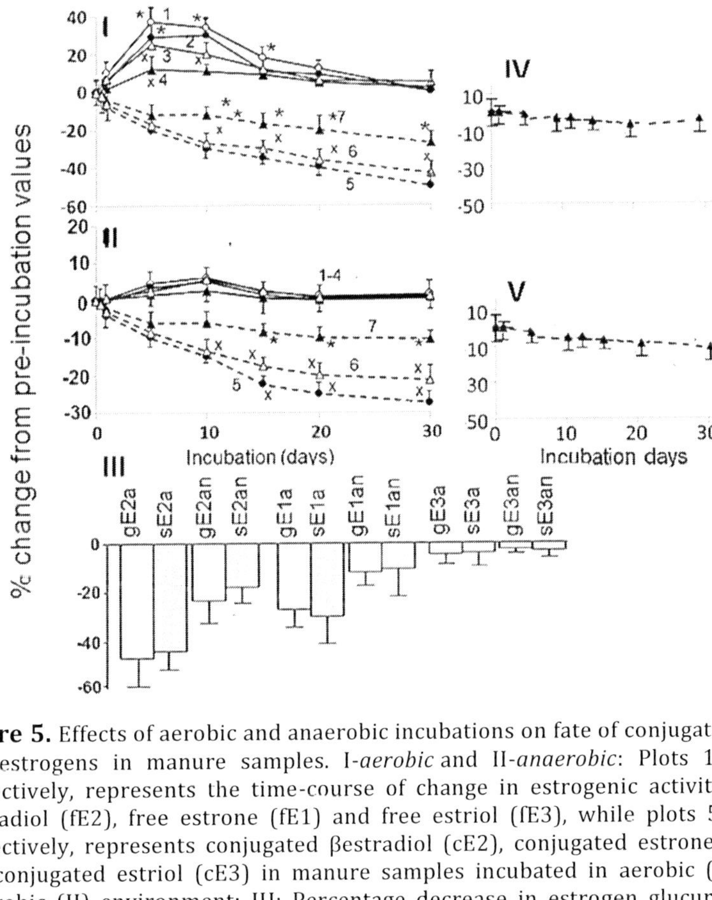

Figure 5. Effects of aerobic and anaerobic incubations on fate of conjugated and free estrogens in manure samples. I-*aerobic* and II-*anaerobic*: Plots 1 to 4, respectively, represents the time-course of change in estrogenic activity, free βestradiol (fE2), free estrone (fE1) and free estriol (fE3), while plots 5 to 7, respectively, represents conjugated βestradiol (cE2), conjugated estrone (cE1) and conjugated estriol (cE3) in manure samples incubated in aerobic (I) and anaerobic (II) environment; III: Percentage decrease in estrogen glucuronides (gE2, gE1 and gE3) and estrogen sulfates (sE2, sE1 and sE3) in manure at 30-days after aerobic (a) or anaerobic (an) incubation. Both conjugates were comparably sensitive to manure processing; IV-*aerobic* and V-*anaerobic:* Time-course of change in cEs in sterilized-manure following aerobic or anaerobic incubations; values are mean ± SD, n = 27, * significant ($p < 0.05$) compared with control and x: significant when compared with corresponding plot 1 values for free estrogens and plot 7 values for conjugated estrogens.

Manure Estrogens: Anaerobic digestion caused <10% (Figure 5II plot 7), 30% a (Figure 5II plot 6) and 32% (Figure 5II plot 7) decrease in cE3, cE2 and cE1 concentrations, respectively, without increasing fE concentrations and E2Eq activity levels (Figure 5II, plot 1–4). In sterilized manure, only 10% decrease in cEs occurred (Figure 5IV and V). Similar to aerobic incubation, gEs and corresponding sEs exhibit comparable sensitivity to anaerobic (an) incubation (for 30 days) in manure.

4. DISCUSSION

4.1. Manure Characteristics

Swine manure contains organic carbon, O-N, mineralized nitrogen compounds such as NH_4^+, NO_2^-, NO_3^-, NH_3, nitrogen gases, soil and other nutrients such as P, S and K, although their proportions may vary depending on its source, moisture content, storage, and handling methods. The C/N ratio for fresh manure was 17:1 that concurs with the values reported earlier [13,14]. A critical C:N ratio of about 15:1 is essential for efficient aerobic transformations of carbon and nitrogen, a higher C:N ratio is associated with (1) a decrease in ammonia emission, possibly due to an increase in nitrogen immobilization [14,15] and (2) an increase in odorous volatile organic acids during anaerobic digestion [16,17]. An earlier study has shown an inverse relationship between C/N ratio and PAN [18,19]. Thus C/N ratio may be a rough indicator of PAN in manure.

Manure contains two forms of nitrogen: (i) plant-unavailable-nitrogen (PUA) such as O-N that accounts for almost 95% of total nitrogen; and (ii) PAN such as NH_4^+ and NO_3^- that account for <1% of total nitrogen in manure. Contrarily, earlier studies have reported equal proportions of O-N and PAN (NH_4^+ and NO_3^-) in swine manure [2,20]. This discrepancy may be due to differences in the animals' diet or the manures' pH that may range from 7 to 8.4, resulting in 1% to 5% loss of NH_4^+ via volatilization as NH_3. Livestock manure has been identified as a key source of NH_3 emission around the world [21].

In addition to plant nutrients and organic carbon, swine manure also contains conjugated (cE) and free (fE) forms of estrogens originating from animal urine and feces, respectively. Total estrogen concentrations observed in the present study were comparable to those reported by Combalbert *et al.* [7], Raman *et al.* [22] and Zhang *et al.* [23], but lower than those reported by Furuichi *et al.* [24]. The following patterns of estrogen concentrations were observed:

- Manure liquid: cE1G > cE2G > cE1S > cE2S, cE3G, fE2, fE3 > fE1 > cE3S
- Manure Solid: fE1 > fE2 > cE1S > cE1G > cE2G > fE3, cE3G > cE2S, cE3S

Similar to the present study, earlier studies have also shown cE1 and fE1 to be the major estrogens present in swine manure samples [7,23,25]. However, there were several folds differences in estrogen concentrations in swine manure, possibly due to: (1) rapid degradation of fE2 and fE1 bound to the soil [26,27,28]; (2) differential hydrolysis of cEs resulting in an increasing in free estrogens [29]; and (3) potentially higher mobility of conjugated estrogens compared to free estrogens (E2 and E1), because of their high water solubility [26,28,30,31,32,33]. Based on these observations, it was expected that cEs, being water soluble estrogens, will remain in the aqueous phase and leach with water flow, while fEs, being lipophilic, remain bound to the solid particles and leach slowly. Contrarily, the present study showed that fE and cEs both bound to manure solid (although fE binding > cE binding), possibly via different mechanisms: fEs bound to soil via hydrophobic interactions and cEs bound to soil via forming hydrogen bonds.

4.2. Plant Nutrients in Aerobic and Anaerobic Environment

General considerations: This study showed that O-N accounted for most of the nitrogen present in swine manure samples. Since O-N is PUN, manure is processed aerobically or/and anaerobically to convert O-N into PAN such as NH_4^+ and NO_3^-(Figure 1). In general, this is a multi-step process described below [34,35,36,37,38].

1. Heterotrophic bacteria mediated ammonification and/or denitrification of amino acids convert O-N into NH_4^+. Bacteria deamination of amino acids occur via oxidative deamination or anoxic deamination via sehydratase or sulfhydrase enzymes [39].
2. Oxidative nitrification converts NH_3 or NH_4^+ to nitrites (NO_2^-) according to the following equation: $NH_3 + O_2 \rightarrow NO_2^- + 3H^+ + 2e^-$. *Nitrosomonas*, *Nitrosococcus*, and *Nitrosospira* are most frequently associated with this reaction, however, *Nitrosolobus* and *Nitrosovibrio* can also autotrophically oxidize ammonia [40]. This is followed by the oxidation of nitrites to nitrates (NO_3^-)

by *Nitrobacter, Nitrospira, Nitrococcus* [41] according to the equation: $NO_2^- + H_2O \rightarrow NO_3^- + 2H^+ + 2e^-$.

3. Anaerobic oxidation (Anammox) of NH_4^+ to N_2 with NO_2^- as the electron acceptor [42].

4. $NH_4^+ + NO_2^- \rightarrow N_2 + 2H_2O$.

5. Denitrification in the presence of easily available carbon as an energy and carbon source and carried out by numerous facultative bacteria such as Pseudomonas [43]. In this process, nitrates and nitrites are reduced to nitrogen gas, such as N_2O and N_2, when a proton donor (energy source such as carbon) is available. The process involve the following pathway:

6. $NO_3^- \rightarrow NO_2^- \rightarrow NO \rightarrow N_2O \rightarrow N_2$.

In this process microorganisms first reduce nitrates to nitrites and then produce nitric oxide, nitrous oxide and nitrogen gas. A source of organic carbon is an important component of the denitrification process.

This suggests that manure storage conditions may determine whether manure nitrogen will be stabilized with accumulation of PANs or nitrogen will be lost as NH_3, N_2O and N_2.

Aerobic processing of manure nutrients: Aerobic incubation (i) transiently increased pH that converted NH_4^+ into NH_3 gas, resulting in significant decrease in NH_4^+/NH_3 ratio and increase in NH_3 nitrogen loss; and (ii) rapidly increased manure NO_3^- concentrations that peaked at day-10, then decreased gradually, reaching to basal levels at day-30. This suggests that O-N was converted into NO_2^- and NO_3^-, causing transient increase in NO_3^- concentrations and severe decrease in the NH_4^+/NH_3 ratio. Comparable aerobic processing of manure nitrogen has also been reported previously [44,45,46,47,48]. Studies have shown that during complete nitrification, NO_3^- can be reduced to NO by nitrite reductase, which is then reduced to N_2O by nitric oxide reductase [49,50]. During incomplete nitrification, however, ammonia mono-oxygenase may oxidize NH_3 sequentially to hydroxylamine and nitroxyl by hydroxylamine oxidoreductase, which is then converted to N_2O [51]. Approximately10% of O-N may be emitted in the form of N_2O [52,53,54]. This suggests that Aerobic environment may reduce the fertilizer value of swine manure. Since N_2O is considered to be an important factor in ozone depletion [55] aerobic oxidation may pose serious environmental threat.

Anaerobic processing of manure nutrients: Anaerobic digestion involves anoxic degradation of organic materials by microbial organisms that leads to the formation of biogas and stabilization of PAN such as

NH_4^+ and NO_3^- [56].The present study showed that anaerobic incubation converted organic nitrogen (O-N) into NH_4^+ and reduced their pH to less than 7.0, thus stabilizing NH_4^+ in manure. This was confirmed by the observation that, following a rapid increase at day-1, NH_3concentrations did not increase further. Recently Moller [57] has shown that the organic fraction remaining after anaerobic digestion is much more recalcitrant than the input manure leading to a stabilization of the organic matter. The present study also showed that NO_2^- was not detected, but NO_3^+concentrations began to increase at day-10, peaked at day-20 (about 270% of basal) and remained elevated thereafter. The time-course of increase in NO_3^+ and decrease in NH_4^+/NH_3ratio were comparable. The decrease in NH_4^+/NH_3 ratio may be due to conversion of NH_4^+ to NO_3^- via bacterial thermophilic anaerobic deamination that also yields corresponding fatty acid [40,58]. In field applications, however, contradictory results regarding the effects of anaerobic digestion on NH_3 volatilization has been revealed: a decrease, an increase and no change have all been reported [59,60,61,62,63,64,65]. This may be due to differences in experimental design, soil composition and influx nitrogen composition.

4.3. Effects of Aerobic and Anaerobic Digestions on Estrogen Concentrations and Estrogenic Activity

Aerobic incubation: The present study showed that aerobic incubation (1) degraded cEs (cE2, cE1 > cE3); and (2) transiently increased fEs concentrations (fE1, fE2 > fE3) and ensuing estrogenic activity in manure samples, possibly due to diverse groups of aerobic bacteria that activated (resulted in an increase in estrogenic activity) and/or hydrolyzed (resulted in a decrease in estrogenic activity) estrogens [66,67,68,69]. In aerobic environment, aerobic bacteria were several folds more active than anaerobic ones [70]. A recent study has isolated bacteria that transformed fE1 into fE2, resulting in further increase in manures' estrogenic activity since a transformation of fE1 to fE2 results in two folds increase in estrogenic activity [71]. In addition, estrogen glucuronates and sulfates both exhibited comparable de-conjugation of conjugated estrogens in manure in oxidative and reductive environment, although D'Ascenzo *et al.* [72] and Hutchins *et al.* [73] have shown that sulfate-conjugated estrogen are more recalcitrant to biodegradation. Taken together, these observations suggest that an increase in estrogenic activity during aerobic incubation may be due to de-conjugation of estrogens and/or transformation of E1 to E2, while, the subsequent decrease in estrogenic activity may be due to hydroxylation of estrogens by mono- and di-oxygenases and ensuing decarboxylation of the phenolic groups [74]. Chemical hydrolysis contributes about 10% of total estrogen degradation without a decrease in estrogenic activity.

Anaerobic Incubation: Unlike the effects of aerobic environment, effects of anaerobic environment on estrogen biotransformation are not fully characterized: an increased, a decrease and no change in estrogen levels have all been reported [69,75,76]. In the present study, anaerobic incubation for 30 days caused about 20% to 30% decrease in cE concentrations without any increase in fE concentrations in manure samples. These observations concur with previous studies concerning estrogen degradation in manure incubated under anaerobic conditions [77] or in anaerobic plants processing active sludge [78,79]. Furuichi *et al.* [24] have shown that fE2 degraded more rapidly than fE1 or fE3, while de-conjugation of conjugated estrogens did not occur in up-flow anaerobic sludge biological (UABS) effluent. However, Shappel *et al.* [80] have shown comparative removal (approximately 90%) of all three estrogens from swine primary lagoon. Pholchan *et al.* [81] have also shown differential degradation of E1 and E2 (E3 was not reported) in different nitrite accumulating reactors. Unlike the earlier studies, the present study showed differential degradation of conjugated (glucoronate and sulfate conjugates) estrogens in manure exposed to anaerobic environment (cE2 > cE1 > cE3).

4.4. Environmental Consequences

Swine manure has been shown to contain both fEs and cEs (glucuronide/sulfate), and cEs, in the presence of manure glucuronidase and sulfatase, release fEs. Conjugated hormones, therefore, can be considered hormone reserve. Manure also contains two types of nitrogen compounds, O-N that is PUN and NH_4^+ nitrogen that is considered PAN. Untreated manure contained mostly O-N, thus nutrition value of untreated manure is limited. The present study showed that (1) aerobic processing of manure degraded estrogens, transiently hydrolyzed cEs into fEs, and then degraded fEs, but severely reduced PAN (fertilizer value of manure) and (2) anaerobic processing of manure poorly degraded the hormones, did not induce hydrolysis of cEs, but increased PAN, thus increasing the fertilized value of manure. This distinct effect of aerobic and anaerobic manure processing on their estrogenic activity and fertilized capacity presents an interesting dilemma: Anaerobic incubation that increases manures' PAN does not degrade estrogens, while aerobic incubation that effectively degrades estrogens also decreases PAN, thus making manure less profitable. New techniques are need to fully use manure as organic fertilizer.

5. CONCLUSIONS

This study showed that anaerobic incubation stabilized both PAN that increased the fertilizer value and estrogens that increased the toxicity of manure samples, while aerobic processing decreased PAN thus decreasing the manure's fertilizer value, but increased estrogen degradation thus in decreasing the manure's toxicity. This suggests that aerobic or anaerobic processing may not convert manure into effective and safe fertilizer. A new approach such as an aerobic-anaerobic hybrid may be needed.

ACKNOWLEDGEMENTS

This project was partially funded by grants from the National Pork Producers, Consortium for Water Quality and the University of Minnesota, Graduate School—faculty seed grant. Technical support was provided by Gupta Shveta, Gupta Arun and Jiang Yin.

REFERENCES

1. Hribar, C. Understanding Concentrated Environmental Health Animal Feeding Operations and Their Impact on Communities. Available online: http://www.cdc.gov/nceh/ehs/Docs/Understanding CAFOs_NALBOH.pdf (accessed on 29 July 2015)
2. Hamilton, D.W.; Luce, W.G.; Heald, A.D. Production and Characteristics of Swine Manure. Available online: http://agrienvarchive.ca/bioenergy/download/F-1735_swine _man_char_OK.pdf (accessed on 28 July 2015).
3. Khanal, S.K.; Xie, B.; Thompson, M.L.; Sung, S.; Ong, S-K.; van Leeuwen, J. Fate, transport, and biodegradation of natural estrogens in the environment and engineered systems. *Environ. Sci. Technol.* **2006**, *40*, 6537–6546.
4. Weida, W.J. Economic Implications of Confined Animal Feeding Operations. Available online: http://www.sraproject.org/wp-content/uploads/2007/12/economicimplicationsofcafos.pdf (accessed on 28 July 2015).
5. Liney, K.E.; Jobling, S.; Shears, J.A.; Simpson, P.; Tyler, C.R. Assessing the sensitivity of different life stages for sexual disruption in roach (*Rutilus rutilus*) exposed to effluents from wastewater treatment works. *Environ. Health Perspect.* **2005**, *113*, 1299–1307.
6. Hutchinson, T.H.; Ankley, G.T.; Segner, H.; Tyler, C.R. Screening and testing for endocrine disruption in fish—Biomarkers as "signposts" not "traffic lights" in risk assessment. *Environ. Health Perspect.* **2006**, *114*, 106–114.
7. Combalbert, S.; Hernandez-Raquet, G. Occurrence, fate, and biodegradation of estrogens in sewage and manure. *Appl. Microbiol. Biotechnol.* **2010**, *86*, 1671–1692.
8. Singh, A.K.; Gupta, S.; Kumar, K.; Gupta, S.K.; Chander, Y.; Gupta, A.; Saxena, R. Quantitative analysis of conjugated and free estrogens in swine manure:

Solutions to overcome analytical problems due to matrix effects. *J. Chromatogr. A***2013**, *1305*, 203–212.

9. Panter, G.H.; Thompson, R.S.; Beresford, N.; Sumpter, J.P. Transformation of a non-oestrogenic steroid metabolite to an oestrogenically active substance by minimal bacterial activity. *Chemosphere* **1999**, *38*, 3579–3596.
10. Ritter, J.K. Roles of glucuronidation and UDP-glucuronosyltransferases in xenobiotic bioactivation reactions. *Chem. Biol. Interact.* **2000**, *129*, 201–208.
11. Masse, D.I.; Croteaua, F.; Masses, L. The fate of crop nutrients during digestion of swine manure in psychrophilic anaerobic sequencing batch reactor. *Bioresource Technol.* **2006**, *98*, 2819–2823.
12. Krom, M.D. Spectrophotometric determination of ammonia: A study of a modified Berthelot reaction using salicylate and dichloroisocyanurate. *Analyst* **1980**, *105*, 305–316.
13. Zeng, Y.; de Guardia, A.; Daumoin, M.; Benoist, J.-C. Characterizing the transformation and transfer of nitrogen during the aerobic treatment of organic wastes and digestates. *Waste Manag.* **2012**, *32*, 2239–2247.
14. Kirchmann, H.; Witter, E. Ammonia volatilization during aerobic and anaerobic manure decomposition. *Plant Soil***1989**, *115*, 35–41.
15. Amberger, A.; Huber, J.; Rank, M. Gulleausbringung: Vorsicht, Ammonia kerverluste. *DLG Miteilungen* **1987**, *20*, 1084–1086.
16. Hobbs, P.J.; Misselbrook, T.H.; Cumby, T.R. Production and emission of odours and gases from ageing pig waste. *J. Agric. Engng. Res.* **1999**, *72*, 291–298.
17. Zahn, J.A.; Hatfield, J.L.; Do, Y.S.; Pfeiffer, R.L. Characterization of volatile organic emissions and wastes from a swine production facility. *J. Environ. Qual.* **1997**, *26*, 1687–1696.
18. Gale, E.S.; Sullivan, D.M.; Cogger, C.G.; Bary, A.I.; Hemphill, D.D.; Myhre, E.A. Estimating plant-available nitrogen release from manures, composts, and specialty products. *J. Environ. Qual.* **2006**, *35*, 2321–2332.
19. Van Kessel, J.S.; Reeves, J.B.; Meisinger, J.J. Nitrogen and carbon mineralization of potential manure components. *J. Environ. Qual.* **2000**, *29*, 1669–1677.
20. Westerman, P.W.; Huffman, R.L.; Feng, J.S. Swine-lagoon seepage in sandy soil. *Trans. ASAE* **1995**, *38*, 1749–1760.
21. Buijsman, E.; Mass, H.J.; Asman, W.A. Anthropogenic NH_3 emissions in Europe. *Atmos. Epyiron.* **1987**, *11*, 1009–1022.
22. Raman, D.R.; Williams, E.L.; Layton, A.C.; Burns, R.T.; Easter, J.P.; Daugherty, A.S.; Mullen, D.M.; Sayler, G.S. Estrogen content of dairy and swine wastes. *Environ. Sci. Technol.* **2004**, *38*, 3567–3573.
23. Zhang, H.; Shi, J.; Liu, X.; Xinmin, Z.; Qingcai, C. Occurrence of free estrogens, conjugated estrogens, and bisphenol A in fresh livestock excreta and their removal by composting in North China. *Environ. Sci. Pol. Res.* **2014**, *21*, 9939–9947.
24. Furuichi, T.; Kannan, K.; Suzuki, K.; Tanaka, S.; Giesy, J.P.; Masunaga, S. Occurrence of estrogenic compounds in and removal by a swine farm waste treatment plant. *Environ. Sci. Technol.* **2006**, *40*, 7896–7902.
25. Johnson, A.C.; Williams, R.J.; Matthiessen, P. The potential steroid hormone contribution of farm animals to freshwaters, the United Kingdom as a case study. *Sci. Total Environ.* **2006**, *362*, 166–178.
26. Lee, L.S.; Strock, T.J.; Sarmah, A.K.; Rao, P.S. Sorption and dissipation of testosterone, estrogens, and their primary transformation products in soils and sediment. *Environ. Sci. Technol.* **2003**, *37*, 4098–4105.
27. Hakk, H.; Sikora, L. Dissipation of 17β-estradiol in composted poultry litter. *J. Environ. Qual.* **2011**, *40*, 1560–1566.
28. Stumpe, B.; Marschner, B. Long-term sewage sludge application and wastewater irrigation on the mineralization and sorption of 17β-estradiol and testosterone in soils. *Sci. Total Environ.* **2007**, *374*, 282–291.

29. Shrestha, S.L.; Francis, X.M.; Casey, F.X.M.; Hakk, H.; Smith, D.J.; Padmanabhan, G. Fate and transformation of an estrogen conjugate and its metabolites in agricultural soils. *Environ. Sci. Technol.* **2012**, *46*, 11047–11053.
30. Shore, L.S.; Reichmann, O.; Shemesh, M.; Wenzel, A.; Litaor, M.I. Washout of accumulated testosterone in a watershed. *Sci. Total Environ.* **2004**, *332*, 193–202.
31. Sarmah, A.K.; Scherr, F. Mediator saolution influence on the sorption potential of sulfo-conjugated esterogenic steroid hormone and its metabolite in New Zealand dairy farm soil. In Proceedings of the 19th World Congress of Soil Science, Soil Solutions for a Changing World, Brisbane, Australia, 1–6 August 2010; International Union of Soil Sciences: Rome, Italy, 2010; pp. 102–105.
32. Zitnick, K.K.; Shappel, N.W.; Hakk, H.; DeSutter, T.M.; Khan, E.E.; Casey, FXM. Effects of liquid swine manure on dissipation of 17β-estradiol in soil. *J. Hazard. Mat.* **2011**, *186*, 1111–1117.
33. Dinova, M.V. Transformation of steroids by actinobacteria: A review. *Appl. Biochem. Microbiol.* **2001**, *43*, 1–14.
34. Nikaeen, M.; Nafez, A.H.; Bina, B.; Nabavi, B.F.; Hassanzadeh, A. Respiration and enzymatic activities as indicators of stabilization of sewage sludge composting. *Waste Manag.* **2015**, *39*, 104–110.
35. Jiang, J.; Liu, X.; Huang, Y.; Huang, H. Inoculation with nitrogen turnover bacterial agent appropriately increasing nitrogen and promoting maturity in pig manure composting. *Waste Manag.* **2015**, *39*, 78–85.
36. Kulikowska, D.; Gusiatin, Z.M. Sewage sludge composting in a two-stage system: Carbon and nitrogen transformations and potential ecological risk assessment. *Waste Manag.* **2015**, *38*, 312–320.
37. Kuypers, M.M.; Sliekers, A.O.; Lavik, G.; Schmid, M.; Jørgensen, B.B.; Kuenen, J.G.; Sinninghe, D.J.S.; Strous, M.; Jetten, M.S. Anaerobic ammonium oxidation by anammox bacteria in the Black Sea. *Nature* **2003**, *422*, 608–611.
38. Oleszkiewicz, J.A. A comparison of anaerobic treatment of low concentration piggery wastewaters. *Agric. Wastes***1983**, *8*, 215–231.
39. Orlygsson, J.; Houwen, F.P.; Svensson, B.H. Thermophilic anaerobic amino acid degradation: deamination rates and end-product formation.
40. Watson, S.W.; Valos, F.W.; Waterbury, J.B. The family nitrobacteraceae. In *The Prokaryotes*; Starr, M.P., Ed.; Springer-Verlag: Berlin, Germany, 1981.
41. Anthonisen, A.C., Loehr, R.C.; Prakasam, T.B.S. Inhibition of nitrification by ammonia and nitrous-acid. *J. Water Pol. Control Fed.* **1976**, *48*, 835–852.
42. Yamamoto, T.; Wakamatsu, S.; Qiao, S.; Hira, D.; Fujii, T.; Furukawa, K. Partial nitritation and anammox of a livestock manure digester liquor and analysis of its microbial community. *Bioresour. Technol.* **2011**, *102*, 2348–2351.
43. Naik, S.; Setty, Y.P. Effect of Carbon Sources on Biological Denitrification of Wastewater by Immobilized *Pseudomonas Stutzeri* Bacteria in a Fluidized Bed Bio Reactor (FBBR). Available online: http://ipcbee.com/vol23/23-CCEA2011-BEM30014.pdf (accessed on 24 August 2015).
44. Lqynachatn, E.; Bartholomeww, V.; Wolluma, G. Nitrogen transformations in aerated swine manure slurries. *J. Environ. Qual.* **1976**, *5*, 293–297.
45. Evans, M.R.; Hissetert, R.; Smith, M.P.W.; Ellamd, D.F.; Baines, S. Effect of microorganism residence time on aerobic treatment of piggery waste. *Agric. Wastes* **1979**, *1*, 67–85.
46. Evans, M.R.; Smith, M.P.W.; Deans, E.A.; Svoboda, I.F.; Thacker, F.E. Nitrogen and aerobic treatment of slurry. *Agric. Wastes* **1986**, *15*, 205–213.
47. Garrawaj, Y.L. Investigations on the aerobic treatment of pig slurry. *Agric. Wastes* **1982**, *4*, 131–142.
48. Blouin, M.; Bisailloj, N.-G.; Beaudet, R.; Ishaque, M. Nitrification of swine waste. *Can. J. Microbiol.* **1990**, *36*, 273–278.

49. Hino, T.; M atsumoto, Y.; Nagano, S.; Sugimoto, H.; Fukumori, Y.; Murata, T.; Iwata, S.; Shiro, Y. Structural basis of biological N_2O generation by bacterial nitric oxide reductase. *Science* **2010**, *330*, 1666–1670.
50. Moenne-Loccoz, P.; Fee, J.A. Catalyzing NO to N_2O in the nitrogen cycle. *Science* **2010**, *330*, 1632–1633.
51. Canfield, D.E.; Glazer, A.N.; Falkowski, P.G. The evolution and future of Earth's nitrogen cycle. *Science* **2010**, *330*, 192–196.
52. Sommer, S.G. Effect of composting on nutrient loss and nitrogen availability of cattle deep litter *Eur. J. Agron.* **2001**, *14*, 123–133.
53. Fukumoto, Y.; Osada, T.; Hanajima, D.; Haga, K. Patterns and quantities of NH_3, N_2O and CH_4 emissions during swine manure composting without forced aeration—Effect of compost pile scale. *Bioresour. Technol.* **2003**, *89*, 109–114.
54. Szanto, G.L.; Hamelers, H.M.; Rulkens, W.H.; Veeken, A.H.M. NH_3, N_2O and CH_4 emissions during passively aerated composting of straw-rich pig manure. *Bioresour. Technol.* **2007**, *98*, 2659–2670.
55. Ravishankara, A.R.; Daniel, J.S.; Portmann, R.W. Nitrous oxide (N_2O): The dominant ozone-depleting substance emitted in the 21st century. *Science* **2009**, *326*, 123–125.
56. Kelleher, B.P.; Leahy, J.J.; Henihan, A.M.; O'dwyer, T.F.; Sutton, D.; Leahy, M.J. Advances in poultry litter disposal technology—A review. *Bioresource Technol.* **2002**, *83*, 27–36.
57. Moller, K. Effects of anaerobic digestion on soil carbon and nitrogen turnover, N emissions, and soil biological activity. A review. *Agron. Sustain. Dev.* **2014**, *35*, 1021–1041.
58. Seto, B. The stickland reaction. In *Diversity of Bacterial Respiratory Systems*; Knowles, C.J., Ed.; CRC Press: Boca Raton, FL, USA, 1980; Volume 2, pp. 49–64.
59. Pain, B.F.; Misselbrook, T.H.; Clarkson, C.R.; Rees, Y.J. Odor and ammonia emissions following the spreading of anaerobically digested pig slurry on grassland. *Agric. Wastes* **1990**, *34*, 259–267.
60. Chantigny, M.H.; Rochette, P.; Angers, D.A.; Masse, D.; Cote, D. Ammonia volatilization and selected soil characteristics following application of anaerobically digested pig slurry. *Soil Sci. Soc. Am. J.* **2004**, *68*, 306–312.
61. Chantigny, M.H.; Angers, D.A.; Rochette, P.; Belanger, G.; Masse, D.I.; Cote, D. Gaseous nitrogen emissions and forage nitrogen uptake on soils fertilized with raw and treated swine manure. *J. Environ. Qual.* **2007**, *36*, 1864–1872.
62. Chantigny, M.H.; Angers, D.A.; Belanger, G.; Rochette, P.; Eriksen-Hamel, N.; Bittman, S.; Buckley, K.; Masse, D.; Gasser, M.O. Yield and nutrient export of grain corn fertilized with raw and treated liquid swine manure. *Agron. J.***2008**, *100*, 1303–1309.
63. Chantigny, M.H.; MacDonald, J.D.; Beaupre, C.; Rochette, P.; Angers, D.A.; Masse, D.; Parent, L.E. Ammonia volatilization following surface application of raw and treated liquid swine manure. *Nutr. Cycl. Agroecosyst.* **2009**, *85*, 275–286.
64. Möller, K.; Stinner, W. Effects of different manuring systems with and without biogas digestion on soil mineral nitrogen content and on gaseous nitrogen losses (ammonia, nitrous oxides). *Eur. J. Agron.* **2009**, *30*, 1–16.
65. Ni, K.; Pacholski, A.; Kage, H. Analysis of ammonia losses after field application of biogas slurries by an empirical model. *J. Plant Nutr. Soil Sci.* **2012**, *175*, 253–264.
66. Yoshimoto, T.; Nagi, F.; Fujimoto, J.; Watanabe, K.; Mizukoshi, H.; Makino, T.; Kimora, K.; Saino, H.; Sawada, H.; Omura, H. Degradation of estrogens by *Rhodococcus zopfi* and *Rhodococcus equi* isolates from activated sludge in wastewater treatment plan. *Appl. Environ. Microbiol.* **2004**, *70*, 5283–5289.
67. Yu, C.-P.; Roh, H.; Chu, K.-H. 17β-Estradiol-Degrading bacterial isolated from activated sludge. *Environ. Sci. Technol.***2007**, *41*, 486–492.
68. Ivanov, V.; Lim, J.-W.; Stabnikova, O.; Gin, K.Y.-H. Biodegradation of estrogens by facultative anaerobic iron-reducing bacteria. *Process. Biochem.* **2010**, *45*, 284–287.

69. Lee, H.B.; Liu, D. Degradation of 17β-Estradiol and its metabolites by sewage bacteria. *Water Air Soil Pollut.* **2002**, *134*, 351–366.
70. Isabelle, M.; Villemur, R.; Juteau, P.; Lepine, F. Isolation of estrogen-degrading bacteria from an activated sludge bioreactor treating swine waste, including a strain that converts estrone to β-estradiol. *Can. J. Microbiol.* **2011**, *57*, 550–568.
71. Barbier, O.; Albert, C.; Martineau, I.; Vallee, M.; High, K.; Labrie, F.; Hum, D.W.; Labrie, C.; Belanger, A. Glucuronidation of the nonsteroidal antiestrogen EM-652 (SCH 57068), by human and monkey steroid conjugating UDP-glucuronosyltransferase enzymes. *Mol. Pharmacol.* **2001**, *59*, 636–645.
72. D'Ascenzo, G.; di Corcia, A.; Gentili, A.; Mancin, R.; Mastropasqua, R.; Nazzari, M.; Samperi, R. Fate of natural estrogen conjugates in municipal sewage transport and treatment facilities. *Sci. Total Environ.* **2003**, *302*, 199–209.
73. Hutchins, S.R.; White, M.V.; Hudson, F.M.; Fine, D.D. Analysis of lagoon samples from different concentrated animal feeding operations for estrogens and estrogen conjugates. *Environ. Sci. Technol.* **2007**, *41*, 738–744.
74. Furimsky, A.M.; Green, G.A.; Sharp, L.E.H.; Catz, P.; Adjei, A.A.; Parman, T.; Kapetanovic, I.M.; Weinshilboum, R.M.; Iyer, L.V. Effect of resveratrol on 17β-estradiol sulfation by human hepatic and jejunal S9 and recombinant sulfotransferase 1E1. *Drug Metab. Disp.* **2008**, *36*, 129–136.
75. Czajka, C.P.; Londry, K.L. Anaerobic biotransformation of estrogens. *Sci. Total Environ.* **2006**, *367*, 932–941.
76. Yost, E.E.; Meyer, M.T.; Dietze, J.E.; Meissner, B.M.; Davis, L.W.; Williams, C.M.; Lee, B.; Kullman, S.W. Comprehensive assessment of hormones, phytoestrogens, and estrogenic activity in an anaerobic swine waste lagoon. *Environ. Sci. Technol.* **2013**, *47*, 13781–13790.
77. De Mes, T.; Kujawa-Roeleveld, K.; Zeeman, G.; Lettinga, G. Anaerobic biodegradation of estrogens—Hard to digest. *Water Sci. Technol.* **2008**, *57*, 1177–1182.
78. Andersen, H.; Siegrist, H.; Halling-Sorensen, B.; Ternes, T.A. Fate of estrogens in a municipal sewage treatment plant. *Environ. Sci. Technol.* **2003**, *37*, 4021–4026.
79. Ternes, T.A.; Kreckel, P.; Mueller, J. Behaviour and occurrence of estrogens in municipal sewage treatment plants—II. Aerobic batch experiments with activated sludge. *Sci. Total Environ.* **1999**, *225*, 91–99.
80. Shappell, N.W.; Billey, L.; Forbes, D.; Matheny, T.A.; Poach, M.E.; Reddy, G.B.; Hunt, P.G. Estrogenic activity and steroid hormones in swine wastewater through a lagoon constructed-wetland system. *Environ. Sci. Technol.* **2007**, *41*, 444–450.
81. Pholchan, P.; Mnes, J.; Donnelly, T.; Sallis, P.J. Fate of estrogens during the biological treatment of synthetic wastewater in a nitrite-accumulating sequencing batch reactor. *Environ. Sci. Technol.* **2008**, *42*, 6141–6147.

CHAPTER 14

Relationship between Mineral Soil Surface Area and the Biological Degradation of Biosolids Added to Soil

Dongqi Wen [1], Wenjuan Zhai [1], Demetrios Moschandreas [1], Guanglong Tian [2] and Kenneth E. Noll [1,*]

[1] Civil, Architecture and Environmental Engineering, Illinois Institute of Technology, 3201 S Dearborn St., Chicago, IL 60616, USA;

[2] Environmental Monitoring and Research Division, Monitoring and Research Department, Metropolitan Water Reclamation District of Greater Chicago, 6001 W. Pershing Road, Cicero, IL 60804, USA

ABSTRACT

Geochemical and biological processes that operate in the soil matrix and on the soil surface are important to the degradation of biosolids in soil. Due to the large surface area of soils it is assumed that the microbial ecology is associated with mineral soil surface area. The total mineral surface areas were determined for soils from eight different fields selected from a long term study (1972–2006) of annual biosolids application to 41 fields in central Illinois varying in size from 3.6 to 66 ha. The surface areas for the soils varied from 1 to 9 m^2/g of soil. The biological degradation rates for the eight soils were determined using a biological degradation rate model (DRM) and varied from 0.02 to

0.20/year^{-1}. Regression analysis revealed that the degradation rate was positively associated with mineral soil surface area (1 m^2/g produces 0.018 year^{-1} increase in the degradation rate). The annual soil sequestration rate was calculated to increase from 1% to 6% when the soil total surface area increased from 1 to 9 m^2/g of soil. Therefore, land application of biosolids is an effective way to enhance carbon sequestration in soils and reduce greenhouse gas emissions.

Keywords: soil carbon; carbon sequestration; biosolids; biological degradation; mineral soil surface area

1. INTRODUCTION

The ability for soils to biologically degrade biosolids and sequester carbon (C) is recognized as one method to mitigate greenhouse gas emissions [1,2]. Due to the large surface area of soils, biological processes that operate on the soil surface are potentially important to the degradation rate of biosolids in soil. However detailed knowledge of where microorganisms reside is very difficult to obtain. In this study the relationship between the mineral soil surface area and the degradation rate of biosolids added to soil is evaluated to provide a better understanding of the important variables that control sequestration and lead to application of technologies based on their ability to increase the rate of sequestration.

Biosolids are nutrient-rich organic materials resulting from anaerobic digestion of primary and secondary sludge from wastewater treatment plants. Jarecki and Lal [3] suggested that application of biosolids is an important management practice to increase soil C sequestration in agricultural soils. Net C sequestration rates from biosolids applied to soil have been reported to be between 1 to 3 Mg ha^{-1}·year^{-1} with biosolids application rate between 56 to 71 Mg ha^{-1}·year^{-1} [4].

The rate of biosolids decomposition is one of the key variables governing the dynamic of C sequestration and therefore numerous investigators have studied degradation rates of biosolids applied to soil [5,6,7,8,9,10,11,12,13,14]. One of the most comprehensive models of C sequestration in soils is CENTURY [10], which simulates the flux of C through plant litter using three soil organic fractions (active, slow, and passive), each of which has a significantly different degradation rate above and below ground litter pools. Parton *et al.* [10] applied the CENTURY model to estimate soil organic matter decomposition rates and pointed out that decomposition rates are affected by climate (*i.e.*, soil temperature, soil moisture content) and soil texture. The average slow soil organic fraction decomposition rate (0.12 $year^{-1}$) was similar to biosolids degradation rates presented in this study under similar climate conditions. Sorenson [15] applied ^{14}C-labiled cellulose to seven different soils containing different silt and clay amounts to determine the effect of soil texture on cellulose degradation decomposition rates. The study reported that clay content influents soil organic carbon content (biomass) with higher clay concentrations related to higher cellulose degradation rates. Gilmour and Gilmour [9] developed a model for the rate of decomposition of biosolids based on experimental results from Terry *et al.* [8] to compute the rate of conversion to carbon dioxide (CO_2) for a single biosolids application to soil. The model indicates that the biosolids degradation process follows the exponential first-order decay model. Zhai *et al.* [14] formulated a degradation rate model (DRM) using data from a long term field study of biosolids applied to soil [4]. The DRM is based on quantification of the degradation rate for biosolids and yield for residual microbial biomass from repeated application of biosolids to soil and provides an easy quantitative method for evaluating C sequestration.

The objective of this paper is to examine the effect of mineral soil surface area on the biosolids degradation rate and C sequestration rate. The degradation rate and surface area distribution are determined for soils from eight fields from the Tian *et al.* [4] database. Tian *et al.* [4] reported on biosolids degradation and C sequestration from a long term study (1972–2006) of annual biosolids application to 41 fields varying size from 3.6 to 66 ha. The biosolids applied to the fields were in the liquid phase with an average organic carbon content of 23.2% [4].

2. MATERIALS AND METHODS

This paper explores the effect of mineral soil surface area on the biosolids degradation rate and C sequestration. We hypothesize that soils with more surface area have the capability of biodegrading biosolids at a faster rate. To evaluate this hypothesis, the total surface areas (1 to 9 m^2/g) and degradation rates (0.02 to 0.20 $year^{-1}$) were compared for soils from eight fields. The fields were selected with care from the Tian *et al.* [4] database. Application of the DRM model determined the required degradation rates. The fields were selected strategically from the 41 fields to represent different mineral surface areas. This approach facilitates the objective of establishing an association between total mineral surface area and degradation rate.

Tian divided 41 fields into three groups based on soil type. Group I consisted of 20 fields of "coarse" mine spoil soils, primarily Lenzburg

(Fine-loamy, mixed, active, calcareous, mesic Haplic Udarents) and Lenzwheel soil series [16]. Group II consisted of nine fields of "fine" mine spoil soils, primarily Rapatee (fine-silty, mixed, superactive, non-acid, mesic Mollic Udarents) soil series. Group III consisted of 12 fields of various non-mined soils that were degraded by intensive cultivation or overgrazing. Of the eight fields selected for evaluation in this study, three were from group I (F32, F39 and F15), three from group II (F43, F45, and F47) and two from group III (F10 and F37), see Table 1 [17]. Six of the soils were selected because they represented pairs with similar annual biosolids application rates but different soil types (F37 and F39, F45 and F10, and F43 and F15). Soils from fields 47 and 32 were selected because they represent large variations in degradation rate and application rate (See Table 1). A sieve analysis was performed to determine the size distribution of the coarse particles, and the hydrometer method was used to determine the size distribution of the fine particles [18] for the eight soils. The degradation rates and surface area were also calculated for three soils evaluated by Terry *et al.* [8] based on laboratory conditions and compared to results from soils from the eight fields.

2.1. Application of DRM to Determine Degradation Rate

The DRM [14] was used in this study to determine the biological degradation rate and C sequestration amounts for biosolids applied to eight fields from 1972 to 2006 in Fulton County, Illinois [2]. 1. The biosolids decomposition in the DRM is described by first order kinetics:

$$\frac{dC}{dt} = -kC \tag{1}$$

where:

- C = the carbon concentration present in the biosolids (Mg/Ha)
- k = the first-order rate ($year^{-1}$)
- t = biosolids decomposition time (year).

Equation (1) can be quantified as Equation (2) below:

$$C_t = C_o \cdot e^{-kt} \tag{2}$$

where:

- C_o = initial concentration of C in biosolids (Mg/Ha)
- C_t = the concentration at time *t* (Mg/Ha).

With the same annually biosolids C added (C_o); the accumulated residue of biosolids C (r) after time *t* (year.) is given by Equation (3) [19]:

$$r = (1 + f^1 + f^2 + f^3 + \cdots + f^t) \cdot Co \tag{3}$$

where:

- f = the fraction left after 1 year. decay, or $f = Ct / C_o = e^{-kt}$, with t=1.

The DRM was developed by curve-fitting the measured values of SOC for each year (1972–2006) with model-generated values of SOC for each of the 41 fields using trial and error to produce a best-fit average degradation rate for biosolids degradation and biomass yield (Figure 1). When the appropriate degradation rate and biomass yield are estimated from curve fitting, it is possible to use the DRM to estimate (1) the degradation of biosolids as a function of time; (2) the portion of the SOC due to biosolids remaining and (3) the portion due to residual microbial biomass (C sequestered). The average estimated and measured SOC are 3.94 and 3.34 Mg/Ha/year, respectively, for Group I fields. For Group II fields, the average estimated and measured SOC amounts are 3.17 and 2.59 Mg/Ha/year, respectively, while the average estimated and measured SOC for Group III fields are 2.15 and 1.68 Mg/Ha/year [14]. Figure 1 provides examples of the measured and modeled SOC concentrations and amount of C sequestered over the 34 years for two of the eight fields that have long application time (15 to 22 year). The coefficient of determinations between SOC measured and estimated with the DRM were very good (average coefficient of determination is 0.94) indicating that the curve-fitting technique was able to provide degradation rate and yield information that allows the DRM model to provide acceptable estimates of the measured SOC values and sequestration amounts.

2.2. Soil Particulate Size Distribution

Tian *et al.* [4] classified the eight fields evaluated in this study as coarse (group I), fine (group II), and mixed (group III) but provided no additional information concerning difference in the type of soils in each group. Based on the assumption that biological degradation of biosolids takes place on the surface of soils [20], the mass and surface area for the eight soil samples were measured and compared to the degradation rates for the soils to see if higher degradation rates occurred for soils with greater total surface area.

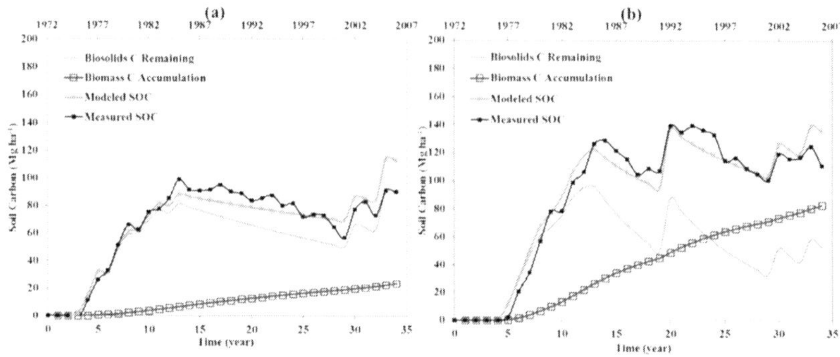

Figure 1. Example of the variation in biosolids remaining, measured and modeled soil organic carbon (SOC); (**a**) field 32 and (**b**) field 47.

2.2.1. Particle Mass Size Distribution Analysis

As suggested by Parton *et al.* [10,21], who developed the CENTURY model and determined that organic decomposition rates are affected by soil texture, a soil texture analysis was used to determine the physical characteristics of the soils [22]. The soil textures of fields investigated in this study are shown in Table 2. Quantitatively, soil texture denotes the proportion of sand (0.05 to 2 mm diameter), silt (0.002 to 0.05 mm diameter) and clay (less than 0.002 mm diameter) that occur in a given soil.

2.2.2. Particulate Surface Area Analysis

In order to explore the effect of surface area on biosolids degradation, particulate surface area size distributions were estimated from the mass size distributions. For this analysis, the soil particles were assumed to be spherical with a smooth surface [23] and the number of soil particles was estimated by Equation (4):

$$N = M/(\rho \cdot \frac{3}{4} \cdot \pi \cdot r^3) \qquad (4)$$

where:

- N = the number of soil particles
- r = soil particle radius (cm)
- M = the mass of the soil particle (g)
- ρ = soil density [17] (g/cm^3)

The soil particulate surface area (S) was estimated by Equation (5):

$$S = 4\pi r^2 \cdot N \quad (5)$$

where:

- N = the number of soil particles
- r = soil particle radius (m)
- S = particle surface area (m^2).

2.3. Reference Soils from Terry

Terry *et al.* [8] evaluated the biological degradation of synthetic biosolids using three different soil types under controlled conditions in the laboratory over a one year period to determine the effect of soil type, moisture, pH, temperature, and application rate on the decomposition rate of biosolids in soil. The synthetic sludge applied to Terry's experiment was in liquid phase with volatile solids similar to biosolids applied to 41 fields in Illinois. The syntactic biosolids had organic carbon percent (22.3%) similar to biosolids used in this study. Results from Terry's experiments were compared with results from this study. Terry's paper reports on the emissions from biosolids added to soil with little analysis of the results. We have reanalyzed the data in detail to demonstrate that the mineral soil surface area and degradation rate are related. This analysis provides new data for comparison of laboratory and field degradation rate (Biological degradation rates under laboratory and field conditions are different due to differences in average soil temperature and moisture content [9]).

3. RESULTS AND DISCUSSION

3.1. Biosolids Degradation Rate

3.1.1. DRM Estimated Rate of Biosolids Degradation

Analyses of DRM simulation results for the eight fields (Table 1) indicated that the degradation rate varied between 0.02 and 0.20 year^{-1}. It can be seen from Figure 1 that the slow rates of biosolids degradation and the large application rates resulted in large increases in SOC due to the presence of biosolids that have not reacted. The SOC concentration peaked in the mid-1980s after 12 years of annual biosolids application. The application of biosolids declined after 1985 and the accumulation of biosolids did not continue to increase. During the early stages of biosolids addition, the biosolids accumulation exceeded degradation, resulting in the accumulation of stored biosolids and, therefore, an increase in SOC. Eventually the amount of decomposition converged on the amount of biosolids applied and the biosolids approached a steady state. This result is supported by Hamake [19], who developed a mathematical model to predict the cumulative levels of pesticides in soil. The study indicated that when pesticides application rates equal the decomposition rates, a steady state is approached. Also, Jastrow *et al.* [24] suggested carbon sequestration occurs when a positive disequilibrium sustained between C input and C degraded over some period of time. A new steady-state system would eventually be achieved when the amount of degradation converge on the amount of application. These studies explain the increasing of SOC during the early stages of biosolids addition and the achievement of a steady state in this study.

Application of biosolids increased the sequestered C concentration up to 1985 in all eight fields, as suggested by Figure 2. There was a marked increase in the ratio of C sequestered to C application rate up to year 1985 due to accumulation of the C from the conversion of biosolids to new biomass. Beyond 2025, the trend in Figure 2 indicates almost the same ratio of C sequestration to C application rate. In the short term, lower biosolids degradation rates result in less microbial production and produce a smaller increase in C sequestration, i.e. F32 . In the long term, all of the applied biosolids may undergo degradation and the amount of C sequestered is determined by the total amount of biosolids applied. Under aerobic conditions, approximately 35% to 40% of the C in biosolids may eventually be sequestered, although this may take a long time (between 20 and 100 years) with k between 0.20 and 0.02 year^{-1}, and 95% biosolids conversion (see Table 1 and Figure 2).

3.1.2. Rate of Degradation Related to Coarse and Fine Soil Type

Analyses of DRM simulation results for the 41 fields in the Tian *et al.* database [4] indicated higher biosolids degradation rates occur with finer soils. This effect can be seen in Figure 3 that shows that the microbial degradation rate was smaller for coarse (group I) than for fine (group II) soil type (separation of the regression lines). The error bars in Figure 3 represent one standard deviation for each average biosolids application rate for related fields in Tian's database [4]. The Tian *et al.* database [4] has included only one data point for fine soils for each application rate in Group II. Figure 3 also identified the eight fields from the database [4] selected for evaluation in this paper. Four of the fields represent fine soils and four represent coarse soils.

Table 1. Variation in the first order biosolids degradation rate (from Equation (1)* in DRM), annual application rate, soil density, and surface area for soils evaluated.

Soil Type	Fields No.	Group No.	Biosolids Degradation Rate ($year^{-1}$)	Annual Application Rate (Mg/ha)	Soil Density [17] (g/cm^3)	Surface Area (m^2/g)
Coarse	F32	I	0.02	8.93	2.6	1.29
	F39	I	0.05	11.95	2.6	1.73
	F10	III	0.06	12.85	2.6	2.29
	F15	I	0.11	15.44	2.6	2.42
Fine	F37	III	0.14	11.75	2.6	6.55
	F47	II	0.15	18.02	2.6	8.54
	F45	II	0.15	12.95	2.6	7.63
	F43	II	0.20	15.63	2.6	7.65
Other soil samples [6]	Fincastle	N/A	0.13	5.15	2.6	4.81
	Chalmers	N/A	0.10	5.15	2.6	4.66
	Tracy	N/A	0.07	5.15	2.6	2.13
*			$dC/dt = -kC$			

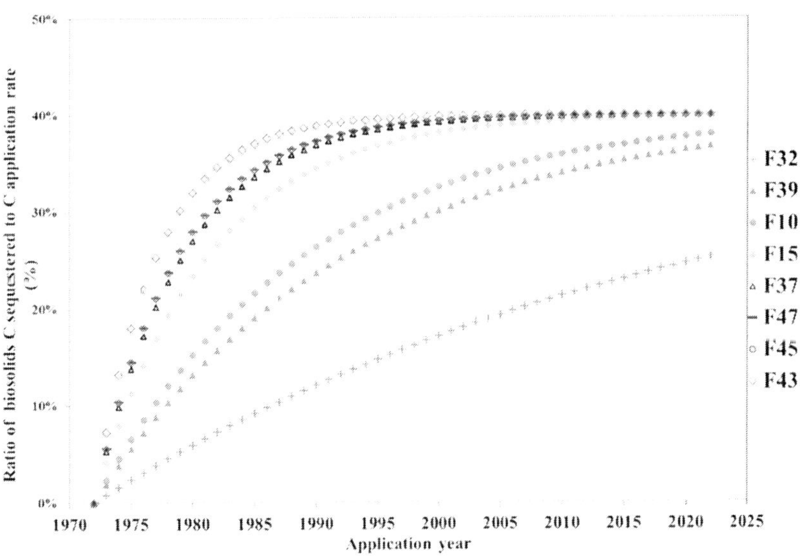

Figure 2. Percent of biosolids converted to biomass (carbon sequestration) between 1972 and 2025 for the eight fields. The variation between fields is due to difference in the degradation rate (See Table 1).

The average difference in the degradation rate between the linear regression lines was near 0.10 $year^{-1}$. This represents the different in the average degradation rate of biosolids when applied on coarse and fine soils. This result is supported by other researchers who used soil texture rather than soil surface area to relate to biological decomposition [15,25]. Sorenson [15] conducted an experiment to investigate the relationship between clay content concentrations and organic material decomposition rate. He suggested that higher clay concentrations (finer texture) causes a more rapid biological decomposition,. Parton [25] determined that active organic carbon turnover time (1/k) decreased with increase in clay and silt content. Both of the results agree with the findings of this study.

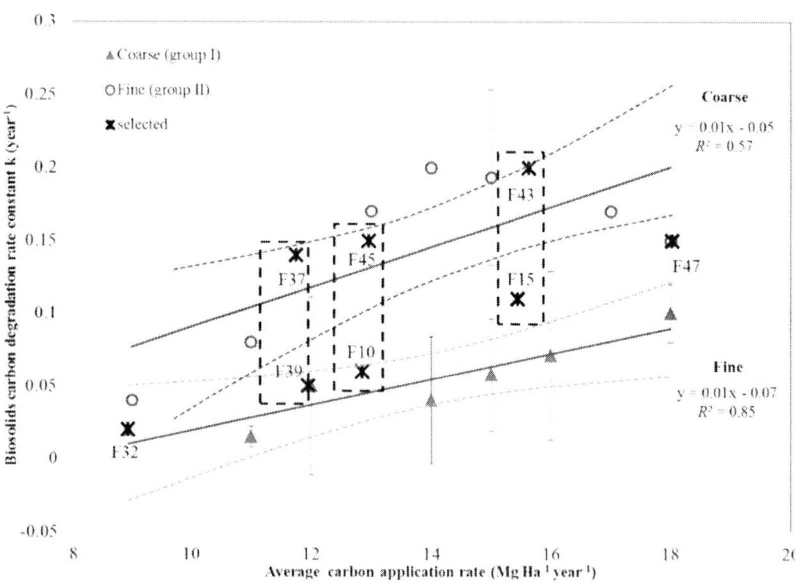

Figure 3. Biosolids degradation rate as a function of average annual carbon application rate separated into coarse and fine soil types (group I and II). Also identified are the eight fields used in this study including fields with similar application rates (paired field) and different soil types with 95% confidence interval shown for each soil group. The soil groups are from Tian *et al.* database [4].

3.1.3. Degradation Rate of Synthetic Biosolids

Decomposition was initially very rapid in the three soils evaluated by Terry [8] with 54% to 63% of the total C in the biosolids removed during the first 28 days followed by a slow decomposition for the period from 28 to 336 days. Gilmour and Gilmour [9] developed a model for the rate of decomposition of biosolids based on the experiment results from Terry *et al.* [8]. The model indicated that the biosolids degradation process follows first-order exponential decay with rapid and slow decomposition terms. The model provided corrections for field conditions of temperature and moisture that may be different than the laboratory conditions. Based on the study of Gilmour and Gilmour [9], Tian *et al.* [4] determined that the decomposition of biosolids at a site similar to the location of the 41 fields in Illinois (average temperature near 10 °C and 10% moisture) was only 37% of that under Terry's [8] laboratory conditions (temperature 21 °C and 20% moisture).

The laboratory degradation rates, shown in Figure 4, were determined for the slow decomposition phase for the biosolids added to the three soils based on first order kinetics using Terry's decomposition data for synthetic biosolids incubated for 28 to 336 days at 21 °C and 20% moisture. The degradation rates need to be reduced to 37% of the laboratory values for comparison to field data presented in Table 1. The reduced rates varied from 0.07 to 0.13 year^{-1} for the three soils. It is generally assumed that the rapid fraction is consumed completely before the decomposition of the slow fraction starts [26]. The slow decomposition phase was used in this study because the biosolids have already undergone short term volatile losses and has no remaining rapid fraction. This was similar to the field conditions where the long term repeated annual application of biosolids results in the accumulation of biosolids over long time periods with only slow decomposition remaining (see Figure 1).

3.2. Particulate Mass and Surface Area Distribution Analysis

Soil texture for each of the eight soil samples is presented in Table 2. The eight fields can be divided into two types of soil, with the coarse samples from fields 10, 15, 32, and 39 classified as soils with a sandy loam texture; and the fine samples from fields 37, 43, 45, and 47 classified as soils with a clay loam texture.

The soil texture for the 11 soils analyzed in this study and mineral surface areas are closely related (See Table 2), and therefore literature relating soil texture to degradation rate can also apply to mineral soil surface area [15,25,27]. Both Parton [25] and Paul [27] indicated that active organic carbon decomposition rates increased with increased clay plus silt contents, while increased sand content caused decreased active organic carbon decomposition rates. The results from this study correspond to those findings and demonstrate that soil texture is closely related to mineral soil surface area. However, mineral soil surface area

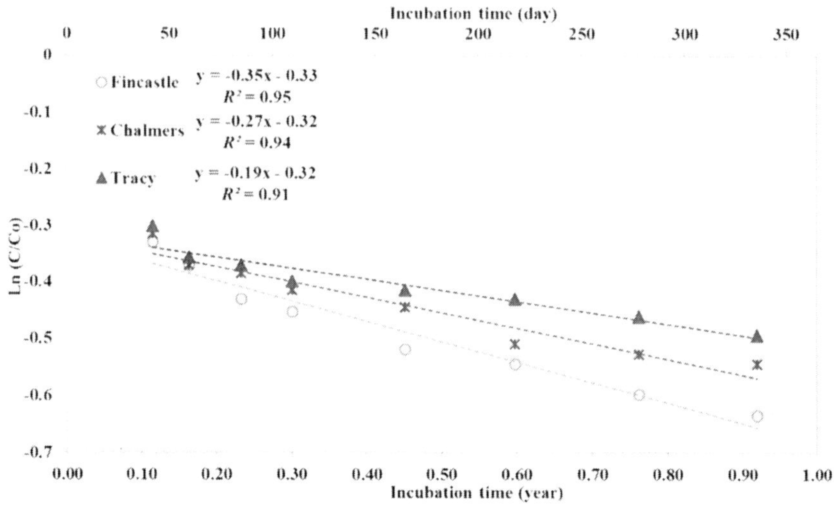

Figure 4. Variation in degradation rate ($year^{-1}$, slope of trend line) for three soils based on first order kinetics ($\ln(C/C9)=-kt$) and Terry et.al [8] laboratory decomposition data (incubated for 28 to 336 days at a temperature of 21 °C and moisture content of 20%).

provides a more definitive relationship to biological degradation of biosolids added to soil than soil texture. This is because surface area measurements are related to individual soil mass percentage while soil texture represents a range of mass percentages.

Figure 5 provides the surface area distribution for soils evaluated in this study. The total surface area for the coarse soils varied from a low of 1.3 to a high of 2.4 m^2/g of soil. The total surface area for the fine soils varied from a low of 6.5 to a high of 8.5 m^2/g of soil. The average difference between the two groups was 5.2 m^2/g of soil. As mentioned in Section 3.1.2, the average difference in biosolids degradation rates between these two groups was 0.1 $year^{-1}$ (see Figure 3). This leads to the conclusion that the average difference in total mineral soil surface area of 5.2 m^2/g causes the average difference in degradation rate of 0.1 $year^{-1}$. Please notice that the average difference in biosolids degradation rate is caused by soil surface area and not biosolids application rate, see Section 3.3.

Table 2. Mass and mineral soil surface area percentage (clay, silt, and sand) and soil texture of selected soil samples.

Field	Clay		Silt		Sand		Soil Texture
	Mass (%)	Surface Area (%)	Mass (%)	Surface Area (%)	Mass (%)	Surface Area (%)	
F32	10.00	23.00	21.10	70.00	68.90	7.00	Sandy loam
F39	10.00	50.50	23.01	46.30	67.00	3.21	Sandy loam
F10	14.80	39.10	23.20	56.10	62.00	4.79	Sandy loam
F15	10.00	54.20	24.10	43.30	65.90	3.42	Sandy loam
F37	28.70	66.10	40.10	33.10	31.20	0.72	Clay loam
F47	31.02	69.30	49.10	31.40	19.90	0.31	Clay loam
F45	30.20	65.50	47.90	34.10	21.90	0.36	Clay loam
F43	30.10	66.50	45.05	33.20	25.00	0.35	Clay loam
Fincastle	16.80	55.80	69.70	43.60	13.50	0.30	Silt loam
Chalmers	24.80	61.40	60.90	37.90	14.30	0.36	Silt loam
Tracy	8.90	53.80	37.20	43.00	53.90	3.10	Sandy Loam

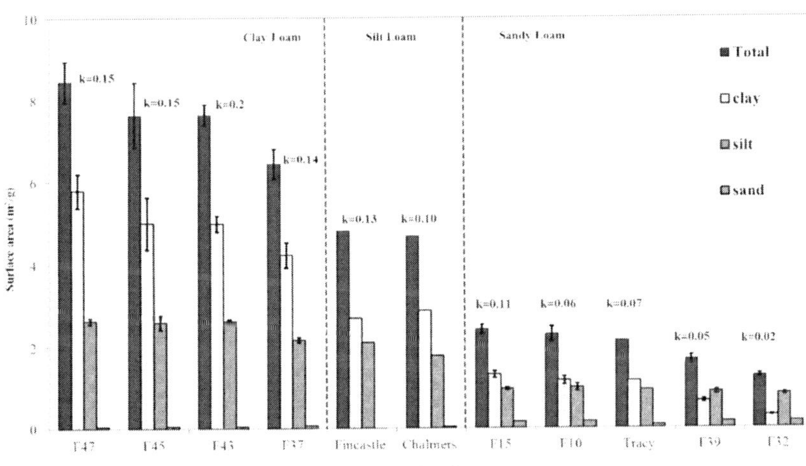

Figure 5. Overall total surface area and surface area for clay, silt, and sand components for soils [8]. The average and deviation of mineral soil surface area are shown.

The total surface areas for Terry's soil samples were calculated using Equations (4) and (5); see Table 2 and Figure 5. The Fincastle and Chambers soils had total surface areas (average 4.7 m^2/g) between Tian *et al.* fine and coarse soil groups [4] and the Tracy soil was comparable to the coarse soil group (2.1 m^2/g). One can observe from Figure 5 that biosolids degradation rates increase with finer soil texture which indicated strong dependent between degradation rates and texture. This result is observed by other researchers [15,27,28,29,30]. Schimel [28,29] pointed out that clay content influents soil organic carbon content with higher clay concentration (finer soil texture) related to higher organic compounds degradation rates. Allison *et al.* [30] showed that the soil texture of mineral soil colloid affects the rate of organic carbon loss and the quantity of humus formed from readily decomposable plant materials.

3.3. Relationship between Mineral Soil Surface Area and Biosolids Degradation

Figure 6 shows the relationship between the total mineral soil surface area and the degradation rate for the eight fields. The figure indicates that the degradation rate increased when there was an increase in the total surface area. The degradation rate varied from 0.02 to 0.20 $year^{-1}$ when the soil total surface area varied from 1.3 to 8.5 m^2/g of soil.

The values from Table 1 for the three reference soils from Terry *et al.* [8] are also given in Figure 6 for comparison but the coefficient of determination between the variables presented in the figure is based only on the eight samples from the Tian *et al.* [4] database. The reference soils provide results similar to the relationship developed for the eight Tian soils when they are corrected for field temperature and moisture conditions as suggested by Gilmour and Gilmour [9] and Tian *et al.* [4].

Table 3 provides information on the rate of change in the degradation rate between the coarse and fine soil pairs based on the same application rates. The average difference in total mineral soil surface area of 5.13 m^2/g of soil produces an average difference in the rate of degradation near 0.09 $year^{-1}$. Therefore, there is an average increase of 0.018 $year^{-1}$ in the degradation rate due to an increase of 1 m^2/g of soil when the application rates are the same. This indicates that changes in total surface area produce a greater change in the degradation rate than changes in application rate.

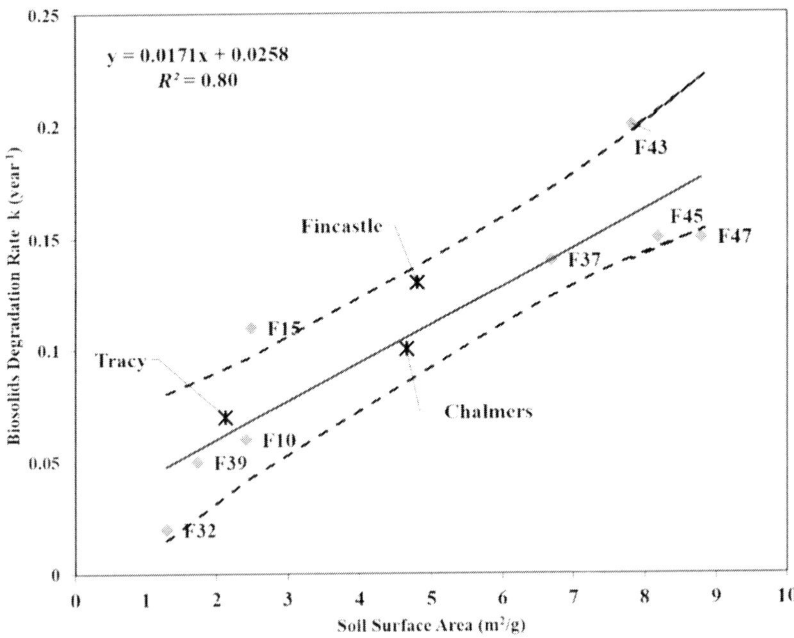

Figure 6. Biosolids degradation rate as a function of surface area for eight soils from different fields with 95% confidence interval. The three reference soils from Terry *et al.* [8] laboratory degradation experiments are also identified but not included in the correlation calculations.

Table 3. Comparison of changes in biosolids degradation rate and mineral soil surface area for paired soils (same biosolids application rate, but different soil texture).

Paired Samples	Average Biosolids Application Rate (Mg ha^{-1}·year^{-1})	Difference of k (Δk) (year^{-1})	Difference of Surface Area (ΔSF) (m^2/g)	Δk/ΔSF (g·year^{-1}·m^{-2})
F39 *vs.* F37	11.85	0.09	4.82	0.019
F10 *vs.* F45	12.90	0.09	5.34	0.017
F15 *vs.* F43	15.54	0.09	5.23	0.017
Average	N/A	0.09	5.13	0.018

To further explore the effects of surface area and application rate on the degradation rate, we formulated a multiple linear regression analysis between the degradation rate for the eight soils and the variables of total mineral soil surface area and the annual biosolids application rate. The analysis performed by SPSS (version 22.0, SPSS Inc., Chicago, IL, USA), indicates that the total mineral soil surface area is a significant indicator of degradation rate with a high coefficient of determination (R2=0.86) but annual application rate is not statistically significant ($p=0.32>0.05$). Therefore the regression model relating degradation rate and total mineral soil surface area is given by Equation (6) as indicated in Figure 6:

$$y = 0.017x + 0.026 \tag{6}$$

where:

- y = the degradation rate ($year^{-1}$)
- x = the total mineral soil surface area (m^2/g)

Equation (6) was applied to the decomposition of biosolids for the three reference soils from Terry *et al.* [8] since these observations (see Figure 6) were not used to generate Equation (6). The estimated biosolids degradation rates from Terry *et al.* [8] were related to the rate predicted by the simple regression model shown in Equation (6). The agreement between values of estimated three values) from Terry *et al.* data and predicted by Equation (6) is statistically significant.

3.4. C Sequestration

Figure 7 shows the increase in the annual percentage of applied biosolids that can be sequestered as a function of total mineral soil surface area. The annual percentage of biosolids converted to biomass is determined based on a mass balance between gases C emitted to the atmosphere and stored as biomass C sequestered in the soil and are computed using the DRM. The C sequestration values were determined based on the assumption of first order degradation using Equations (1) to (3) with a biomass yield of 40%. The total mineral soil surface area for different degradation rates was determined from Equation (6). The sequestration rates for Terry *et al.* [8] soils were based on degradation rates determined from Figure 4 (0.19 to 0.35 $year^{-1}$, 21 °C and 20% moisture) and for the eight fields from the DRM (Table 1). For Terry's [8] soils the increase in total surface area was 2.7 m^2/g and produced an increase in the annual sequestration rate from 6% to 11%. For the soils from the eight fields (10 °C and 10% moisture), the increase in total surface area was 7.2 m^2/g and this produced an increase in the annual sequestration rate from 1% to 6%. Importantly, the present study indicates that land application of biosolids is an appropriate way to enhance C sequestration in soils [1], which was recommended by United Nations Framework Convention on Climate Change (UNFCCC) as a necessary step toward reducing greenhouse gas (GHG) [2]. This leads to the conclusion that applying biosolids to soils with fine texture contributes to the reduction of GHG more effectively than applying to coarser soils, since higher biosolids degradation rates are strongly related to finer soil texture and will generate higher soil organic C sequestration [14]. Pertinent literature also shows that soil with finer texture has a higher capacity [25] to hold water which is positive for both plant growth and ensuring biosolids degradation rates. Increasing soil C sequestration is an important option not only to mitigate climate change but also to enhance soil fertility and the productivity of agroecosystems [31]. Applying biosolids are much more effective in enhancing C sequestration than other agriculture method such as applying animal manure or plant materials [30,32]. Soil total surface area as well as temperature and moisture affect the rate of biosolids degradation and the rate of C sequestration.

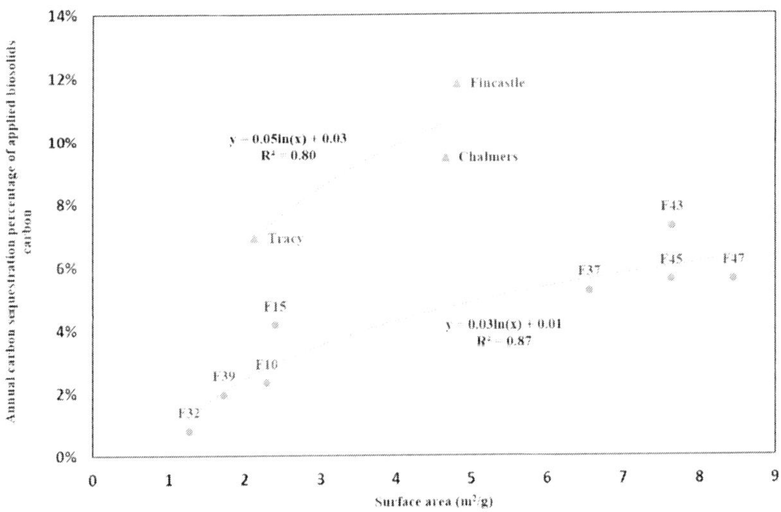

Figure 7. Annual percentages of applied biosolids converted to sequestered carbon as a function of mineral soil surface area for field (10 °C and 10% moisture) and laboratory conditions (21 °C and 20% moisture). Sequestered % is based on first order kinetics and a biomass yield of 40%.

4. CONCLUSIONS

This study has reanalyzed data on biosolids application to soil from a field study and a laboratory study to determine the relationship between biosolids mineral soil surface area and biosolids degradation. Regression analysis indicates that there is an association between total mineral soil surface area (of sand, silt, and clay) and biosolids degradation rate. There was an average increase of 0.018 $year^{-1}$ in the degradation rate due to an increase of 1 m^2/g of soil when the application rates were the same. Biological degradation of biosolids added to soil was significantly associated (R2=0.80) with the total mineral soil surface area. Mineral soil surface area (related to soil texture) has been shown to influence other soil characteristics such as soil organic matter content, soil aggregation and biomass (also defined as soil humus or sequestered carbon) accumulation that play an important role in soil degradation rates. It was not possible to include these additional variables in the regression analysis based on the data available; however, it was shown that application rate does not have a statistically significant effect on the degradation rate.

Applying biosolids to soils with fine texture (higher mineral soil surface area) contributes to the reduction of GHG more effectively than application to coarse soils since higher biosolids degradation rates are related to finer soil texture and generate higher soil organic C sequestration. Increasing soil C sequestration is an important option not only to mitigate climate change but also to enhance soil fertility and the productivity of agroecosystems.

One limitation of this study is that only eight different soils were evaluated. In the future a wider range of soil types will be investigated to evaluate the effect of mineral soil surface areas on biosolids degradation rates

ACKNOWLEDGMENTS

We recognize Metropolitan Water Reclamation District of Greater Chicago (MWRDGC) for collecting the data used in this study and Cecil Lu-Hing who was responsible for initiating and supervising the data collection program.

REFERENCES

1. Post, W.M.; Peng, T.H.; Emauel, W.R.; King, A.W.; Dale, V.H.; de Angelis, D.L. The global carbon cycle. *Am. Sci.* **1990**, *78*, 310–326.
2. Brown, S.; Leonard, P. Building carbon credits with biosolids recycling. *BioCycle* **2004**, *45*, 25–29.
3. Jarecki, M.K.; Lal, R. Crop management for soil carbon sequestration. *Crit. Rev. Plant. Sci.* **2003**, *22*, 471–502.
4. Tian, G.; Granato, T.C.; Cox, A.E.; Pietz, R.I.; Carlson, C.R.; Abedin, Z. Soil carbon sequestration resulting from long-term application of biosolids for land reclamation. *J. Environ. Qual.* **2009**, *38*, 61–74.
5. Conant, R.T.; Paustian, K. Potential soil carbon sequestration in overgrazed grassland ecosystems. *Glob. Biogeochem. Cycles* **2002**, *16*, 901–909.
6. West, T.Q.; Post, W.M. Soil organic carbon sequestration rates by tillage and crop rotation: A global data analysis. *Soil Sci. Soc. Am. J.* **2002**, *66*, 1930–1946.
7. Miller, A., III; Morgan, M.E.; Libbey, L.M. Lactobacillus maltaromicus, a new species producing a malty aroma. *Int. J. Syst. Bacterial.* **1974**, *24*, 346–354.
8. Terry, R.E.; Nelson, D.W.; Sommers, L.E. Decomposition of anaerobically digested sewage sludge as affected by soil environmental conditions. *J. Environ. Qual.* **1979**, *8*, 342–347.
9. Gilmour, J.T.; Gilmour, C.M. A simulation model for sludge decomposition in soil. *J. Environ. Qual.* **1980**, *9*, 194–199.
10. Parton, W.J.; Schimel, D.S.; Cole, C.V.; Ojima, D.S. Analysis of factors controlling soil organic matter levels in great plains grasslands. *Soil Sci. Soc. of Am. J.* **1987**, *51*, 1173–1179.
11. Brown, S.; Beecher, N.; Carpenter, A. Calculator tool for determining greenhouse gas emissions for biosolids processing and end use. *Environ. Sci. Technol.* **2010**, *44*, 9509–9515.
12. Bolan, N.S.; Kunhikrishnan, A.; Naidu, R. Carbon storage in a heavy clay soil landfill site after biosolids application. *Sci. Total Environ.* **2013**, *465*, 216–225.
13. Thangarajan, R.; Bolan, N.S.; Tian, G.; Naidu, R.; Kunhikrishnan, A. Role of organic amendment application on greenhouse gas emission from soil. *Sci. Total Environ.* **2013**, *465*, 72–96.
14. Zhai, W.; Moschandreas, D.J., Tian, G.; Venkatesan, D.; Noll, K.E. Degradation rate model formulation to estimate soil carbon sequestration from biosolids application. *Soil Sci. Soc. Am. J.* **2013**, *78*, 238–247.
15. Sorenson, L.H. Carbon-nitrogen relationships during the humiliation of cellulose in soils containing different amounts of clay. *Soil Biol. Bio chem.* **1981**, *13*, 313–321.
16. USDA. *Soil Survey of Fulton County, Illinois*; U.S. Department of Agriculture: Washington, DC, USA, 1997.
17. Hillel, D. *Fundamentals of Soil Physics*, 2nd ed.; Academic Press Inc.: Orlando, FL, USA, 1980; pp. 123–124.
18. Liu, C.; Evett, J. Grain-size analysis of soil (including both mechanical and hydrometer analyses). In *Soil Properties: Testing, Measurement, and Evaluation*, 6th ed.; Prentice Gall: Upper Saddle River, NJ, USA, 2008; pp. 117–128.
19. Hamaker, J. Mathematical prediction of cumulative levels of pesticides in soil. In *Organic Pesticides in the Environment*, 1st ed.; American Chemical Society: Washington, DC, USA, 1966; pp. 124–131.
20. Loosdrecht, M.C.; Lyklema, J.; Norde, W.; Zehnder, A.J. Influence of interfaces on microbial activity. *Microbiol. Mol. Biol. Rev.* **1990**, *54*, 75–87.
21. Parton, W.J.; Ojima, D.S.; Cole, C.V.; Schimel, D.S. A general model for soil organic matter dynamics: Sensitivity to litter chemistry, texture and management. *Soil Sci. Am. J.* **1994**, *39*, 147–167.

22. USDA. *Textural Soil Classification*; U.S. Department of Agriculture: Washington, DC, USA, 1987.
23. Borkovec, M.; Wu, Q.; Degovics, G.; Laggner, P.; Sticher, H. Surface area and size distributions of soil particles. *Colloid Surface A* **1993**, *73*, 65–76.
24. Jastrow, J.D.; Amonette, J.E.; Bailey, V.L. Mechanisms controlling soil carbon turnover and their potential application for enhancing carbon sequestration. *Clim. Change* **2007**, *80*, 5–23.
25. Parton, W.J.; Flach, K.; Burke, I.C.; Cole, C.V.; Yonker, C.M. Texture, climate and cultivation effects on soil organic matter content in US grassland soils. *Soil Sci. Am. J.* **1989**, *53*, 800–805.
26. Gilmour, J.T.; Clark, M.D. Predicting long-term decomposition of biosolids with a seven-day test. *J. Environ. Qual.* **1996**, *25*, 766–770.
27. Paul, E.A. Dynamics of organic matter soils. *Plant Soil.* **1984**, *76*, 275–285.
28. Schimel, D.S.; Coleman, D.C.; Horton, K.C. Soil Organic matter dynamics in paired rangeland and cropland topo sequences in North Dakota. *Geoderma* **1985**, *36*, 201–214.
29. Schimel, D.S.; Parton, W.J. Microclimatic controls of nitrogen mineralization and nitrification in shortgrass steppe soils. *Plant Soil.* **1986**, *93*, 347–357.
30. Allison, F.E.; Sherman, M.S.; Pinck, L.A. Maintenance of soil organic matter. *Soil Sci.* **1949**, *68*, 463–478.
31. Manlay, R.J.; Feller, C.; Swift, M.J. Historical evolution of soil organic matter concepts and their relationships with the fertility and sustainability of cropping systems. *Agric. Ecosyst. Environ.* **2007**, *119*, 217–233.
32. Gerzabek, M.H.; Haberhauer, G.; Kirchmann, H. Soil organic matter pools and carbon-13 natural abundances in particle-size fraction of a long-term agricultural field experiment receiving organic amendments. *Soil Sci. Soc. Am. J.* **2001**, *64*, 352–358.

CHAPER 15

Dynamics of Soil Nitrogen Availability Indices in a Sandy Clay Loam Soil Amended with Animal Manures

Habeebah Adewunmi Saka[1], Jamiu Oladipupo Azeez[1], Joy Nwakaego Odedina[1] Shade John Akinsete[1]

[1]*Federal University of Agriculture Abeokuta, Abeokuta, Nigeria*

ABSTRACT

Purpose

Transformation of nitrogen (N) from different manure sources to available forms to promote food security in context of rising climate change is vital. Effect of manure (types, nutrients, high rates and application time) on soil N transformation requires further investigation. This study evaluated effects of three animal manures on soil N mineralization of sandy clay loam.

Methods

A 22-week field study in a Randomized Complete Block Design (three replicates) was conducted using dried Cattle, Goat and Poultry Manures (CGPM), applied at 5, 10, 20, 40, 60, 80, 120 and 150 t ha^{-1} once at onset of the study. Also, $N_{15}P_{15}K_{15}$ (0.4 t ha^{-1}) was incorporated in three splits of 2, 8 and 14 weeks after incorporation (WAI) of manures as reference. Soil NH_4^+–N and NO_3^-–N were determined bi-weekly.

Results

The results showed N release peaked at 10 WAI with highest NH_4^+–N (830, 400, 253 mg kg^{-1}) and NO_3^-–N (316, 398, 250 mg kg^{-1}) at 150, 60 and 60 t ha^{-1} for CGPM, respectively. Initial rapid N release phase occurred at 0–4 WAI, NH_4^+–N and NO_3^-–N increased averagely by 182, 183, 139% and 131, 175, 144% for CGPM, respectively. Declines occurred at 8, 14 and 22 WAI but reduction observed at 22 WAI tripled 8 and 14 WAI.

Conclusions

Application of NPK and all the manures provided highest soil NH_4^+–N and NO_3^-–N at 12 WAI. High rates of CGPM were not injurious to these soil properties; hence this has implication for N to maximize plants uptake as well as decrease losses to environment.

Keywords: Animal manures Rates NPK fertilizer and nitrogen mineralization

INTRODUCTION

Agricultural growth as a result of population increase had contributed to intensification of land cultivation; this causes severe soil fertility depletion. However, losses of nutrients in the soil have been replenished through the use of animal manures (Ramsasa 2010). Hence, the potential of animal manure use in soil replenishment has received the global attention. Currently, to many soil fertility difficulties, animal manures offer an affordable and readily available solution (Spore 2006) and are important source of Ca, Mg, S and micronutrients (Gomez-Brandon et al. 2013); they contain low and highly variable amounts of N, P and K. Aside from being source of plant nutrients, the physical properties of soil are also improved (Akanni and Ojeniyi 2008). However, the use of animal manures can pose a major problem of excess nitrate in soil (Azeez and Van Averbeke 2010b; Navarro Pedreno et al. 1996). The difficulty in the prediction of manure nutrients to plants is due to turnover processes and losses in manures (Azeez and Van Averbeke 2010a; Sorenson 2001). Therefore, incorporation of animal manures to supply nitrogen in proper dose at the proper time without posing any injury to soil and plants is very important in soil replenishment to support crop productivity. Hence, in order to predict the net mineralization of nitrogen in animal manures, transformation of nitrogen from organic into inorganic form and immobilization processes need to be understood. These processes are mediated through microbial activities and naturally, they are biochemical (Vel and Swamam 2013; Bartholomow 1965). Large variability in manure qualities (Chadwick et al. 2000) or types had posed difficulties for researchers in concluding on a specific rate of manure to support development of different crops. The consequence of these led to differences in the recommendations of 60 t ha^{-1} of cattle kraal manure for *Corchorus olitorius* L. (Masarirambi et al. 2012b), goat manure rate of 20 t ha^{-1} for sweet maize (Uwah and Eyo 2014); and poultry manure at the levels of 60 t ha^{-1} (Masarirambi et al. 2012a) and 30 t ha^{-1}(Enujeke et al. 2013) for lettuce and maize yield, respectively.

Animal manures at proper application time, adequate rates and proper nutrient contents when expanding the uses of organic matter in place of inorganic fertilizers for nitrogen mineralization to meet crop N requirement without adverse problem on soil, crop and the environment need to be investigated. This necessitated the study to (1) examine the release patterns of NH_4^+–N and NO_3^-–N from different rates of three animal manures and recommended inorganic fertilizer rate to soil, (2) evaluate the potential rates of mineralization of the three animal manures on soil and (3) determine the influence of high application rates of the three animal manures on soil.

MATERIALS AND METHODS

Location of the experiment

This research was conducted at the Teaching and Research Farm of the Federal University of Agriculture, Abeokuta, Ogun State, Nigeria, located between latitude 7°12′ N and longitude 3°20′ E (FUNAAB 2013). Irregularities of rainfall occurred during the period of the study (Table 1).

Table 1 Agrometeorological data for the period of the experiment

	2 WAI	4 WAI	6 WAI	8 WAI	10 WAI	12 WAI	14 WAI	16 WAI	18 WAI	20 WAI	22 WAI
Agrometeorological data for the period of the experiment											
Rainfall (mm)	119.4	33.3	8.8	15.1	62.5	147.8	3.5	NA	87.3	35.6	43.1
Mean Temp./day (°C)	27.9	28.1	26.9	27.5	47.3	54.6	39.1	24.9	26.1	26	26.9
Soil Temp./day (°C) (20 cm)	28.7	28.9	28.1	29.9	27.3	26.8	36.3	27.4	26.2	27.9	27.4
Average RH/day (%)	79.8	77.6	73.6	73.1	81.1	89.1	80.6	80.6	80.9	84.1	77.5
Mean sunshine (h/day)	5.5	6.7	5.2	5.4	3.7	2.7	2.9	3.4	3.1	4.1	5.2
Evaporation/day (mm)	3.1	4	3	2.9	1.5	NA	0.8	3.2	2.5	2.8	3.4

Source: Federal University of Agriculture, Abeokuta AGROMET STATION, May–October, 2013

WAI weeks after incorporation of manures, Soil temperature at the depth of 20 cm, *NA* not available, *Temp.* temperature, *mm* millimeter, *RH* relative humidity

Experimental design

The experiment was a randomized complete block design (RCBD) with three replicates. Cattle, goat and poultry manures were applied at the rates of 5, 10, 20, 40, 60, 80, 120 and 150 t ha^{-1}separately; these were used to test the pattern of manure mineralization irrespective of the rates, to test the manure-induced soil factors that affect plant growth and also evaluate the residual effect of manure rates on soil properties. NPK 15–15–15 fertilizer was used as check for each manure treatment and applied at the rate of 400 kg ha^{-1} (Schippers 2000). Starting from 2 weeks after incorporation of manures but a day after application of NPK fertilizer, soil samples were collected at every 2-week intervals from the net portion of each bed at a depth of 0–15 cm with the exception of 6 and 16 WAI due to rainfall failure. The samples were air dried, prepared for laboratory analysis and analyzed for NH_4^+–N and NO_3^-–N.

Soil and manures analyses

Initial soil sample and those collected bi-weekly were analyzed using the following methods. Soil EC and pH were extracted with 1:2 soil:water ratio and measured potentiometrically using HANNA 215 electrical conductivity meter (Richards 1954) and glass pH meter electrodes (HANNA pHep), respectively, (Van Reeuwijk 1993). The NH_4^+–N and NO_3^-–N were extracted in 2 Molar K_2SO_4 and determined colorimetrically using the method of Okalebo et al. (1993). Available P was extracted by Bray P-1 extraction (Bray and Kurtz 1965), obtained colorimetrically (Murphy and Riley 1962). Exchangeable bases were extracted with 1 N Ammonium-acetate solution in 1:10 soil solution ratio, K^+ and Na^+ were analyzed with flame photometer, while Ca^{2+} and Mg^{2+} with Atomic Absorption Spectrophotometer (AAS) (Anderson and Ingram 1993). Organic carbon (OC) content was determined by Walkley–Black method (Nelson and Sommers 1990) and particle size by hydrometer method (Bouyoucos 1965).

The cattle, goat and poultry manures used were digested with nitric per chloric acid (2:1) (Watanabe et al. 2013; Silva and QueirÓz 2002). The digests were analyzed for macronutrients and micronutrients using standard procedures (Kaira and Maynard 1991; Cater 1993). Distilled water was used for extraction to determine the electrical conductivity and pH of the manures.

Statistical analysis

Data collected were subjected to Analysis of Variance (ANOVA) using SAS (1999). Duncan Multiple Range Test at 5% level of probability was used to determine differences in the treatment rates means and also to show the significance effects on parameters measured in relation to the control.

RESULTS

Agrometeorological data for the experimental period

Agrometeorological average data for rainfall, relative humidity, mean temperature, soil temperature, mean sunshine hour and evaporation at 2-week intervals during the experiment were presented in Table 1.

Soil characteristics

The result revealed that the soil used for the research had EC value of 0.69 dS m^{-1}, pH of 7.6 and total N value of 0.8 g kg^{-1}. The NH_4^+–N, NO_3^-–N, OC and available P values were 0.013, 0.014, 14.2 g kg^{-1} and 7.5 mg kg^{-1}, respectively. Exchangeable bases; K^+, Na^+, Ca^{2+} and Mg^{2+}had the values of (0.42, 0.82, 10.96 and 1.34) $cmol^+$ kg^{-1}, respectively. The carbon to nitrogen ratio (C: N) was 11.83. The micro-nutrients; Mn^{2+}, Fe^{2+}, Cu^{2+} and Zn^{2+} had the values of (55.1, 9.85, 1.2 and 5.1) mg kg^{-1}, respectively. The percentages of sand, silt and clay were 770, 68 and 162 g kg^{-1}, respectively.

Manures characteristics

The EC values of the cattle, goat and poultry manures were 12.4, 8.8 and 16.4 dS m^{-1}, respectively, while the pH values of the cattle, goat and poultry manures were 7.9, 7.7 and 7.8, respectively. The equivalent amount of total N, NH_4^+–N, NO_3^-–N, P, K, Na, Ca, Mg and OC in cattle manure added to the soil between the rates of 5–150 t ha^{-1} were 225.5–6765, 2.1–63, 2.2–66, 16–480, 37–1110, 46.5–1395, 148–4440, 47.5–1425 and 894–26,820 kg, respectively. However, addition of goat manure between the rates of 5 and 150 t ha^{-1} is equivalent to adding total N, NH_4^+–N, NO_3^-–N, P, K, Na, Ca, Mg and OC of 241–7230, 1.6–48, 1.8–54, 3.5–105, 18.5–555, 23–690, 90.5–2715, 50–1500 and 1115–33,450 kg, respectively, to the soil. Hence, the equivalent amount of total N, NH_4^+–N, NO_3^-–N, P, K, Na, Ca, Mg and OC in the poultry manure applied to the soil at the rates ranged from 5 to 150 t ha^{-1} estimated to the amount of 139–

4170, 7.4–220.5, 7.5–223.5, 81–2430, 19.5–585, 24.5–735, 88.5–2655, 26.5–795 and 296.5–8895.0 kg, respectively (Tables 2, 3, 4).

Table 2 Equivalent amount of nutrients present in cattle manure rates added to the soil

Rates (t ha^{-1})	Total N	NH_4^+–N	NO_3^-–N	P	K	Na	Ca	Mg	OC
Cattle manure nutrients (kg)									
5	225.5	2.1	2.2	16.0	37.0	46.5	148.0	47.5	894.0
10	451.0	4.2	4.4	32.0	74.0	93.0	296.0	95.0	1788.0
20	902.0	8.4	8.8	64.0	148.0	186.0	592.0	190.0	3576.0
40	1804.0	16.8	17.6	128.0	296.0	372.0	1184.0	380.0	7152.0
60	2706.0	25.2	26.4	192.0	444.0	558.0	1776.0	570.0	10,728.0
80	3608.0	33.6	35.2	256.0	592.0	744.0	2368.0	760.0	14,304.0
120	5412.0	50.4	52.8	384.0	888.0	1116.0	3552.0	1140.0	21,456.0
150	6765.0	63.0	66.0	480.0	1110.0	1395.0	4440.0	1425.0	26,820.0

Total N total nitrogen, *K* total potassium, *NH_4^+–N* ammonium–nitrogen, *Na* total sodium, *NO_3N* nitrate–nitrogen, *Ca* total calcium, *P* total phosphorus, *Mg* total magnesium, *OC* organic carbon

Table 3 Equivalent amount of nutrients present in goat manure rates added to the soil

Rates (t ha^{-1})	Total N	NH_4^+–N	NO_3^-–N	P	K	Na	Ca	Mg	OC
Goat manure nutrients (kg)									
5	241.0	1.6	1.8	3.5	18.5	23.0	90.5	50.0	1115.0
10	482.0	3.2	3.6	7.0	37.0	46.0	181.0	100.0	2230.0
20	964.0	6.4	7.2	14.0	74.0	92.0	362.0	200.0	4460.0
40	1928.0	12.6	14.4	28.0	148.0	184.0	724.0	400.0	8920.0
60	2892.0	19.2	21.6	42.0	222.0	276.0	1086.0	600.0	13,380.0
80	3856.0	25.1	28.8	56.0	296.0	368.0	1448.0	800.0	17,840.0
120	5784.0	38.4	43.2	84.0	444.0	552.0	2172.0	1200.0	26,760.0
150	7230.0	48.0	54.0	105.0	555.0	690.0	2715.0	1500.0	33,450.0

Total N total nitrogen, *K* total potassium, *NH_4^+–N* ammonium–nitrogen, *Na* total sodium, NO_3^-–*N* nitrate–nitrogen, *Ca* total calcium, *P* total phosphorus, *Mg* total magnesium, *OC* organic carbon

Table 4 Equivalent amount of nutrients present in poultry manure rates added to the soil

Rates (t ha^{-1})	Total N	NH_4^+–N	NO_3^-–N	P	K	Na	Ca	Mg	OC
Poultry manure nutrients (kg)									
5	139.0	7.4	7.5	81.0	19.5	24.5	88.5	26.5	296.5
10	278.0	14.7	14.9	162.0	39.0	49.0	177.0	53.0	593.0
20	556.0	29.4	29.8	324.0	78.0	98.0	354.0	106.0	1186.0
40	1112.0	58.9	59.6	648.0	156.0	196.0	708.0	212.0	2372.0
60	1668.0	88.2	89.4	972.0	234.0	294.0	1062.0	318.0	3558.0
80	2224.0	117.6	119.2	1296.0	312.0	392.0	1416.0	424.0	4744.0
120	3336.0	176.4	178.8	1944.0	468.0	588.0	2124.0	636.0	7116.0
150	4170.0	220.5	223.5	2430.0	585.0	735.0	2655.0	795.0	8895.0

Total N total nitrogen, *K* total potassium, *NH_4^+–N* ammonium–nitrogen, *Na* total sodium, NO_3^-–*N* nitrate–nitrogen, *Ca* total calcium, *P* total phosphorus, *Mg* total magnesium, *OC* organic carbon

Effect of cattle, goat and poultry manures on soil NH_4^+–N and NO_3^-–N (mg kg^{-1})

Effect of cattle manure on soil NH_4^+–N (mg kg^{-1})

Table 5 reveals that at 2 WAI, soil applied with cattle manure rate of 60 t ha^{-1} gave highest value of 62.06 mg kg^{-1} in NH_4^+–N and lowest of 27.69 mg kg^{-1} at 80 t ha^{-1} by 171 and 21%, respectively, above the control, whereas percentage decreases of 13, 9 and 0.04% were recorded in soil with cattle manure rates of 120, 150 t ha^{-1} and NPK, respectively. But there were no significant differences in control and the amendment rates at 4 WAI. While at 8 WAI, soil with cattle manure rate of 120 t ha^{-1} had NH_4^+–N of 51.66 mg kg^{-1} with the highest increase of 98%, soil applied with 20 t ha^{-1} of cattle manure had 36.50 mg kg^{-1} of NH_4^+–N with the lowest increase of 40% in relation to control. The soil with cattle manure rate of 150 t ha^{-1}gave highest NH_4^+–N value of 830.4 mg kg^{-1} by 636% but 5 t ha^{-1} resulted in the least value of 221.4 mg kg^{-1}by 96% relative to control at 10 WAI. However, the amendment rates were not significantly different from control soil at 12 WAI. At 14 WAI, highest increase of 61% and lowest increase of 14% above the control were obtained from soil applied with cattle manure rates of 150 and 40 t ha^{-1} with NH_4^+–N values of 29.32 and 20.88 mg kg^{-1}, respectively. While 5, 10, 20 t ha^{-1}and NPK decreased by 38, 29, 31 and 72%, respectively relative to control. At 18 WAI, soil with cattle manure rate of 150 t ha^{-1} with NH_4^+–N value of 86.55 mg kg^{-1} recorded the highest increase of 290%, but 10 t ha^{-1} had 23.33 mg kg^{-1} and recorded the least increase of 5% over the control, while 5 t ha^{-1} with 21.11 mg kg^{-1} of NH_4^+–

N decreased by 5%. At 20 WAI, soil with cattle manure rate of 150 t ha^{-1} gave 49.53 mg kg^{-1} of NH_4^+–N and had the highest increase of 111% while the rate of 5 t ha^{-1} had 25.03 mg kg^{-1} of NH_4^+–N and recorded the lowest increase of 7% relative to control. But the rate of 60 t ha^{-1} with NH_4^+–N value of 5.26 mg kg^{-1} recorded the highest increase of 149%, while 20 t ha^{-1} and 120 t ha^{-1} with NH_4^+–N value of 2.99 mg kg^{-1}each had the least increase of 42% compared to control at 22 WAI. However, the level of soil NH_4^+–N decreased in relation to the corresponding rates at 20 WAI.

Table 5 Influence of cattle manure rates and NPK fertilizer (t ha^{-1}) on mean soil NH_4^+–N (mg kg^{-1}) at 2 weeks intervals

Rates (t ha^{-1})	Means of bi-weekly analyses of soil NH_4^+–N (mg kg^{-1}) with applied cattle manure								
	2 WAI	4 WAI	8 WAI	10 WAI	12 WAI	14 WAI	18 WAI	20 WAI	22 WAI
0	22.88^{b}	68.11^{a}	26.10^{f}	112.90^{f}	79.90^{a}	18.245^{bc}	22.19^{g}	23.50^{f}	2.11^{e}
5	21.60^{b}	69.31^{a}	38.30^{b}	221.40^{d}	93.50^{a}	11.315^{de}	21.11^{g}	25.03^{e}	3.81^{c}
10	31.96^{b}	74.26^{a}	47.40^{bc}	270.30^{c}	118.80^{a}	14.43^{cd}	23.33^{f}	26.69^{d}	4.81^{b}
20	29.61^{b}	78.80^{a}	36.50^{bc}	246.80^{d}	133.60^{a}	12.63^{cde}	24.44^{f}	25.06^{d}	2.99^{d}
40	37.45^{ab}	75.52^{a}	38.29^{bc}	246.90^{d}	121.00^{a}	20.88^{b}	26.61^{f}	38.14^{c}	3.53^{c}
60	62.06^{a}	71.32^{a}	49.10^{bc}	219.20^{c}	107.90^{a}	23.94^{b}	36.66^{e}	39.53^{c}	5.26^{a}
80	27.69^{b}	73.13^{a}	49.10^{bc}	202.30^{d}	122.60^{a}	22.67^{b}	42.16^{d}	40.00^{c}	3.49^{c}
120	20.03^{b}	88.39^{a}	51.66^{a}	408.60^{b}	142.40^{a}	25.20^{ab}	66.58^{b}	40.59^{c}	2.99^{d}
150	20.79^{b}	75.35^{a}	40.50^{bc}	830.40^{a}	149.90^{a}	29.32^{a}	86.55^{a}	49.53^{a}	3.39^{c}
0.4	22.87^{b}	73.25^{a}	37.90^{bc}	270.40^{c}	149.40^{a}	5.2^{e}	50.30^{c}	46.52^{b}	3.59^{c}

Means followed by the same letter(s) within the same columns do not differ significantly at $P < 0.05$ according to Duncan multiple range test

WAI weeks after incorporation of manures

Effect of goat manure on soil NH_4^+–N

Table 6 showed that aside the soil applied with goat manure rate of 150 t ha^{-1} that was significantly higher by 114% with NH_4^+–N of 48.27 mg kg^{-1} at 2 WAI, soil with other amendment rates were not statistically different from control. The soil with goat amendment rates followed similar trend as cattle treatment at 4 WAI. The soil NH_4^+–N increased in comparison with the corresponding rates at 2 WAI. At 8 WAI, soil with goat manure rate of 150 t ha^{-1} gave NH_4^+–N value of 44.17 mg kg^{-1} and recorded the highest increase of 53% while 20 t ha^{-1} had NH_4^+–N value of 34.60 mg kg^{-1} and with the least increase of 20% above the control. But reductions of 18 and 28% were obtained from 5 to 10 t ha^{-1}, respectively. The level of NH_4^+–N increased relative to corresponding rates at 4 WAI. The soil applied with goat manure rate of 60 t ha^{-1} resulted in NH_4^+–N value of 400.3 mg kg^{-1} with highest increase of 300% while 10 t ha^{-1} had 225 mg kg^{-1} and recorded the least increase of 126% relative to control at 10 WAI. But 12 WAI followed the same trend as soil applied with cattle amendment. Furthermore, soil NH_4^+–N decreased in comparison with the corresponding rates at 10 WAI. Highest increase of 24% was recorded from soil with goat manure rate of 150 t ha^{-1} which had NH_4^+–N value of 22.88 mg kg^{-1} at 14 WAI. The soil with rate of 60 t ha^{-1} resulted in 18.67 mg kg^{-1} and gave the lowest increase of 1% more than the control. However, 5, 10 t ha^{-1}and NPK reduced by 29, 13 and 70%, respectively. The level of soil NH_4^+–N decreased in relation to the corresponding rates at 12 WAI. However, soil applied with goat manure rate of 150 t ha^{-1} had highest NH_4^+–N value of 88.77 mg kg^{-1} by 300% while 5 t ha^{-1} resulted in 26.63 mg kg^{-1} and recorded the least increase of 20% at 18 WAI, compared with control. At 20 WAI, soil with goat manure rate of 120 t ha^{-1} recorded NH_4^+–N value of 47.76 mg kg^{-1} and had the highest increase of 105%. But the soil with goat manure rate of 5 and 10 t ha^{-1} with the NH_4^+–N value of 24.45 mg kg^{-1} each recorded the lowest increase of 5%. However, there was a decrease in soil NH_4^+–N level in relation to the corresponding rates at 18 WAI. Consequently, the soil with goat manure rate of 120 t ha^{-1} which gave NH_4^+–N value of 4.99 mg kg^{-1} recorded the highest increase of 141% at 22 WAI while 5 t ha^{-1} resulted in 3.04 mg kg^{-1} with the least increase of 47% relative to control. There were decreases in soil NH_4^+–N level compared with the corresponding rates at 20 WAI.

Table 6 Influence of goat manure rates and NPK fertilizer (t ha^{-1}) on mean soil NH_4^+–N (mg kg^{-1}) at 2 weeks intervals

Rates (t ha^{-1})	Means of bi-weekly analyses of soil NH_4^+–N (mg kg^{-1}) with applied goat manure								
	2 WAI	4 WAI	8 WAI	10 WAI	12 WAI	14 WAI	18 WAI	20 WAI	22 WAI
0	22.54^{b}	66.92^{a}	28.93^{d}	100.10^{i}	76.70^{a}	18.52^{b}	22.19^{j}	23.35^{f}	2.07^{c}
5	20.89^{b}	67.43^{a}	23.85^{f}	212.10^{g}	85.00^{a}	13.00^{cd}	26.63^{i}	24.45^{e}	3.04^{c}
10	21.50^{b}	75.99^{a}	20.75^{e}	225.70^{h}	93.50^{a}	16.12^{bc}	37.77^{h}	24.45^{e}	3.29^{c}
20	25.77^{b}	77.00^{a}	34.60^{c}	234.00^{f}	95.00^{a}	19.57^{ab}	47.71^{g}	29.79^{d}	3.95^{c}
40	27.57^{ab}	70.55^{a}	39.46^{b}	308.70^{c}	111.10^{a}	19.51^{ab}	49.93^{f}	30.89^{d}	3.28^{c}
60	27.75^{b}	77.68^{a}	37.87^{b}	400.30^{a}	112.20^{a}	18.67^{b}	68.80^{d}	30.89^{d}	4.12^{b}
80	27.98^{b}	80.27^{a}	37.58^{b}	298.00^{d}	134.60^{a}	19.42^{ab}	72.12^{c}	40.00^{c}	4.28^{b}
120	30.74^{b}	74.52^{a}	34.80^{c}	374.60^{b}	123.50^{a}	18.94^{b}	79.62^{b}	47.76^{a}	4.99^{a}
150	48.27^{a}	92.08^{a}	44.17^{a}	308.70^{c}	141.30^{a}	22.88^{a}	88.77^{a}	43.35^{c}	3.69^{c}
0.4	23.98^{b}	72.70^{a}	37.14^{b}	241.32^{e}	145.50^{a}	5.49^{e}	54.00^{e}	46.52^{b}	3.78^{c}

Means followed by the same letter(s) within the same columns do not differ significantly at $P < 0.05$ according to Duncan multiple range test

WAI weeks after incorporation of manures

Effect of poultry manure on soil NH_4^+–N

At 2 WAI, soil applied with poultry manure rate of 60 t ha^{-1} gave NH_4^+–N value of 50.86 mg kg^{-1} and recorded the highest increase of 153%. The rate of 10 t ha^{-1} had soil NH_4^+–N value of 20.18 mg kg^{-1} and had the least increment of 0.5% more than the control. But the values at 4 WAI followed similar trend with soil on which cattle and goat manures were incorporated. The soil NH_4^+–N increased compared to the corresponding rates at 2 WAI. Meanwhile, soil applied with poultry manure rate of 150 t ha^{-1} recorded highest soil NH_4^+–N value of 47.85 mg kg^{-1} by 90% at 8 WAI. However, soil on which NPK was added with NH_4^+–N value of 37.21 mg kg^{-1} recorded the least increase of 47% over the control. The level of soil NH_4^+–N decreased in relation to the corresponding rates at 4 WAI. More so, soil with poultry manure rate of 60 t ha^{-1} resulted in 253.4 mg kg^{-1} of NH_4^+–N and had the highest increase of 138%, while the rate of 5 t ha^{-1} with soil NH_4^+–N value of 110.7 mg kg^{-1} recorded the least increase of 4% relative to control at 10 WAI. The soil applied with poultry amendment rates were not differerent significantly from control at 12 WAI but soil NH_4^+–N level decreased relative to corresponding rates at 10 WAI. At 14 WAI, soil with poultry manure rate of 150 t ha^{-1}had NH_4^+–N value of 28.06 mg kg^{-1} and recorded the highest increase of 49% but 40 t ha^{-1}resulted in 19.86 mg kg^{-1} of NH_4^+–N with lowest increase of 6% more than the control. However, at 14 WAI the level of soil NH_4^+–N decreased in relation to the corresponding rates at 12 WAI. At 18 WAI, soil with poultry manure rate of 150 t ha^{-1} gave highest NH_4^+–N value of 73.23 mg kg^{-1} by 230% increase while 5 t ha^{-1} had least NH_4^+–N value of 23.27 mg kg^{-1} by 5% increase compared with control, whereas, 80 t ha^{-1} resulted in soil NH_4^+–N value of 52.44 mg kg^{-1} and had the highest increase of 127% more than the control at 20 WAI while 20 t ha^{-1} with soil NH_4^+–N value of 27.41 mg kg^{-1} recorded the least increase of 19%. At 22 WAI, soil with poultry manure at the rate of 150 t ha^{-1} with NH_4^+–N value of 4.45 mg kg^{-1}recorded the highest increase of 119% while 5 t ha^{-1} gave soil NH_4^+–N value of 2.08 mg kg^{-1}and recorded the least increase of 3% above the control. In addition, soil NH_4^+–N values decreased in comparison with the corresponding rates at 20 WAI, Table 7.

Table 7 Influence of poultry manure rates and NPK fertilizer (t ha^{-1}) on mean soil NH_4^+–N (mg kg^{-1}) at 2 weeks intervals

Rates (t ha^{-1})	Means of bi-weekly analyses of soil NH_4^+–N (mg kg^{-1}) with applied poultry manure								
	2 WAI	4 WAI	8 WAI	10 WAI	12 WAI	14 WAI	18 WAI	20 WAI	22 WAI
0	20.08^{b}	67.22^{a}	25.25^{d}	106.50^{i}	71.50^{b}	18.79^{b}	22.19^{j}	23.13^{g}	2.03^{e}
5	19.94^{b}	79.79^{a}	23.50^{d}	110.72^{h}	99.17^{ab}	12.89^{cd}	23.27^{i}	28.60^{e}	2.08^{e}
10	20.18^{b}	77.08^{a}	24.72^{d}	244.86^{b}	120.20^{ab}	15.29^{bcd}	28.82^{h}	28.60^{e}	2.64^{d}
20	29.40^{ab}	76.61^{a}	23.57^{d}	219.16^{d}	127.43^{ab}	16.54^{bc}	32.18^{g}	27.41^{f}	2.89^{cd}
40	34.26^{ab}	74.74^{a}	35.25^{c}	217.03^{e}	121.00^{ab}	19.86^{b}	35.51^{f}	27.44^{f}	3.01^{c}
60	50.86^{a}	71.19^{a}	47.35^{a}	253.37^{a}	144.01^{a}	20.45^{b}	45.59^{d}	33.23^{d}	2.23^{de}
80	41.82^{ab}	80.70^{a}	43.37^{b}	227.75^{c}	156.33^{a}	21.95^{ab}	38.84^{e}	52.44^{a}	2.44^{d}
120	46.06^{a}	87.39^{a}	46.43^{a}	212.78^{f}	140.80^{a}	26.54^{a}	65.55^{b}	48.97^{b}	3.28^{b}
150	43.58^{ab}	75.44^{a}	47.85^{a}	200.07^{g}	153.40^{a}	28.11^{a}	73.23^{a}	47.44^{c}	4.45^{a}
0.4	21.75^{b}	72.98^{a}	37.21^{c}	212.24^{f}	144.50^{a}	5.88^{e}	57.70^{c}	46.52^{c}	3.02^{c}

Means followed by the same letter(s) within the same columns do not differ significantly at $P < 0.05$ according to Duncan multiple range test

WAI weeks after incorporation of manures

Effect of cattle manure on soil NO_3^-–N

The soil applied with cattle manure rate of 60 t ha^{-1} resulted in NO_3^-–N value of 54.09 mg kg^{-1}at 2 WAI by 119% while the soil with NPK had the NO_3^-–N value of 25.98 mg kg^{-1} and recorded the least increase of 5% more than the control. The 4 WAI was in line with soil NH_4^+–N but the soil NO_3^-–N values increased compared with the corresponding rates at 2 WAI. However, at 8 WAI, soil with cattle manure at the rate of 120 t ha^{-1} gave NO_3^-–N value of 50.42 mg kg^{-1} with highest increase of 117% but 10 t ha^{-1} with soil NO_3^-–N value of 25.56 mg kg^{-1} recorded the least increase of 10% in relation to control. However, the soil NO_3^-–N values reduced in comparison with the corresponding rates at 4 WAI, Table 8. Moreover, soil with cattle manure rate of 150 t ha^{-1} with soil NO_3^-–N of 316.68 mg kg^{-1} resulted in highest increase of 201% at 10 WAI while 5 t ha^{-1} with 133.78 mg kg^{-1} had the least increase of 27% in relation to control. At 12 WAI, 80 t ha^{-1} with soil NO_3^-–N value of 139.60 mg kg^{-1} gave highest increase of 52% but 5 t ha^{-1} had soil NO_3^-–N value of 97.70 mg kg^{-1} and had lowest increase of 7% relative to control. However, soil NO_3^-–N decreased relative to the corresponding rates at 10 WAI. The soil with cattle manure rate of 120 t ha^{-1} gave highest NO_3^-–N value of 26.42 mg kg^{-1} by 32% at 14 WAI while 80 t ha^{-1} with the value of 24.28 mg kg^{-1} resulted in least increase of 21% relative to control. But in comparison with the corresponding rates at 10 WAI, soil NO_3^-–N decreased, Table 8. At 18 WAI, soil applied with cattle manure at the rate of 120 t ha^{-1} gave NO_3^-–N value of 58.28 mg kg^{-1} by 106%, but 5 t ha^{-1} with NO_3^-–N value of 31.4 mg kg^{-1} recorded the least increase of 11%, relative to control. Consequently, at 20 WAI, soil applied with NPK had NO_3^-–N value of 43.28 mg kg^{-1} and recorded the greatest increment in soil NO_3^-–N by 116% followed by 80 t ha^{-1} with NO_3^-–N value of 40.00 mg kg^{-1} that had 100% increases, while 10 t ha^{-1} with soil NO_3^-–N value of 29.24 mg kg^{-1} had the lowest increase of 44% more than control. There was a decrease in soil NO_3^-–N for cattle treated soil at 20 WAI compared to the corresponding rates at 18 WAI. However, at 22 WAI, soil appiled with cattle manure rates of 10 and 60 t $ha^{-1}$with the same NO_3^-–N value of 6.58 mg kg^{-1} gave the highest increase of 107% while 150 t $ha^{-1}$with NO_3^-–N value of 3.31 mg kg^{-1} gave the least increase of 4% in comparison with control. The soil NO_3^-–N decreased in relation to the corresponding rates at 20 WAI.

Table 8 Influence of cattle manure rates and NPK fertilizer (t ha^{-1}) on mean soil NO_3^-–N (mg kg^{-1}) at 2 weeks intervals

Rates (t ha^{-1})	Means of bi-weekly analyses of soil NO_3^-–N (mg kg^{-1}) with applied cattle manure								
	2 WAI	4 WAI	8 WAI	10 WAI	12 WAI	14 WAI	18 WAI	20 WAI	22 WAI
0	24.75^{b}	68.70^{a}	23.28^{d}	105.07^{f}	91.70^{b}	20.09^{bc}	28.25^{g}	20.00^{j}	3.18^{d}
5	21.81^{b}	75.61^{a}	29.17^{b}	133.78^{ef}	97.70^{b}	12.68^{d}	31.40^{fg}	29.24^{h}	4.50^{c}
10	35.95^{ab}	75.63^{a}	25.56^{cd}	179.56^{d}	99.28^{b}	13.89^{d}	38.51^{f}	28.74^{i}	6.58^{a}
20	31.29^{b}	71.55^{a}	26.00^{cd}	151.61^{e}	133.95^{a}	14.76^{d}	42.41^{e}	30.00^{g}	3.90^{c}
40	40.31^{ab}	73.86^{a}	27.24^{cd}	140.80^{e}	126.00^{a}	18.76^{cd}	45.31^{d}	37.66^{e}	4.64^{c}
60	54.00^{a}	69.75^{a}	48.69^{ab}	148.78^{e}	108.30^{ab}	24.78^{c}	48.39^{b}	38.95^{d}	6.58^{a}
80	30.96^{b}	76.65^{a}	47.09^{ab}	206.51^{c}	139.60^{a}	24.28^{c}	58.26^{a}	40.00^{b}	3.96^{c}
120	22.78^{b}	83.65^{a}	50.42^{a}	208.28^{c}	135.60^{a}	26.42^{a}	58.28^{a}	37.31^{f}	3.31^{d}
150	24.70^{b}	81.01^{a}	45.74^{ab}	316.68^{a}	123.40^{a}	25.74^{b}	54.19^{ab}	39.27^{c}	5.20^{b}
0.4	25.98^{b}	82.56^{a}	26.91^{cd}	267.77^{b}	145.50^{a}	10.76^{e}	46.82^{fg}	43.28^{a}	4.38^{c}

Means followed by the same letter(s) within the same columns do not differ significantly at $P < 0.05$ according to Duncan multiple range test

WAI weeks after incorporation of manures

Effect of goat manure on soil NO_3^-–N

At 2 WAI, soil applied with goat manure rate of 150 t ha^{-1} had NO_3^-–N value of 50.29 mg kg^{-1}which recorded the highest increase of 133% while 5 t ha^{-1} with soil NO_3^-–N value of 23.13 mg kg^{-1} recorded the least increase of 7% relative to control. At 4 WAI, the trend was in line with cattle manure treated soil as NH_4^+–N. However, there were increases of soil NO_3^-–N compared with the corresponding rates at 2 WAI. At 8 WAI, goat manure rate of 40 t ha^{-1} with NO_3^-–N value of 45.22 mg kg^{-1} recorded the highest increase of 102% but 10 t ha^{-1} with NO_3^-–N value of 24.95 mg kg^{-1} recorded the least increase of 11% above the control. In comparison of 8 WAI values with the corresponding rates at 4 WAI, soil NO_3^-–N decreased. The soil with goat manure rate of 60 t ha^{-1} which had NO_3^-–N value of 398.54 mg kg^{-1} recorded highest increase of 293%, 5 t ha^{-1} gave NO_3^-–N value of 227.02 mg kg^{-1} with the least increase of 124% in relation to control at 10 WAI. However, at 12 WAI, 80 t ha^{-1} resulted in soil NO_3^-–N value of 137.85 mg kg^{-1} and had highest increase of 51% but 5 t ha^{-1} with soil NO_3^-–N value of 100.70 mg kg^{-1} had lowest increase of 11% more than the control. The result revealed that soil NO_3^-–N decreased compared to the corresponding rates at 10 WAI. More so, at 14 WAI, 150 t ha^{-1} with soil NO_3^-–N value of 23.22 mg kg^{-1} was significantly higher while other goat manure treated soil rates decreased in soil NO_3^-–N compared with control. Consequently, the soil NO_3^-–N decreased relative to corresponding rates at 12 WAI. The soil applied with goat manure at the rate of 150 t ha^{-1} had NO_3^-–N value of 82 mg kg^{-1} and highest increase of 183%. The least increase of 4% was recorded from 5 t ha^{-1} with soil NO_3^-–N value of 30 mg kg^{-1}compared with control, at 18 WAI, while 120 t ha^{-1} with soil NO_3^-–N value of 44.69 mg kg^{-1}resulted in highest increment of 111% at 20 WAI. But 5 t ha^{-1} gave 28.75 mg kg^{-1} of NO_3^-–N and recorded lowest increase of 36% in relation to control. In comparison with the corresponding rates at 18 WAI, there were decreases in soil NO_3^-–N level. At 22 WAI, soil with goat manure rate of 120 t ha^{-1} had NO_3^-–N value of 5.79 mg kg^{-1} and gave highest increase of 78% but 5 t ha^{-1} with NO_3^-–N value of 3.34 mg kg^{-1} recorded least increase of 3% more than the control. The values also decreased compared to corresponding rates at 20 WAI, Table 9.

Table 9 Influence of goat manure rates and NPK fertilizer (t ha^{-1}) on mean soil NO_3^-–N (mg kg^{-1}) at 2 weeks intervals

Rates (t ha^{-1})	Means of bi-weekly analyses of soil NO_3^-–N (mg kg^{-1}) with applied goat manure								
	2 WAI	4 WAI	8 WAI	10 WAI	12 WAI	14 WAI	18 WAI	20 WAI	22 WAI
0	21.61^{b}	70.33^{a}	22.39^{d}	101.35^{f}	91.10^{d}	19.00^{ab}	29.00^{i}	21.22^{g}	3.26^{d}
5	23.43^{b}	78.52^{a}	25.49^{bcd}	227.02^{e}	100.70^{d}	14.68^{cde}	30.00^{h}	28.75^{f}	3.34^{d}
10	35.21^{ab}	70.40^{a}	24.95^{cd}	232.24^{e}	114.10^{b}	15.92^{cde}	31.00^{g}	30.00^{f}	4.27^{c}
20	30.42^{ab}	81.69^{a}	28.25^{c}	270.24^{d}	112.10^{bc}	18.14^{bcd}	49.00^{d}	30.08^{e}	5.34^{b}
40	27.64^{b}	74.28^{a}	45.22^{a}	309.51^{b}	113.50^{bc}	19.25^{ab}	36.00^{f}	33.55^{d}	5.55^{b}
60	27.11^{b}	83.51^{a}	33.84^{b}	398.54^{a}	115.80^{b}	18.95^{bc}	72.00^{b}	41.76^{c}	5.34^{b}
80	24.49^{b}	83.83^{a}	28.46^{c}	313.51^{c}	137.80^{a}	19.85^{ab}	70.00^{c}	40.00^{c}	5.41^{b}
120	31.26^{ab}	80.55^{a}	31.81^{bc}	317.41^{c}	132.70^{b}	17.56^{bcd}	72.00^{b}	44.69^{a}	5.79^{a}
150	50.29^{a}	84.04^{a}	31.47^{bc}	374.18^{b}	117.80^{b}	23.22^{a}	82.00^{a}	40.77^{c}	5.51^{b}
0.4	26.32^{b}	78.15^{a}	27.59^{c}	261.51^{d}	141.05^{a}	10.82^{e}	46.68^{e}	42.22^{b}	4.53^{c}

Means followed by the same letter(s) within the same columns do not differ significantly at $P < 0.05$ according to Duncan multiple range test

WAI weeks after incorporation of manures

Effect of poultry manure on soil NO_3^-–N

Table 10 presented that at 2 WAI, soil applied with poultry manure rate of 60 t ha^{-1} had NO_3^-–N value of 51.22 mg kg^{-1} and recorded the highest increase of 127%. The rate of 10 t ha^{-1} gave NO_3^-–N value of 23.54 mg kg^{-1} and recorded the least increase of 7% more than the control. The 4 WAI was also similar to others above. The soil with poultry manure rate of 120 t ha^{-1} with the value of 35.42 mg kg^{-1} recorded the highest increase of 59% in soil NO_3^-–N while 20 t ha^{-1}with NO_3^-–N value of 23.74 mg kg^{-1} had the least increase of 7% relative to control at 8 WAI. Soil NO_3^-–N decreased in comparison with the corresponding rates at 4 WAI. Furthermore, soil applied with NPK gave highest NO_3^-–N value of 261.76 mg kg^{-1} with increment of 151% while 5 t ha^{-1} with NO_3^-–N value of 114.46 mg kg^{-1} presented least increase of 10% compared with control at 10 WAI. However, soil with poultry manure rate of 80 t ha^{-1} with 153.75 mg kg^{-1}of NO_3^-–N resulted to highest increment of 68% but 5 t ha^{-1} with NO_3^-–N value of 99.49 mg kg^{-1} had lowest increase of 9% relative to control at 12 WAI. There were decrements at 12 WAI in relation to the corresponding rates at 10 WAI. More so, soil with poultry manure rate of 150 t ha^{-1} recorded NO_3^-–N value of 25.87 mg kg^{-1} and gave the highest increase of 35% at 14 WAI. But soil NO_3^-–N decreased in relation to the corresponding rates at 12 WAI. The rate of 120 t ha^{-1} recorded highest NO_3^-–N value of 46.55 mg kg^{-1} by 62% and the least increase of 8% was recorded from 5 t ha^{-1} with NO_3^-–N value of 30.42 mg kg^{-1} relative to control, at 18 WAI. Consequently, at 20 WAI, 80 t ha^{-1} presented soil NO_3^-–N value of 56.38 mg kg^{-1} and this resulted in highest increment of 178% while 5 t ha^{-1} had NO_3^-–N value of 30 mg kg^{-1} by 88% over the control. Increments in relation to 18 WAI were erratic. At 22 WAI, the rates of soil amended with poultry manure were not significantly different from control but NO_3^-–N level reduced when compared to the corresponding rates at 20 WAI.

Table 10 Influence of poultry manure rates and NPK fertilizer (t ha^{-1}) on mean soil NO_3^-–N (mg kg^{-1}) at 2 weeks intervals

Rates (t ha^{-1})	Mean of bi-weekly analyses of soil NO_3^-–N (mg kg^{-1}) with applied poultry manure								
	2 WAI	4 WAI	8 WAI	10 WAI	12 WAI	14 WAI	18 WAI	20 WAI	22 WAI
0	22.59^{d}	68.51^{a}	22.46^{d}	104.19^{g}	91.40^{g}	19.11^{b}	28.18^{j}	20.28^{j}	3.19^{a}
5	22.38^{d}	72.95^{a}	24.23^{cd}	114.46^{f}	99.49^{f}	12.08^{e}	30.42^{i}	30.00^{i}	5.59^{a}
10	23.54^{d}	82.25^{a}	24.88^{cd}	239.48^{b}	145.95^{bc}	12.12^{d}	30.96^{h}	30.48^{h}	4.16^{a}
20	28.91^{bcd}	76.99^{a}	23.94^{c}	230.91^{b}	123.60^{de}	13.79^{d}	35.63^{g}	38.77^{f}	4.12^{a}
40	46.00^{abc}	68.69^{a}	25.55^{c}	240.08^{b}	131.15^{d}	17.87^{cd}	37.37^{f}	42.08^{e}	3.55^{a}
60	51.22^{a}	65.63^{a}	24.98^{cd}	250.14^{a}	144.35^{c}	16.50^{cd}	44.12^{d}	49.64^{c}	2.96^{a}
80	43.11^{abcd}	77.76^{a}	28.96^{b}	228.89^{c}	153.75^{a}	16.28^{cd}	40.22^{e}	56.38^{a}	3.18^{a}
120	47.37^{ab}	84.95^{a}	35.42^{a}	217.42^{d}	121.90^{e}	17.19^{cd}	46.55^{a}	54.19^{b}	4.73^{a}
150	37.65^{abcd}	75.35^{a}	28.05^{b}	211.43^{e}	145.95^{bc}	25.87^{a}	45.66^{b}	31.32^{g}	2.37^{a}
0.4	24.76^{cd}	84.32^{a}	28.96^{b}	261.76^{a}	145.40^{bc}	10.34^{f}	45.23^{c}	42.81^{d}	4.35^{a}

Means followed by the same letter(s) within the same columns do not differ significantly at $P < 0.05$ according to Duncan multiple range test

WAI weeks after incorporation of manures

DISCUSSION

Soil characteristics

The soil used for the research was sandy clay loam (USDA 2010), slightly saline (LAS 2014), and slightly alkaline (Pam and Brian 2007) and these could support plant performances. More so, the soil was deficient in total nitrogen (McBride 2015; USDA-SCS 1974) and had low available P content (ENDMEMO 2015; Mallarino 2000) which could retard crop development. It contained moderate OC (McBride 2015; USDA-SCS 1974), exchangeable K^+ and Mg^{2+} (Pam and Brian 2007); these were optimal for performance of the crop. The high exchangeable $Na^+$and Ca^{2+} contents (Pam and Brian 2007) could be attributed to the decomposition of the organic matter content of the soil due to high temperature, evaporation rate, sunshine hours/day and low rainfall during the research. This confirmed the studies of some researchers (Davidson and Jannssens 2006; Friedlingstein et al. 2006). The NH_4^+–N and NO_3^-–N contents were low and C: N ratio was considered normal (Hill 2001) and could be as a result of low rainfall during this period. The copper, iron and zinc contents were very high (Enwezor et al. 1989).

Manure characteristics

The EC of the three manures applied were very strongly saline (LAS 2014). This could be as a result of high concentration of cationic salts in the manures. More so, the highest EC value of poultry manure compared with the cattle and goat manures support the findings of Azeez and Van Averbeke (2010a, b) and could be attributed to the highest NH_4^+–N, NO_3^-–N and total P contents in poultry manure compared to others. The pH of the pure goat and poultry manures was considered mildy alkaline, whereas that of cattle manure was moderately alkaline (Pam and Brian 2007). This would allow for nutrients availability in the soil after mineralization. It was observed from the result that cattle manure had the highest total K, Na and Ca over the goat and poultry manures. However, goat manure was higher in total N, Mg and OC in relative to cattle and poultry manures. Very low value of total P in goat manure could be a result of the nutrient concentrations of the feed given to the animals. This is because they were fed with elephant grasses and wheat offal under intensive system although the nutrient contents of the feed were not measured. Equivalent amount of total N contents present in goat manure applied was higher than cattle and poultry manures. High P content in poultry manure could also contribute to the yield quality of crops at lower rates than cattle and goat manures. Cattle manure recorded highest equivalent amount of pH, total K, Na and Ca than goat and poultry manures; this could be attributed to the higher EC value in cattle manure than goat manure and confirmed the report of Monica (2013). Highest contents of total N, Mg and OC in goat manure compared with cattle and poultry manures suggested the reason for lowest value of EC in goat manure applied during the studies. This is because Mg was the only salt forming cation that was highest in goat manure. Highest total N could be a result of N level in the feed given to the goat animals, although the nutrient contents of the feed were not determined.

Manure mineralization processes in the soil

Incorporation of cattle, goat and poultry manures increased the concentration of NH_4^+–N; this supports the findings of Sajal and Abul Kashem (2014) and NO_3^-–N depending on the rates (Eghball et al. 2002) compared with control. However, the low values of NH_4^+–N between 0 and 8 WAI of the manures could be attributed to conversion of NH_4^+–N to NO_3^-–N as supported in the study of (Hoskins 2015) that, between 2 and 4 weeks, NH_4^+–N converted relatively quickly to NO_3^-–N in a soil applied with broiler manure. However, mineralization of organic N into inorganic form of NH_4^+–N and NO_3^-–N occurred between 8 and 14 WAI of manures. Consequently, the low NH_4^+–N between 18 and 22 WAI could be attributed to ammonia volatilization (Meissinger and Jokela 2000) while low NO_3^-–N contents between 18 and 22 WAI could be attributed to leaching (Tom 2002; Marco et al. 2002) and plant root uptake. The low NO_3^-–N contents in soil applied with NPK fertilizer at 18 WAI compared with 8 and 14 WAI could be attributed to acidic properties of NPK fertilizer, since acidity reduces soil microbial activity and this would in turn reduce nutrient mineralization (Xu et al. 2002).

Mineralization of NH_4^+–N and NO_3^-–N contents occurred in the soil with the maximum values at 10 WAI for cattle, goat and poultry manures but was contrary to the results of Azeez and Van Averbeke (2010b) that sharp increase in N release was at 120 days; and (Ayeni 2011; Ayeni and Adeleye 2011), that release of poultry manure nutrients proportion is between 1 and 2 months of incubation. The variability of mineralization processes could be attributed to microbial activities as suggested by Vinten et al. (2002); and climatic variations. Hence, the turnover processes and losses have influence on availability of nutrients in manures; this is why the prediction of manure nutrients to plants is problematic (Azeez and Van Averbeke 2010b; Sorenson 2001).

Very low N mineralization at 14 WAI compared to 12 WAI could be attributed to very low rainfall during this period. This corroborated the report of Jonathan (2006) that in dry soils, N mineralization is low because soil micro-organism activity is limited by water availability. More so, low NH_4^+–N suggested to be as a result of volatilization of NH_3 as this is favored by warm temperature wet soils under drying conditions (Tom 2002; James 2001). While low NO_3^-–N could be as a result of leaching that might have occurred at 12 WAI due to very high rainfall at this period, this also corroborated the work of Tom (2002) that NO_3 in the soil can be lost through percolation of water below the active root zone.

Across weeks after incorporation of the three manures, application rates of 20 t ha^{-1} for cattle, 40 t ha^{-1} for goat and poultry manure at the rate of 10 t ha^{-1} were not widely significantly different when compared to higher rates. Therefore, addition of manures more than these rates to the soil could lead to wastage as these did not increase the amount of NH_4^+–N and NO_3^-–N in the soil sequentially according to this study. Although the high rates of cattle, goat and poultry manures were significantly higher relative to lower rates in this research, these suggested not to be encouraged as the increments were not regular across the weeks after manure application. The irregularity could be attributed to volatilization (Meissinger and Jokela 2000) and leaching (Tom 2002; Marco et al. 2002) of the NH_4^+–N and NO_3^-–N in the soil, respectively.

During this work, NPK fertilizer applied was able to supply equal amount of NH_4^+–N to the soil when compared with 10 t ha^{-1} of cattle manure treatment between 2 and 12 WAI. But goat manure treated soil at 60 t ha^{-1} supplied equal amount of NH_4^+–N compared to NPK fertilizer between 2 and 8 WAI, while 40 t ha^{-1} of poultry manure was able to supply equal amount of NH_4^+–N to the soil compared to NPK fertilizer at 0.4 t ha^{-1} between 2 and 8 WAI. This suggested that cattle, goat and poultry manures incorporated to the soil at 10, 60 and 40 t ha^{-1}, respectively, supplied equal amount of NH_4^+–N compared to NPK fertilizer at 0.4 t ha^{-1} within 6 weeks. The low NH_4^+–N and NO_3^-–N contents of the soils applied with NPK fertilizer at 14 WAI compared with cattle, goat and manure treated soils rates could be attributed to acidic properties of NPK fertilizer. This corroborated the study results of Xu et al. (2002) that acidity reduces soil microbial activity resulting in reduced nutrient mineralization. This could also be attributed to depletion of soil organic matter as a result of synthetic fertilization (Jonathan 2006).

CONCLUSION AND RECOMMENDATION

Generally, addition of NPK 15–15–15 fertilizer, cattle, goat and poultry manures improved the soil NH_4^+–N and NO_3^-–N compared with control. The NH_4^+–N and NO_3^-–N contents of the soil between 2 and 8 WAI increased with the application of cattle, goat and poultry manures. However, between 8 and 14 WAI mineralization of NH_4^+–N and NO_3^-–N occurred but peak values were obtained at 10 WAI irrespective of the manures types and rates. Finally, the effect of cattle, goat and poultry manures on NH_4^+–N and NO_3^-–N decreased between 18 and 22 WAI.

To increase soil NH_4^+–N and NO_3^-–N for optimum production of crops yield, application of cattle, goat and poultry manures is recommended. It is recommended that application rate of 20 t ha^{-1} for cattle, 40 t ha^{-1} for goat and poultry manure at the rate of 10 t ha^{-1} would provide optimum and improve the soil nutrients level which will be retained in the soil for longer period. To improve soil with low N contents, poultry manure is recommended. Meanwhile, addition of cattle, goat and poultry manures to supply NH_4^+-N and NO_3^--N to the soil with regard to sowing dates is recommended to be 2 WAI as this would allow equilibration of the manures with the soil prior to planting, while incorporation of these three manures for crop needs is recommended to be 8 WAI.

ACKNOWLEDGEMENTS

The assistance of people that contributed to the success of this research is gratefully appreciated while the expressions, views and conclusions attained are for the authors.

REFERENCES

1. Akanni DI, Ojeniyi SO (2008) Residual effect of goat and poultry manures on soil properties nutrient content and yield of amaranthus in southwestern Nigeria. Res J Agric 2(2):44–47
2. Ayeni LS (2011) Integrated plant nutrition management: a panacea for sustainable crop production in Nigeria. Int J Soil Sci 6:19–24
3. Ayeni LS, Adeleye EO (2011) Soil nutrient status and nutrient interactions as influenced by agro wastes and mineral fertilizer in an incubation study in the South—west Nigeria. Int J Soil Sci 6:60–68. doi:10.3923/ijss.2011.60.68
4. Azeez JO, Van Averbeke W (2010a) Fate of manure phosphorus in a weathered sandy clay loam soil amended with three animal manures. Biores Technol 101:6584–6588. doi:10.1016/j.biortech.2010.03.073
5. Azeez JO, Van Averbeke W (2010b) Nitrogen mineralization potential of three animal manures applied on a sandy clay loam soil. Biores Technol 101:5645–5651. doi:10.1016/j.biortech.2010.01.119
6. Bartholomow WW (1965) Mineralization and immobilization of nitrogen in the decomposition of plant and animal residues. In: Bartholomow WW, Clark FE (eds) Soil nitrogen. American Society of Agronomy, Madison, pp 285–306
7. Bouyoucos GH (1965) A calibration of the hydrometer method for testing mechanical analysis of soils. Agric J 43:434–438
8. Bray RH, Kurtz LT (1965) Determination of total organic and available forms of phosphorus in soil. Soil Sci 59:225–624
9. Cater MR (1993) Soil sampling and methods of analysis. Canadian Society of Soil Science, Ontario
10. Chadwick DR, John F, Pain BF, Chambers BJ, Williams JC (2000) Plant uptake of nitrogen from the organic nitrogen fraction of animal manures: a laboratory experiment. J Agric Sci Camb 134:159–168
11. Eghball B, Wienhold BJ, Gilley JE, Eigenberg RA (2002) Mineralization of manure nutrients. J Soil Water Conserv 57(6):470–473
12. ENDMEMO (2015) Conversion of part per million to milligram per kilogram. Complete concentration percentage unit conversions. (Online). http://www.endmemo.com/sco
13. Enujeke EC, Ojeifo IM, Nnaji GU (2013) Residual effects of organic manure and inorganic fertilizer on maize grain weight and some soil properties in Asaba Area of Delta State. ISSN 2250-3579. Int J Adv Bio Res 3(3):433–442

14. Enwezor WO, Udo EJ, Usoroh NJ, Ayotade KA, Adepetu JA, Chude VO, Udegbe GJ (1989) Fertilizer use and management practices for crop in Nigeria (eds). Federal Ministry of Water Resources and Rural Development (FMAWRRF) Lagos, pp 52–56
15. FUNAAB (2013) Record of daily agrometeorological observation. Department of Water Resources Management and Agrometeorology. College of Environmental Management. Federal University of Agriculture, Abeokuta, pp 1–12
16. Gomez-Brandon M, Fernandez-Delgado MJ, Domiguez J, Insam H (2013) Animal manures: recycling and management technologies. INTECH Open Sci. doi:10.5772/53454
17. Hill laboratories (2001) Soil tests and interpretation. Technical Notes. http://www.hill- laboratories.com, vol 4, 3196, p 9
18. Hoskins B (2015) Nutrient content, availability, and release rates from natural fertilizers. University of Maine. http://www.newengladvfc.org/2015_conference/2, pp 1–3
19. James JC (2001) Nitrogen in soil and fertilizer. First Published in the: Science Turfgrass Foundation News 8(1):6–10
20. Jonathan D (2006) Nitrogen mineralization potential in important agricultural soils of Hawai, i. Cooperative extension service. Soil Crop Managt 15:1–5
21. Kaira YP, Maynard DC (1991) Method manual for forest soil and plant analysis. For, Can., Northwest Reg., North. For, Cent., Edmonton, Albert Inf. Rep. NOR-X-319
22. LAS (2014) soluble salts or electrical conductivity of soils and green house media. Litchfield Analytical Services, pp 1–2
23. Mallarino A (2000) Soil testing and available phosphorus. Integrated crop management. LOWA State University. University Extension, pp 164–166
24. Marco J, Vlado K, Hrvoje P, Jasna S, Mitra R (2002) Influences of Nitrogen fertilizer and irrigation on nitrogen leaching. In proceedings of the second international conference on sustainable agriculture for food, energy and industry, pp 1641–1647
25. Masarirambi MT, Dlamini P, Wahome PK, Oseni TO (2012a) Effects of chicken manure on growth, yield and quality of lettuce (*Lactuca sativa* L.) "Taina" under a lath house in a semi-arid sub-tropical environment. American- Eurasian. J Agric Environ Sci 12(3):399–406
26. Masarirambi MT, Sibandze N, Wahome PK, Oseni TO (2012b) Effects of kraal manureapplication rates on growth and yield of wild okra (*Corchorus olitorius* L.) in a sub-tropical environment. Accepted: November, 29, 2011. ISSN: 2041–3890. Asian J Agric Sci 4(1):89–95
27. McBride C (2015) How to calculate percentages in grams. Demand Media
28. Meissinger JJ, Jokela WE (2000) Ammonium volatilization from dairy and poultry manure. Natural Resource, Agriculture, and Engineering Service. http://www.nraes.org, pp 1–2
29. Monica ZB (2013) Water and soil characterization. pH and electrical conductivity. Microbial life. Educational Research. Montana State University, Bozeman, pp 1–5
30. Murphy J, Riley JP (1962) A modified single solution method for determination of phosphate in natural waters. Anal Chem 27:31–36
31. Navarro Pedreno J, Moral R, Gomez I, Mataix J (1996) Reducing nitrogen losses by decreasing mineral fertilization in horticultural crops in eastern Spain. Agric Ecosyst Environ 59:217–221
32. Okalebo JR, Gathua KW, Woomer PL (1993) Laboratory methods of soil and plant analysis. A working manual. Soil Science Society of East Africa (SSSEA). Technical Publication 1
33. Pam H and Brian M (2007) Interpreting soil test results. What Do all the numbers mean? Pam Hazelton and NSW Department of Natural Resources, pp 1–66
34. Ramsasa C (2010) How to replenish the nutrients in the soil of your garden. Ezine Article. http://EzineArticles.com
35. Sajal R, Abul Kashem M (2014) Effects of organic manures in changes of some soil properties at different incubation periods. Open J. Soil Sci. 4(3):81–86. doi:10.4236/ojss.2014.43011

36. Schippers RR (2000) African indigenous vegetables—an overview of the cultivated species, pp 103–118
37. Silva DJ, QueirÓz AC (2002) Analse de alimentos: metodos quimicos e biologicos, 3 (ed) Vicosa: Imprensa Universitaria/UFV
38. Sorenson P (2001) Short-term nitrogen transformations in soil amended with animal manure. Soil Biol Biochem 33:1211–1216
39. Spore (2006) Issue 123, June 2006, p 13
40. Tom D (2002) Nitrogen Sources. University of Nebraska Lincoln Extensoin. 402. Available through: 441–7180. http://Lancaster.uni.edu, p 288
41. USDA (2010) Soil texture calculator. United State Development of Agriculture Natural Resources Conservation Service Soils. http://www.nrcs.usda.gov/wps
42. USDA–SCS (1974) Total nitrogen and organic carbon ratings and interpretation. Interpretation Guides for Evaluating Analytical Data. USDA–SCS
43. Uwah DF, Eyo VE (2014) Effects of number and rate of goat manure application on soil properties, growth and yield of sweet maize (*Zea mays* L. saccharata strut). Sustainable Agriculture Research; ISSN 1927-050X. Accepted: September 2, 2014. doi:10.5539/sar.v3n4p75
44. Van Reeuwijk LP (1993) Procedure for soil analysis. International Soil Reference and Information Centre Technical paper, pp 9
45. Vel AM, Swamam TP (2013) Nitrogen release pattern from organic manures applied to an acid soil. J Agric Sci 5(6):174–184. ISSN 1916–9760. doi:10.5539/jas.v5n6p174
46. Vinten A, Whitmoore AP, Bloem J, Howard R, Wright F (2002) Factor affecting nitrogen immobilization or mineralization kinetics for cellulose-, glucose-, and straw amended sandy soils. Biol Fert Soils 36:190–199
47. Watanabe PH, Thomas MC, Pascoal LAF, Ruiz, US, Daniel E, Amorim AB (2013) Manure production and mineral excretion in faeces of gilts fed ractopamine. Acta Scientiarum. Animal Sciences. Maringa. Received on September 19, 2012. Accepted on December 14, 2012, v35n3, 267–272. http://www.redalyc.org /articulo.oa?id=303128706007; http://www.ehow.com/how, pp 1–6
48. Xu RK, Coventry DR, Farhoodi A, Schultz JE (2002) Soil acidification as influenced by crop rotations, stubble management, and application of nitrogenous fertiliser, Tarlee, South Australia. Aust J Soil Res 40:483–496

CHAPTER 16

Response of Okra (*Abelmuschus Esculenthus* L. Moench) to Combined Organic and Inorganic Foliar Fertilizers

A. I. Afe[1], F. Oluleye[1]

[1] *Department of Crop Production, Kwara State University,Malete, Nigeria*

ABSTRACT

Aim

This study investigated the effects of combined organic poultry manure (PM) and inorganic foliar fertilizer (FF) on growth and yield of okra (*Abelmuschus esculenthus* L. Moench).

Materials and methods

Field trials were carried out at Shao and Teaching and Research Farm of Kwara State University, Malete, respectively, in 2013 and 2014 cropping seasons. The treatment consisted of PM at 10.0 t/ha, combinations of FF with 10.0, 7.5, 5.0, or 2.5 t/ha, FF alone and control without fertilizer arranged in a randomized complete block design and replicated three times.

Results

Combinations of PM and FF influenced the growth and fruit yield of okra. PM at 10.0 t/ha plus FF produced the highest plant height of 48.40 and 58.50 cm in 2013 and 2014, respectively. Also, highest number of fruits per plant and fruit yield was obtained at the same treatment over the two years. The control, without fertilizer, showed significantly longer days to flower compared with other treatments in the two years.

Conclusion

The highest yield obtained at PM 10.0 t/ha +FF was not significantly different from PM 7.5 t/ha+FF. Hence, poultry manure applied at 7.5t/ha+FF is therefore recommended.

Keywords: Foliar fertilizer Okra fruit yield.Poultry manureSoil-appliedYield component

INTRODUCTION

Sustainable soil productivity is one of the major constraints of tropical agriculture. Continuous cultivation is a common practice by poor resource farmers in the tropics. This has resulted in rapid decline of soil nutrients and unstable soil microbial population (ECA 2001). Consequently, there is decline productivity, low farmers income, and increasing poverty (FAO 2006). Hence, the use of organic manure as alternative soil amendment strategy for soil nutrient management has been advocated (Shehu et al. 1997).

Fertilizer plays a major role among cultural practices for increased crop production. However, blanket application of inorganic fertilizer to farmland soils without adequate knowledge of the nutrient status, often leads to increased soil acidity, particularly when nitrogen fertilizers are applied (Akande et al. 2010). Although, inorganic fertilizers, contain high concentrations of nutrients that are rapidly available and released for plant uptake, their use is limited due to scarcity, high cost, nutrient imbalance, and soil acidity (Akande et al. 2010).

Manure provides necessary macro- and micro-nutrients in available form, and improves the physical and biological properties of the soil (Abou El-Magd et al. 2006). Poultry manure (PM) is an excellent source of organic manure. It supplies both macro and micro-nutrients during mineralization, increases the organic matter content of the soil, and consequently enhances the texture, structure, aeration, moisture holding capacity, nutrient retention and water infiltration in the soil (Akinrinde et al. 2006; Dekissa et al. 2008). According to Garge and Bahla (2008), PM supplies phosphorous more readily to plants than other organic sources. In another study, Abd El-Kader et al. (2010) reported that manure from poultry increased okra yield and water use efficiency than composted plant residue. Additionally, abattoir waste was found as an excellent organic fertilizer in the cultivation of vegetables and also could promote healthy environment (Roy et al. 2015).

Foliar fertilizers (FF) are applied to the foliage of the plant to boost the nutrient density in the crop. It is a convenient way of applying nutrients to the plant to supplement absorbed soil nutrients. It improves the efficiency of soil-applied nutrients and also acts as a catalyst in the uptake and use of certain macro nutrients (Philips 2004; Fageria et al. 2009). Besides being environmentally safe, it increases crop yield and quality (Fageria et al. (2009). As such, El-Aal et al. (2010), Zodape et al. (2011) have recommended the technique in integrated nutrient management. Paliwal et al. (1999) had earlier reported enhanced development of branches per plant in okra due to foliar fertilization. Similarly, Abasi et al. (2010) reported improved days to 50% flowering, plant height number of branches and fruits per plant and fruit yield in okra using combinations of foliar fertilizer and soil-applied inorganic fertilizer in okra.

Boost Xtra is one of the foliar fertilizers that have gained popularity among farmers in Nigeria. It is formulated as soluble granules containing 20% each of nitrogen, phosphorous and potassium along with certain micro-nutrients. Farmers use this fertilizer alone or in combination with soil-applied inorganic fertilizer on various crops to improve growth and yield. However, there is a paucity of information on the combinations of foliar fertilizer and organic manure usage for vegetable crop production. Therefore, this study was initiated to investigate the effects of combined PM and FF on growth and yield of okra.

MATERIALS AND METHODS

Field trials were carried at Shao and the Teaching and Research Farm of Kwara State University, Malete, respectively, in 2013 and 2014 cropping seasons. The two locations are in Moro Local Government Area of Kwara

State, Nigeria. The PM was collected from deep liter system of poultry unit in the University Farm and the boost Xtra purchased from the agro chemical outlet in Ilorin, Nigeria. The treatments consisted of PM at 10.0 t/ha, combinations of FF with 10.0, 7.5, 5.0, or 2.5 t/ha, FF alone and control without fertilizer arranged in a randomized complete block design and replicated three times. The nutrient compositions of the PM and FF are shown in Tables 1 and 2.

Table 1 Nutrient composition of foliar fertilizer (Boost Xtra)

Analysis (wt/vol.)		
Nitrogen	N	20%
Phosphorous	P	20%
Potassium	K	20%
Magnesium	Mg	1.5%
Iron (EDTA)	Fe	0.35%
Manganese EDTA	Mn	0.075%
Copper EDTA	Cu	0.075%
Zinc EDTA	Zn	0.075%
Boron	B	0.0315%
Cobalt EDTA	Co	0.0012%
Molybdenum	Mo	0.0012%
pH (10% solution)	4.0–4.5	
Density	1.5SG@18C	

Table 2Physio-chemical properties of the experimental field (0–15 cm) and nutrient composition of the poultry manure

Soil composition			Poultry manure
Physical properties	Shao	Malete	
Sand (%)	55.28	53.55	
Silt (%)	18.45	19.64	
Clay (%)	28.44	30.23	

Textural class	Sandy-loam		
Chemical properties			
pH (H_2O)	5.88	5.72	5.83
Organic carbon	1.05	1.22	7.10
Available P mg kg^{-1}	6.76	6.85	5.21
Total N	0.18	0.92	8.05
Exchangeable bases			
Ca (mg kg^{-1})	2.62	2.48	0.13
Na (mg kg^{-1})	0.55	0.39	0.22
Mg (mg kg^{-1})	1.15	2.01	0.88
Mn (mg kg^{-1})	0.92	0.88	36.58
Zn (mg kg^{-1})	2.36	2.61	28.94
Cu (mg kg^{-1})	0.33	0.29	4.10
Fe (mg kg^{-1})	3.24	2.98	3.72
K (mg kg^{-1})	0.92	0.88	

Cultural practices

The land was plowed once and harrowed twice. Planting was carried out at the two locations on the flat. Each plot size measured 2.5 m × 3.5 m (8.75 m^2) with 0.5 m between plots and 1.0 m between blocks. Four seeds were planted at inter and intra row spacings of 0.6 and 0.3 m, respectively. Crops were thinned at two weeks after planting (WAP) to two plants per stand. Air-dried PM was applied uniformly at two weeks before planting. Foliar fertilizer was split applied at four and five WAP using knapsack sprayer. Pendimethalin [N-(1-ethylpropyl)-3, 4dimethyl-2, 6 dinitrobenzene amine] was applied two days after planting at the rate of 1.5 kg active ingredients per hectare. This was followed by manual weeding at 5 WAP to keep the experimental plot weed free. Insect pests were controlled with cyalothrin [3-(2-chloro-3,3,3,-trifluoropropenyl)-2.2, dimethylclopropanecarboxylate] at the rate of 1 ml of the product per liter of water using knapsack sprayer.

Data collection

Plant height at fruiting, stem girth, number of fruits per plant, fruit length, fruit circumference, and fresh fruit weight at each harvest were measured and days to 50% flowering was observed. Fruits were harvested at four days interval for eight and nine consecutive periods in 2013 and 2014, respectively.

Statistical analysis

Data were subjected to analysis of variance using SAS (1989) package and treatment means compared using Duncan's Multiple Range Test. This was done where results were significant at 5% level of probability.

RESULTS

Soil and poultry manure analysis

Soils at the two experimental sites were sandy-clay loam, slightly acidic with pH 5.88 and 5.72, total nitrogen, 0.18 and 0.92, available phosphorous, 6.76 and 6.85 and exchangeable potassium 0.92 and 0.88, respectively, for Shao and Malete (Tables 1, 2). The PM used was slightly richer in the plant nutrients compared to un-amended experimental soils of the study locations.

Plant growth

Growth of okra was significantly affected with the application of FF in combination with PM (Table 3). The highest plant height of 48.40 cm and 58.50, respectively, in 2013 and 2014 were observed when PM at 10.0 t/ha was combined with FF. Application of PM at 7.5 t/ha combined with FF had similar plant height compared to when PM was applied at 10 t/ha alone. Compared with other treatments, the control without fertilizer had significantly reduced plant height. Days to 50% flowering and fruiting followed similar trends in all the treatments. Where neither of the fertilizer was applied, took longer days to flower and fruit compared with other treatments. It flowered and fruit at 40.03 and 55.00 days in 2013 and 39.10 and 54.32 days in 2014. Although not statistically significant, combination of PM with FF had shorter days to flower and fruit compared with control treatment.

Table 3 Growth of okra as affected by the combined application poultry manure and foliar fertilizer in 2013 and 2014 cropping seasons

Treatments	2013				2014			
	Plant height (cm)	Days to 50% flowering	Days to 50% fruiting	Stem girth (cm)	Plant height (cm)	Days to 50% flowering	Days to 50% fruiting	Stem girth (cm)
PM10T +foliar	48.40 a	32.77c	43.43b	4.20	58.50a	34.85b	45.62b	4.40
PM7.5t +foliar	46.67 b	33.00c	43.40b	3.95	56.10a	34.72b	46.18b	4.18
PM 5.0t +foliar	42.33 c	33.47c	43.57b	3.80	55.75ab	35.10b	45.74b	3.60
PM 2.5t +foliar	43,33 c	32.50c	41.97b	2.95	48.03c	33.90b	43.39b	3.20
PM 10t alone	45.43 b	33.50c	44.00b	3.85	53.50b	34.62b	46.10b	4.25
Foliar alone	38.80 d	34.50b	43.80b	2.68	36.36d	35.20b	44.98b	2.62
No fertilizer	30.63 e	40.03a	55.00a	2.59 NS	27.50e	39.10a	54.32a	2.55 NS

Values with the same letter(s) under the same column are not significantly different at 5% level of probability by Duncan's Multiple Range Test

Yield and yield components

Okra fruit yield and yield components were significantly affected by different treatments (Table 4). Plots treated 10.0 t/ha PM mixed with FF produced the highest number of fruits per plant in 2014. Similar number of fruits were produced with application of 10.0 t/ha PM alone and 7.5 t/ha PM mixed with FF in 2014. The least number of fruits were recorded at the control treatments in the two years.

Application of 10.0 t/ha PM mixed with FF significantly produced longer fruits than other treatments, except combinations of 7.5 t/ha PM with FF. There was no difference in fruit length between 10.0 t/ha PM alone, as well as 7.5, 5.0, and 2.5 t/ha PM combined with FF. Shorter fruit length was observed at the control treatments.

Table 4 Effects of combined application of poultry manure and foliar fertilizer (boost Xtra) on yield and yield components of okra in 2013 and 2014 cropping seasons

Treatments	2013					2014		
	No of fruits/plant	Fruit length (cm)	Fruit circum (cm)	Yield kg/ha	No of fruits/plant	Fruit length (cm)	Fruit circum (cm)	Yield (kg/ha^{-1})
PM10t +foliar	23.10a	11.00a	11.81a	1,892.94a	23.65a	10.76a	12.00a	1825.92a
PM7.5t +foliar	20.20a	11.10a	11.60a	1,140.00a	19.37b	9.58ab	11.33a	1549.31ab
PM5.0t +foliar	11.80b	9.30bc	10.95a	796.62b	13.22d	8.13b	10.40a	1057.70cd
PM2.5t +foliar	11.50b	8.50cd	9.61a	720.62bc	11.98d	8.15b	10.20a	958.52cd
PM10t alone	13.20b	10.90ab	11.42a	1092.12a	16.35c	8.23b	11.10a	1307.78bc
Foliar alone	10.00b	8.64cd	9.63a	632.22c	9.57e	8.06b	9.45a	958.33cd
No fertilizer	4.63c	7.10d	9.10a	283.38d	7.30f	6.38c	8.45a	637.40d

Values with the same letter(s) under the same column are not significantly different at 5% level of probability by Duncan's Multiple Range Test

Okra fruit yield was highest (1182.94 kg/ha in 2013 and 1825.92 kg/ha in 2014) with the integrated application of PM at 10.0 t/ha plus FF. Recorded fruit yield in this treatment was not statistically different from those obtained with application of 7.5 t/ha PM mixed with FF in the two years. Application of 7.5 t/ha PM integrated with FF had similar fresh fruit weight with applied 10.0 t/ha PM alone in the two cropping seasons. Over the two years, there was no significant different between combinations of FF with PM at 2.5 t/ha and when boost Xtra was applied alone. The yield was minimum, 283.38 kg/ha and 637.40 kg/ha in 2013 and 2014, respectively, at the control treatment without fertilizer. There was no significant difference in fruit yield between application of FF alone and the control.

DISCUSSION

The inherent low nutrient status of the two locations suggests the need for external soil amendment for optimum crop yield. In overall, result of the study indicated that growth and yield attributes of okra were enhanced with combinations of PM and FF. The improved growth and yield observed in this study was due to balanced nutrient in the PM that was synergistically complemented with nutrient present in the FF. This was evident by the earlier commencement of flowering and fruiting in the treatments where FF was combined with high tonnage of PM. The higher values in all the parameters with the combinations of FF with PM at 7.5 t/ha compared to 10.0 t/ha PM alone also corroborated this observation. Earlier commencement of flowering and fruiting, growth, and yield parameters as observed in the study could also be due to excellent macro- and micro-nutrients in the poultry manure that was complemented with the nutrient that was supplied through the foliage. Garge and Bahla (2008) reported that poultry manure readily supply phosphorous to plants than other organic sources. Similarly, Tu et al. (2006) observed that microbial biomass and activity were higher in organically managed soils compared to conventional soil amendment with synthetic inorganic fertilizer. In a more recent study, Roy et al. (2013) reported rapid released of available plant nitrogen in organic fertilizer produced from bovine-rumen-blood-digesta mixture (BBRD).

It is possible that the combined application favored increased production of photosynthates during the growth stages and consequently partitioning and allocation of the dry matter at the developmental stages of okra. This finding is consistent with the observation of earlier researchers on the integration of organic manure with inorganic soil-applied fertilizer for okra (Akande et al. 2010; Abd El-Kader et al. (2010) and other crops such as maize (Makinde and Ayoola 2008), rice (Satyanarayana et al. 2007), sorghum (Bayu et al. 2006), and sweet maize (Uwah et al. 2011). Foliar fertilizers are fast acting because the nutrients are absorbed at the site where they are readily utilized. Apart from supplying micro-nutrients, this method of fertilizer application has also been reported to act as catalyst in the uptake and utilization of certain macronutrients (Philips 2004). In another study, Selvi and Rani (2000), found higher yield, income, and benefit cost ratio in okra when FF was mixed with inorganic NPK fertilizer than when NPK was applied alone.

CONCLUSION

In the present study, combinations of PM applied at 10 t/ha and FF produced the highest fruit yield. However, there was no significant difference in the obtained yield and integration of PM at 7.5 t/ha plus FF. Poultry manure at 7.5 t/ha mixed with foliar fertilizer is therefore recommended.

ACKNOWLEDGEMENT

The authors wish to acknowledge the Centre for Community Development, Kwara State University, for funding part of the project.

REFERENCES

1. Abasi FF, Baloch MA, Zia-ul-hassan Wagan KH, Shah AN, Rajar I (2010) Growth and yield of okra under foliar application of new multinutrients fertilizers products. Pak J Agric Agric Eng Vet Sci 26(2):11–18
2. Abd El-Kader AA, Shaaban SM, Abd El-Fattah MS (2010) Effect of irrigation levels and organic compost on okra plants (*Abelmoschus esculentus.*) grown in sandy calcareous soil. Agric Biol J N Am 1(3):225–231
3. Abou El-Magd MA, El-Bassiony M, Fawzy ZF (2006) Effects of organic manure with or without chemical fertilizer on growth, yield and quality of some varieties of Broccoli plants. J Appl Sci Res 2(10):791–798. http://www.insipub.com /jasr/2006/791-798.pdf
4. Akande MO, Oluwatoyinbo EA, Makinde EA, Adepoju AS, Adepoju IS (2010) Response of okra toorganic and inorganic fertilization. Nat Sci 8(11):261–266
5. Akinrinde EA, Olubakin DA, Omoloso SO, Ahmed AA (2006) Agric J 1(2):96–103
6. Bayu W, Rethman NFG, Hammers PS, Alemu G (2006) Effects of farm yard manure and inorganic fertilizers on sorghum growth, yield and nitrogen use in a semi arid area of Ethiopia. J Plant Nutr 29:391–407. http://www.informatworld. com/smpp/content. doi:10.1080/019041605000320962
7. Dekissa T, Short I, Allen J (2008) In: proceedings of theUCWR/NIWA Annual conference. Intl Water Resources Education; July 22–24, Durhan, NC
8. ECA (2001) State of the Environment in Africa. Economic Commission of Africa. P. O. Box 3001, Addis Ababa, Ethiopia. ECA/FSSDD/01/06. Htt://www.uneca.org /water/State Environ Afri.pdf
9. El-Aal FSA, Shaheen AM, Ahmed AA, Mahmoud AR (2010) Effects of foliar fertilization of urea amino acids mixtures as antioxidants on growth, yield and characteristics of squash. Res J Agric Bio Sci 6:583–588
10. Fageria NK, Filhoa MPB, Moreirab A, Guimarea CM (2009) Foliar fertilization of crop plants. J Plant Nutr 32(6):1044–1064
11. Food and Agricultural Organization (2006) The world sorghum and millet economies: facts trends and outlook. FAO Document repository W1808/E. Friend et al. Robert SD, Schoenholt SH, Mobley JA, Gerard PO
12. Garge S, Bahla GS (2008) Bioresour Technol 99(13):5773–5777
13. Makinde EA, Ayoola AA (2008) Residual influence of early season crop fertilization and cropping systems on growth and yield of cassava. Am J Agric Biol Sci 3(4):715–715. http://www.scipub.org/fulltex/AJAB/34712-715
14. Paliwal R, Naruka IS, Yadav JK (1999) Effect of Nitrogen and Sulphur on growth and yield of Okra (*Abelmoschus esculentus* L. Moench) cv. Pusa Sawani. Prog Hort 31(1–2):94–97
15. Philips M (2004) Economic benefits from using micronutrients for farmer and the fertilizer producer. In: IFA, international symposium on micronutrients, 23-25 Feb 2004
16. Roy M, Kamar S, Debsarcar A, Sen PK, Mukherjee J (2013) Application of rural slaughterhouse waste as an organic fertilizer for pot cultivation of solanaceous vegetables in India. Int J Recycl Org Waste Agric 2:1–11
17. Roy M, Das R, Debsarcar A, Sen PK (2015) Conversion of rural abattoir wastes to an organic fertilizer and its application the field cultivation of tomato in India. Renew Agric Food Syst. doi:10.1017/s1742170515000289
18. Satyanarayana A, Thiyagarajan TM, Uphoff N (2007) Opportunities for water saving with higher yield from the system of rice intensification. Irrig Sci 25(2):99–115
19. Selvi D, Rani P (2000) Effect of integrated nutrient management on yield and economics of okra in inceptisol. Veg Sci 27(2):207–208

20. Shehu YWS, Alhassan WK, Aliyu MA, Phillips CJC (1997) Tropical greenland 32:139–142
21. Statistical Analysis System Institute Inc (1989) SAS/STAT User guide. Version 4th edn.Vol.1, Cary, North Carolina
22. Tu C, Ristaino JB, Hu S (2006) Soil microbial biomass and activity in organic Tomato farming systems: effects of organic inputs and straw mulching. Soil Biol Biochem 38:247–255
23. Uwah DF, Eneji AS, Eshiet UJ (2011) Organic and mineral fertilizers effects on the performance of sweet maize (*Zea mays L.* saccharata strut.) in south eastern rainforest zone of Nigeria. Int J Agric Sci 3(1):54–61
24. Zodape ST, Gupta A, Bhandari SC, Rawat US, Chaudhry DR, Eswaran K, Chikara K (2011) Foliar application of seaweed sap as a biostimulant for enhancement of yield and quality of tomato (*Lycopersicum esculentum.* Mill). J Sci Ind Res 70:215–219

CHAPTER 17

Effects of Co-Composting of Faecal Sludge and Agricultural Wastes on Tomato Transplant and Growth

Eric Gbenatey Nartey[1]• Philip Amoah[1]• Godfred Kwabena Ofosu-Budu[2]•Ashley Muspratt[3]• Surendra kumar Pradhan[1]

[1]*International Water Management Institute (IWMI), PMB CT, 112, Cantonments, Accra, Ghana*

[2]*College of Agricultural and Consumer Sciences, University of Ghana Forest and Horticultural Crops Research Centre- Kade, Legon, Ghana*

[3]*Pivot Ltd, P.O. Box 98394, Mombasa, Kenya*

ABSTRACT

Purpose Faecal sludge (FS) has been co-composted with many organic solid wastes globally. Agricultural wastes, such as oil palm empty fruit bunches (EFB) and cocoa pod husks (CPH), have received very little research attention as far as combining with FS is concerned. This study aimed at co-composting these wastes at different ratios to produce safe compost for use as soilless medium for raising tomato transplants.

Methods Dewatered FS (DFS) was mixed with shredded EFB and CPH at five different ratios: 1DFS:1EFB, 1DFS:1CPH, and DFS:EFB:CPH in ratios of 1:1:1, 2:1:1, and 2:2:1 and composted for 3 months. Select physicochemical parameters and pathogens were monitored every fortnightly and 3 weeks, respectively.

Results Maximum temperatures obtained ranged 46.8–54.5 °C. Though these temperatures were lower than sanitizing temperatures prescribed by USEPA, no *E. coli* was found in any of the piles at the end of composting. The ratio 2DFS:2EFB:1CPH was found to be the safest formulation and hence was used to grow tomato under greenhouse conditions. Tomato seeds were sown in three different growing media: 100% FS-based compost, 100% rice husk biochar, and 50% FS-based compost–50% rice husk biochar mix.

Conclusion Results showed that FS-based compost was a suitable growing medium for tomato. Further studies into the optimal rate and frequency of application of compost teas on tomato are recommended.

Keywords: Faecal sludge Co-composting Cocoa pod husks Transplant Compost tea Tomato

INTRODUCTION

Insufficient waste management has led to growing interests in approaches to safe emptying, transport, and disposal of faecal sludge (FS) (Strauss and Montangero 2002). Numerous technical developments have attempted to address the problems of emptying, transporting, and disposal of FS from urban centres (Boot and Scott 2008). Currently, the focus of research is shifting from FS disposal to reuse, with fertilizer and soil conditioner being among the most common reuse options. FS, like animal manure, is a good soil conditioner and a renewable source of plant nutrients, such as nitrogen (N), phosphorus (P) and potassium (K) (Cofie et al. 2004), and organic matter. These are lacking in arable soils found in the tropics. There is, therefore, need to transform this valuable waste into useful soil amendments for crop production. This will represent an important option for both sustainable waste management and sustainable agriculture (Chang and Davila 2007; Dick and McCoy 1993). Although FS contains these beneficial resources that can be applied to arable soils, the raw sludge contains high pathogen concentrations and direct use without prior treatment puts both farmers and consumers at risks. Composting is one way to sanitize dewatered FS (DFS). On its own, there is too little organic content to achieve thermophilic conditions (i.e., temperatures above 55 °C) during composting, which is necessary for complete pathogen destruction. Co-composting with carbon-rich materials, such as oil palm empty fruit bunches (EFB) and cocoa pod husks (CPH), can impart two key benefits: (a) enabling of thermophilic temperatures and (b) prevention of nitrogen volatilization (Jeong and Kim 2001). Previous composting studies in Ghana experimented with co-composting DFS with municipal solid wastes and sawdust (Kone et al. 2007; Cofie et al. 2009). The use of other agricultural wastes, such as EFB and CPH, has received very little research attention in terms of co-composting with DFS. It is estimated that approximately 400,000 tonnes of cocoa beans are produced annually (Ofosu-Budu et al. 2001, personal communication), and based on this annual production, an estimated 550,750 tonnes of dry cocoa husks are produced (Adamtey 2005). Annual production of EFB in Ghana can be estimated based on annual production of approximately 3,135,000 tonnes of fresh oil palm fruit bunches (calculation based on average yield of 11 tonnes per hectare) (Toledano et al. 2004). Agro-industrial wastes are posing serious challenges for the agro-industries (Abu Qdais and Al-Widyan 2016). Similarly, an estimated amount of about 3.5 million tonnes of human excreta is produced in Ghana based on an average production of 400 g cap^{-1} day^{-1} (Mann 1976). This leaves a lot of human and agricultural wastes in the environment, which are traditionally directly incinerated or deposited in landfills resulting in the production of significant amounts of greenhouse gases (Zhang et al. 2013) or sometimes reapplied to land as a conditioner and could pose health and environmental risks. Many studies have reported that organic wastes composts, such as from sewage sludge (Perez-Murcia et al. 2006), municipal solid waste (Ostos et al. 2008), animal manure (Atiyeh et al. 2001; Eklind et al. 2001), green waste (Grigatti et al. 2007; Ribeiro et al. 2007), and agro-industrial waste (Papafotiou et al. 2004; Bustamante et al. 2008) can be used with very good results as growth medium instead of peat (Kala et al. 2012). These composts and other solid manures are great, but they have one major disadvantage of taking a longer time to breakdown in the soil to release nutrients. Therefore, for successful growing of vegetables and other horticultural crops, such as tomato, which require rapid nitrogen sources, the use of compost tea could play a significant role. Compost tea has evolved from historical horticultural practices, such as steeping manure or plants in water, with the liquid applied to crops for nutritional and plant health reasons. In recent decades, the majority of published studies examining compost tea have focussed on foliar plant disease suppression using liquid materials produced by steeping compost in water for several days to weeks (Scheuerell 2004). It is increasingly being used as an alternative plant disease control measure in commercial horticulture (Diver 2001; Scheuerell and Mahaffee 2002). Compared to the numerous reports on the use of compost tea for managing foliar diseases, little work has been done to assess the nutritional benefits of compost tea on plant growth (Pant 2009). Studies into the nutritional effect of compost tea on horticultural plants are, therefore, necessary. As a liquid fertilizer, its high nutrient value and rapid availability make it a great tonic for the plants. This study, therefore, aims at (i) co-composting DFS, EFB, and CPH at different formulations to determine the appropriate formulation that will produce safe compost for use in soilless medium for tomato transplant production and (ii) to assess the nutritional effect of compost tea brewed from the FS-based compost on the vegetative growth of tomato.

MATERIALS AND METHODS

The study was carried out at the University of Ghana, Forest and Horticultural Crops Research Centre (FOHCREC) (6°05′N; 0°05′W). Climate of the area is humid tropical and monthly average temperature reaches a maximum of 28–29 °C in February and March, and a minimum of 25–26 °C in July/August. Rainfall pattern is bimodal with peaks (150–200 mm) in May/June and September/October (Adamtey 2005). Raw FS coming from household and public toilets at a ratio of 2:1 (*v*/*v*) was dewatered on sand drying beds to obtain DFS. The DFS was characterized and mixed at different formulations with shredded (size 2–3 cm) EFB and CPH, obtained from nearby oil palm processing industries and cocoa farms, respectively. The formulations were as follows: 1 DFS:1 EFB; 1 DFS:1 CPH; and DFS:EFB:CPH in ratio of 1:1:1, 2:1:1, and 2:2:1, respectively. The mixtures were moistened with water to about 60% moisture content following the method described by Estevez-Schwarz et al. (2012) and laid in windrows. Each of the five treatment formulations was replicated three times in a completely randomized design. Average size of a pile was 2 m long, 1.5 m wide, and 0.6 m high. Piles were turned manually every 3 days for the first 2 weeks and then once a week for the rest of the 10 weeks. Throughout the experiment, the moisture contents of the piles were maintained as recommended by Day and Shaw (2000). Daily temperature readings were taken by inserting a thermometer probe into three perforated bamboo pipes of approximately equal sizes mounted in each pile. Samples of the co-composting material were taken every 2 weeks for physicochemical analyses (pH, EC, TOC, NH_4-N, NO_3-N, Total N, P, and K). *E. coli,* total coliform, and helminth egg populations particularly *Ascaris* eggs were determined in samples every 3 weeks.

Feedstock and co-compost analyses

The feedstock and co-compost samples were air dried and milled (<0.5 mm) to homogenize the samples. Total N was determined by the modified Kjeldahl method described by Black (1965). Ammonium (NH_4-N), Nitrate (NO_3-N), Total P, Avail-P, and K were determined by methods, as described in Okalebo et al. (2002). The pH and electrical conductivity (EC) were measured using methods described in USDA and USCC (2001). Total organic carbon (TOC) was determined by the methods described in Walkley and Black (1934). For all samples, total coliform and helminth egg counts were determined using the spread plate method (APHA-AWWA-WEF 2001) and the flotation and sedimentation method following a modified USEPA method (Schwartzbrod and Gaspard 1998), respectively. Compost samples were prepared for total coliform and *E. coli* population estimates as follows: each sample was thoroughly mixed and a subsample of 10 g (dry weight) was weighed into 90 ml of sterile phosphate-buffered saline and shaken vigorously for about 2 min. Further tenfold serial dilutions were made, and Chromocult coliform agar (MERCK1 KgaA 64271 Darmstadt, Germany) was inoculated from each dilution. The inoculated media were incubated at 35–37 °C for 24–48 h (APHA-AWWA-WEF 2001). All colonies with purple-to-blue–black colour were counted as *E. coli.* Temperature, pH, C:N ratio, and CO_2 evolution rates of co-compost were used to determine co-compost stability and maturity. Carbon dioxide (CO_2) evolution rates were determined by methods, as described in Okalebo et al. (2002). Criteria for selecting the best co-compost formulation were based on subjective performance star ratings in terms of pathogen reduction, total nutrient content, and amount of N-mineralised and N-loss. The ratings were as follows: ***** = Excellent, **** = Very Good, *** = Good, ** = Fair and * = Poor. The formulation with the highest number of stars was the best.

Greenhouse experiment

The best co-compost produced above was used as the soilless medium together with rice husk biochar to grow tomato transplants. The matured compost was first milled and sieved through a 2 mm sieve to obtain a uniform texture before mixing with the rice husk biochar to prepare three growing medium types: a 100% FS-based compost (FSC), a 100% rice husk biochar (RHB), and a 50% FS-based compost: 50% rice husks biochar mix (FSC-RHB). Tomato seeds (*Lycopersicon esculentum var. M2*) were sown in the three different growing medium at three seeds/plug and irrigated with tap water until the seeds germinated. Germinated seeds were counted starting from the fourth to tenth days after seeding (DAS). The percentage emergence and mean days to emergence (MDE) were calculated as described by Gerson and Honma (1978). The young transplants were continuously irrigated with tap water until most of them were showing second true leaves after which they were thinned to one plant per plug. Compost tea (CT) was brewed from the same co-compost for fertigation by steeping 2.5 kg of the compost in 15 L of distilled water (1:3 *v/v*) for 6 days following the bucket-fermentation method described by Diver (2001). For application, the CT stock solution was further diluted to 1:5 (CT-1) and 1:10 (CT-2) (stock solution:distilled water). The two compost tea dilutions were prepared in 20L batches and analysed for NPK content before applying to the transplants. To compare with the CT solutions, inorganic fertilizer solutions containing 400 mg N/L:176 mg P/L:332 mg K/L (In-fert-1) and 200 mg N/L:176 mg P/L:332 mg K/L (In-fert-2) were also prepared. After thinning the transplants, irrigation with tap water was terminated. The transplants were then sub-irrigated with the fertilizer solutions (CT-1, CT-2, In-fert-1, or In-fert-2) until the growth medium was saturated. Ordinary tap water served as control solution. Sub-irrigation with the fertilizer solutions was carried out for 21 days after which the experiment was terminated. The experiment was carried out in a complete randomized design with three replications.

Transplant sampling and analyses

Total N in the growth medium and compost tea solutions were determined by the semi–micro-Kjeldahl method (Black 1965). Total phosphorus (P) and potassium (K) were determined using the spectrophotometer (Philip PU8620 UV/VIS/NIR model) and Jenway flame photometer (PFP7), respectively. Plant available N (NH_4 and NO_3) were determined according methods described by Bremmer (1965). Electrical conductivity (EC) and pH of the medium and nutrient solutions were determined following standard procedures, as described in USDA and USCC (2001). Bulk density and water holding capacity of the medium were determined according to methods described in Okalebo et al. (2002) and Vengadaramana and Jashothan (2012), respectively. Transplants were sampled (five plants/treatment rep) every 7 days and the following growth parameters measured: leaf chlorophyll content, number of leaves, plant height, root length, stem diameter, shoot dry matter (DM), root DM, and shoot/root ratio. The leaf chlorophyll content was measured using Apogee chlorophyll content meter (Model CCM- 200), plant height and root length (in cm) were measured with a rule, and stem diameter was measured 1 cm above growth medium surface with a pair of Vernier callipers. The plant samples were then severed and separated into root and shoot parts. They were oven dried at 68 °C for 48 h to obtain dry weight/dry matter.

Statistical analysis

Analysis of variance (ANOVA) was performed on all data using Genstat 9th edition (Release 9.2) statistical package. Data for the microbial analysis were log transformed, while N-mineralised and N- and P-losses were transformed into percentages before ANOVA. Treatment means which were found to be significantly different from each other at ($p < 0.05$) were separated by the Least Significant Differences (LSD) tests. Two tailed correlation analysis between selected characteristics of the compost were determined using SPSS for windows software (version 16). Significant correlations at the 1% level ($p < 0.01$) have been considered.

RESULTS

Co-composting DFS with EFB and CPH

General characterisation of the co-compost feedstock saw total N to be the highest (3.1%) in DFS followed by CPH and then EFB. Ammonium (NH_4-N) in the DFS was about fivefold higher than that in the EFB and CPH (Table 1). The C:N ratios were 43.5, 25.1, and 11.1 for EFB, CPH, and DFS, respectively. Potassium (K) level in CPH was about three and two times higher than that in the DFS and EFB, respectively.

Table 1 Physico-chemical characteristics of feedstock used in co-composting (x ± SD) *(n = 3)*

Parameters	DFS	EFB	CPH
pH (1:5)	7.7 ± 0.0	8.9 ± 0.1	8.6 ± 0.1
TOC (%)	34.3 ± 2.0	64.6 ± 4.4	41.1 ± 0.8
C:N	11.1 ± 0.2	43. 5 ± 1.4	25.1 ± 1.0
Total N (%)	3.1 ± 0.2	1.5 ± 0.1	1.6 ± 0.0
Total P (%)	1.5 ± 0.5	0.1 ± 0.0	0.3 ± 0.0
Total K (%)	0.8 ± 0.0	1.3 ± 0.0	2.2 ± 0.1
NH_4-N (mg kg^{-1})	2062.8 ± 205.2	378.0 ± 5	399.5 ± 3.0
NO_3-N (mg kg^{-1})	662.4 ± 86.4	363.6 ± 126.0	320.4 ± 54.0
Avail-P (%)	NA	NA	NA

EFB empty fruit bunches, *CPH* cocoa pod husks, *DFS* dewatered faecal sludge, *NA* not available

Total coliform populations in the DFS ranged between 2.0×10^3 and 3.0×10^5 CFU g^{-1} with an average of $6.2 \times 10^5 \pm 7.1 \times 10^5$ CFU g^{-1}, while *E. coli* levels ranged between 0 and 6.0×10^3CFU g^{-1}. The average helminth egg population was 1/10 g. The characteristics of the various formulations at the start of co-composting process are shown in Table 2. At the end of co-composting, total N was highest in piles 1 DFS:1 CPH and 2 DFS:2 EFB:1CPH followed by pile 2 DFS:1 EFB:1CPH (Table 2). The differences were, however, not significant. Pile treatment 1 DFS:1 EFB:1 CPH had the lowest total N, but it was not significantly different from pile 1 DFS:1 EFB. Total P did not vary significantly between pile treatments 1 DFS:1 CPH, 1 DFS:1 EFB:1 CPH, and 2 DFS:1 EFB:1 CPH. However, the highest concentration of P was measured in pile 2 DFS:2 EFB:1 CPH and the lowest in pile 1 DFS:1 EFB:1 CPH. Total K was the highest in pile 2 DFS:2 EFB:1CPH and the lowest in pile 1 DFS:1 EFB. Electrical conductivity (EC) was significantly higher in pile 1 DFS:1 CPH compared to the other treatment piles. There were, however, no significant differences between piles 1 DFS:1 EFB, 1 DFS:1 EFB:1 CPH, 2 DFS:1 EFB:1CPH, and 2 DFS:2 EFB:1CPH. In all cases, once the piles were formed, the temperature rose rapidly (Fig. 1). Temperature in all the piles peaked on Day 3, with 1 DFS:1 EFB; 2 DFS:1 EFB:1CPH, and 2 DFS:2 EFB:1 CPH reaching thermophilic conditions (Fig. 1). The maximum temperature attained by each pile ranged between 46.8 and 54.5 °C and the minimum temperature ranged between 31.7 and 32.9 °C.

Table 2 Physico-chemical characteristics of co-compost piles at start and end of co-composting ($p < 0.05$) ($n = 3$)

Piles	EC (μS cm^{-1})	C/N ratio	Total N (%)	NH_4-N (mg kg^{-1})	NO_3-N (mg kg^{-1})	Total P (%)	Avail-P (%)	Total K (%)	*N*-mineralised			*N*- and *P*-losses	
									% Org-N	% NH_4-N	% NO_3-N	% N	% P
At start													
1DFS:1EFB	2150	38.2	2.2	1828.8	284.4	2.6	0.3	1.3	90.5	8.2	1.3	–	–
1DFS:1CPH	2875	32.2	3.1	2368.8	392.4	3.1	0.3	1.9	90.9	7.8	1.3	–	–
1DFS:1EFB:1CPH	2275	45.2	1.8	1569.6	241.2	2.4	0.3	1.6	89.8	8.9	1.4	–	–
2DFS:1EFB:1CPH	1750	61.0	1.5	954.0	252.0	2.4	0.1	1.5	92.1	6.2	1.7	–	–
2DFS:2 EFB:1CPH	2525	51.4	2.0	1785.6	190.8	2.4	0.2	1.5	90.2	8.8	0.9	–	–
LSD	551.4	10.37	0.38	604.2	66.79	0.54	0.05	0.21	1.46	1.57	0.35		
At end													
1DFS:1EFB	1975	20.3	1.2	367.2	849.6	0.8	0.2	0.7	89.4	3.2	7.4	73	84
1DFS:1CPH	2725	12.2	1.8	334.8	1000.8	0.7	0.3	0.9	92.4	1.9	5.4	63	86
1DFS:1EFB:1CPH	2175	19.7	1.1	381.6	561.6	0.6	0.1	0.9	91.0	3.6	5.4	71	88
2DFS:1EFB:1CPH	1925	17.9	1.4	334.8	594.0	0.7	0. 2	0.8	93.5	2.3	4.2	44	82
2DFS:2 EFB:1CPH	2025	19.7	1.8	327.6	568.8	0.9	0.2	1.0	94.9	1.9	3.2	56	82
LSD	417.3	4.20	0.17	30.09	185.36	0.12	0.04	0.14	1.82	0.40	1.55	9.2	3.0

EFB empty fruit bunches, *CPH* cocoa pod husks, *DFS* dewatered faecal sludge, and *LSD* least significant difference

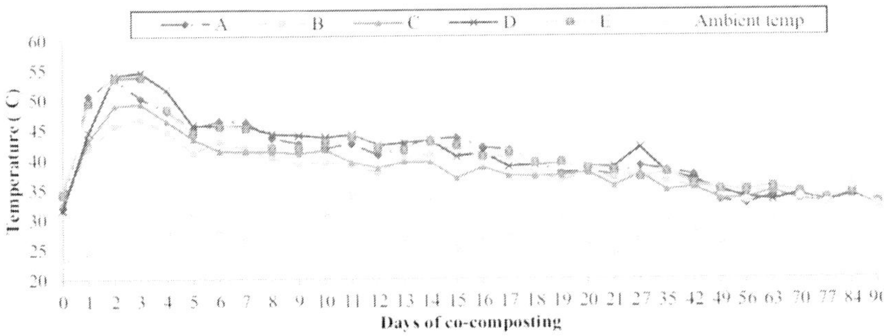

Figure. 1 Changes in temperature in the different piles during the co-composting process (*A* = 1 DFS:1 EFB, *B* = 1 DFS:1 CPH, *C* = 1 DFS:1 EFB:1 CPH, *D* = 2 DFS:1 EFB:1CPH and *E* = 2 DFS:2 EFB:1CPH)

In general, the CO_2 evolution rates indicating microbial respiration decreased with time over the 12-week period for all the piles (Fig. 2). Within the first week, the CO_2 evolution rates were the highest (8.4 mg g^{-1} compost per day) in pile 2 DFS:2 EFB:1CPH and the lowest (2.7 mg g^{-1}compost per day) in pile 1 DFS:1 CPH. By the fifth week, the evolution rates of pile 1 DFS:1 CPH became the highest (1.7 mg/g compost per day) amongst the treatments with pile 2 DFS:2 EFB:1CPH being the lowest (0.8 mg g^{-1} compost day^{-1}). At the end of the 12 week, all the piles recorded stable CO_2 evolutions which were less than 1 mg CO_2 g^{-1} compost day^{-1}.

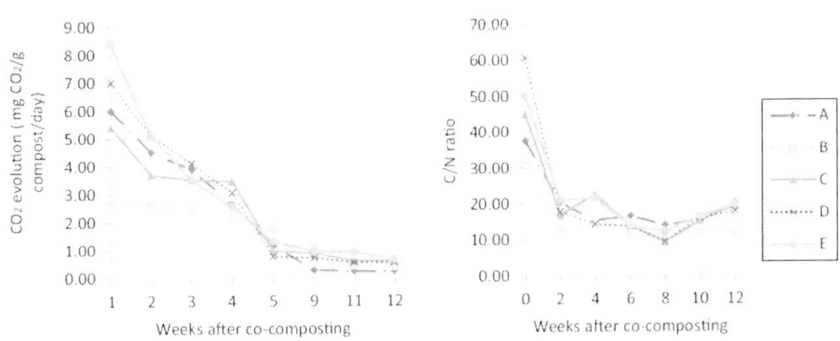

Figure. 2 Changes in CO_2 evolution and C/N in the different piles during the co-composting process (*A* = 1 DFS:1 EFB, *B* = 1 DFS:1 CPH, *C* = 1 DFS:1 EFB:1 CPH, *D* = 2 DFS:1 EFB:1CPH and *E* = 2 DFS:2 EFB:1CPH)

The C:N ratios decreased drastically over the first 2–4 weeks of co-composting for all the piles, and remained low through the week 8 when concentrations began to increase (Fig. 2). Table 2shows the levels of NH_4-N and NO_3-N and org-N after the 12-week composting period. The nitrogen was found predominantly in its organic form, with levels ranging between 89 and 95%, while the inorganic forms (NH_4-N and NO_3-N) ranged between 5 and 11%. Significant ($p < 0.05$) losses of N and P were observed for all the treatment piles. The percentage nitrogen losses ranged from 44 to 73%, while phosphorus losses ranged from 82 to 88% (Table 2). There were, however, no significant differences in N-losses between piles 1 DFS:1 EFB and 1 DFS:1 EFB:1 CPH and then between piles 1 DFS:1 CPH and 1 DFS:1 EFB:1 CPH. The change in levels of *E. coli* is represented in Fig. 3. *E. coli* levels were the highest in co-compost piles DFS:1 EFB:1 CPH and 2 DFS:1 EFB:1CPH (4.2 log units) and this was significantly ($p < 0.05$) higher than that of 1 DFS:1 EFB, 1 DFS:1 CPH, and 2 DFS:2 EFB:1 CPH. Apart from co-compost pile 1 DFS:1 EFB, all the other compost piles recorded *E. coli* concentrations above the USEPA part 503 class A exceptional quality (EG) compost standard (USEPA 1993) at the beginning of the co-composting process. For this standard to be achieved, *E. coli* and faecal coliform concentrations must be <1000 $MPNg^{-1}$ or <3 log units $CFUg^{-1}$. This standard was adopted, because Ghana is yet to have a standard of its own and USA for many years has championed the concept of pathogen reduction as a number one compost quality standard (USEPA 1993). No *E. coli* was, however, recorded in the co-compost piles at the end of the composting period. The total coliform levels, ranging between 3.1×10^4 and 8.9×10^4 $CFUg^{-1}$ at the beginning of the composting process, reduced to levels between 3.1×10^2 and 8.0×10^3 $CFUg^{-1}$ after the period. Helminth eggs were not found in any of the co-compost piles at the end of the co-composting process.

The formulation 2 DFS:2 EFB:1 CPH was found to be the best mixing ratio for co-composting these three organic wastes. This formulation had the highest star ratings (15*), and it contained the highest nutrient (N, P, and K) content and least pathogen content among others, as shown in Table 3, and was used for the greenhouse experiment.

Greenhouse experiment

Electrical conductivity (EC) increased with increasing compost concentration in the growing medium (Table 4). The pH was significantly higher in FSC-RHB (7.8) than in FSC (7.5) and RHB (7.3). Total N, Total P, and Total K were higher in FSC than in FSC-RHB, though the differences were not significant at $p < 0.05$. Nitrate (NO_3) levels showed the following decreasing order: FSC > FSC-RHB > RHB. Available P was significantly higher in FSC-RHB than in the other growing media.

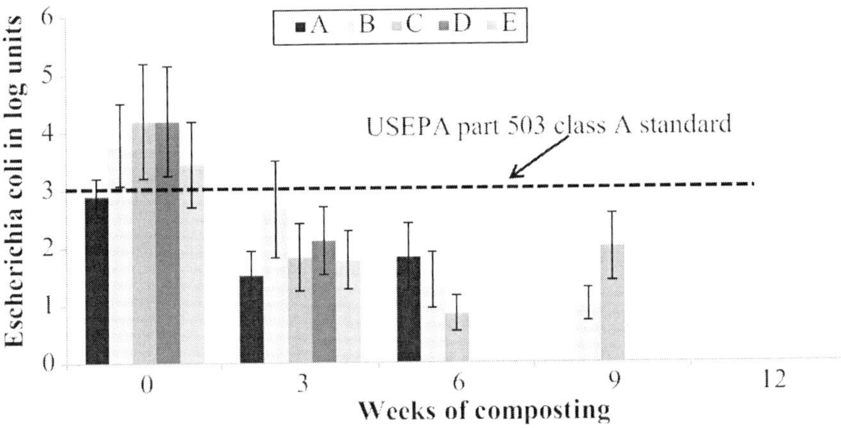

Figure. 3 Changes in *Escherichia coli* population in compost piles during composting process (*A* = 1 DFS:1 EFB, *B* = 1 DFS:1 CPH, *C* = 1 DFS:1 EFB:1 CPH, *D* = 2 DFS:1 EFB:1CPH and *E* = 2 DFS:2 EFB:1CPH)

Table 3 Selecting the best compost ratio based on selected parameters

Compost treatment	Faecal coliform reduction	N- and P-losses	N-mineralization	Total NPK	Total
1DFS:1EFB	*	*	*****	**	9 (*)
1DFS:1CPH	**	***	****	****	13 (*)
1DFS:1EFB:1CPH	***	**	***	*	9 (*)
2DFS:1EFB:1CPH	****	*****	**	***	14 (*)
2DFS:2EFB:1CPH	*****	****	*	*****	15 (*)

The dilution factors did not influence the pH of the compost tea solutions (CT-1 and CT-2) and the inorganic fertilizer solutions (In-fert-1 and In-fert-2) (Table 5). Higher concentrations of Total P and K were recorded in the compost tea solutions than in the inorganic fertilizer solutions.

The different growing media formulation affected tomato seed germination, emergence, and early vegetative growth. Germinated tomato seeds emerged in RHB 24 h earlier than in both FSC and FSC-RHB (Table 6). Percentage emergence was also higher in RHB medium than in the other two growing media.

Table 4 Physico-chemical characteristics of growth medium types ($n = 3$ for each media)

Parameters	FSC	FSC-RHB	RHB	LSD ($p < 0.05$)
pH (1:5)	7.5	7.8	7.3	0.03
EC (1:10) (dS/cm)	2.1	1.6	0.0	0.08
Bulk density (g/cm^3)	0.7	0.6	0.3	0.02
Water holding capacity (per gram)	1.2	1.7	3.1	0.13
Total N (%)	1.8	1.7	1.1	0.28
NH_4 (mg/Kg)	205.0	223.0	158.0	96.10
NO_3 (mg/Kg)	1156.0	423.0	187.0	142.10
Total P (%)	0.5	0.4	0.1	0.16
Avail. P (%)	0.1	0.2	0.0	0.01
Total K (%)	1.1	0.9	0.5	0.05
C/N	19.8	13.3	11.0	3.82

FSC = 100% FS-based compost, FSC-RHB = 50% FS-based compost: 50% rice husk biochar mix and RHB = 100% rice husks biochar

Table 5 Physico-chemical characteristics of compost teas and inorganic fertilizer solutions ($n = 3$)

Parameters	CT-1	CT-2	In-fert-1	In-fert-2	LSD ($p < 0.05$)
pH	8.1	8.1	5.6	5.6	0.16
EC (dS/cm)	0.6	0.1	0.9	0.2	0.10
Total N (mg/L)	74.8	72.8	400.0	200.0	N/A
NH_4 (mg/L)	41.3	30.0	200.0	100.0	N/A
NO_3 (mg/L)	41.4	41.7	200.0	100.0	N/A
Total P (mg/L)	1085.0	560.0	176.0	176.0	N/A
Avail-P (mg/L)	51.5	51.4	176.0	176.0	N/A
Total K (mg/L)	7420.0	3340.0	332.0	332.0	N/A

CT-1 = 1:5 (compost tea stock:water), CT-2 = 1:10 (compost tea stock:water), In-fert-1 = 400 mg N/L:176 mg P/L:332 mg K/L, In-fert-2 = 200 mg N/L:176 mg P/L:332 mg K/L), and N/A = not available

Table 6 Effect of growing medium on tomato seed germination and emergence

Substrate type	Percentage emergence (%)	Mean days to emergence (MDE)
FSC	68.7	6.6
FSC-RHB	78.7	6.6
RHB	84.0	5.6
LSD ($p < 0.05$) ($n = 15$)	10.90	0.59

FSC = 100% FS-based compost, FSC-RHB = 50% FS-based compost: 50% rice husk biochar mix and RHB = 100% rice husks biochar

Young transplants were taller in FSC than in RHB medium (Table 7); however, plant height was not significantly different between FSC and FSC-RHB media. Transplants in the RHB growth medium were significantly thinner than transplants growing in both FSC and FSC-RHB media. Transplants in the FSC medium recorded the highest shoot/root ratio, whereas those growing in the RHB medium recorded the least shoot/root ratio.

Table 7 Early vegetative growth of tomato transplant before treatment (sub-irrigation)

Substrate type	Plant height (cm)	Stem diameter (mm)	Root length (cm)	No. of true leaves	Shoot DM (mg)	Root DM (mg)	Shoot/root ratio
FSC	7.4	0.2	2.3	1.9	12.0	3.0	4.0
FSC-RHB	6.8	0.2	2.8	1.9	11.0	3.0	3.7
RHB	5.7	0.1	3.4	0.9	7.9	3.0	2.7
LSD ($p < 0.05$) ($n = 15$)	1.19	0.03	0.57	0.37	0.70	0.63	0.62

FSC = 100% FS-based compost, FSC-RHB = 50% FS-based compost: 50% rice husk biochar mix and RHB = 100% rice husks biochar

However, after 21 days of treatment, plant height, stem diameter, total (shoot and root) DM, leaf chlorophyll content, and number of leaves were significantly higher in both FSC and FSC-RHB media than in the RHB medium (Table 8). Shoot/root ratio was higher in FSC than in the FSC-RHB. Leaf chlorophyll content and root DM were significantly higher in FSC-RHB than in FSC.

Table 8 Effect of substrate type on vegetative growth of tomato transplants after 21 days of treatment ($n = 3$)

Vegetative growth parameters	Growing medium			
	FSC	FSC-RHB	RHB	LSD ($p < 0.05$)
Plant height (cm)	27.1	26.7	14.2	0.98
Stem diameter (mm)	0.4	0.4	0.3	0.01
Shoot DM (mg/plant)	441.3	447	118.0	28.48
Root DM (mg/plant)	74.3	101.9	34.7	8.17
Chlorophyll content (CCI)	9.6	11.5	6.3	0.68
No. of leaves	6.5	6.3	4.5	0.23
Shoot/root	6.2	4.5	3.3	0.63

FSC = 100% FS-based compost, FSC-RHB = 50% FS-based compost: 50% rice husk biochar mix and RHB = 100% rice husks biochar

The different nutrient solutions significantly affected growth of the transplants in the greenhouse. Growth response of transplants to the inorganic fertilizer solutions (In-fert-1 and In-fert-2) was significantly higher than that of the compost teas (CT-1 and CT-2) and the control (water) (Table 9). However, there were no significant differences in growth between In-fert-1 and In-fert-2 except in root DM, where it was significantly higher in In-fert-2 than in In-fert-1. Similarly, shoot DM, number of leaves, stem diameter, and plant height were higher in CT-1 than in CT-2. However, there were no significant differences in chlorophyll content, root DM, and shoot/root ratio between the two compost tea solutions. Interestingly, the control solution recorded higher plant height, higher root DM, and higher root volume than in the CT-2 solution.

Table 9 Effect of compost teas and inorganic fertilizer solutions on vegetative growth of tomato transplant after 21 days of treatment (sub-irrigation) ($n = 3$)

Vegetative growth parameters	Nutrient solutions					
	CT-1	CT-2	In-fert-1	In-fert-2	*W*	LSD ($p < 0.05$)
Plant height (cm)	19.9	17.2	29.2	28.5	18.6	1.27
Stem diameter (mm)	0.3	0.3	0.4	0.4	0.3	0.02
Shoot DM (mg/plant)	279.4	213.3	490.0	506.1	221.1	36.77
Root DM (mg/plant)	51.7	44.4	86.1	102.8	66.4	10.55
Chlorophyll content (CCI)	6.3	6.0	14.2	13.2	6.1	0.87
No. of leaves	5.3	4.7	6.8	7.0	4.9	0.30
Shoot/root	4.8	4.2	5.7	5.2	3.3	0.80

CT-1 = 1:5 (compost tea stock:water), CT-2 = 1:10 (compost tea stock:water), In-fert-1 = 400 mg N/L:176 mg P/L:332 mg K/L, In-fert-2 = 200 mg N/L:176 mg P/L:332 mg K/L) and W = control

Increasing concentrations of FS-based compost in the growing medium (from 0 to 100%) and decreasing dilutions of the compost tea solutions (CT-2 to CT-1) resulted in significant increases in tomato shoot DM and shoot/root ratio (Table 10). Stem diameter, root length, leaf chlorophyll content, and root DM reduced with increasing compost concentration but were hardly significant. Similarly, there were no significant differences between FSC and FSC-RHB substrates in terms of root DM and plant height.

Table 10 Combined effect of growing medium and nutrient solutions on vegetative growth of tomato transplants after 21 days of treatment

	Nutrient solutions				
	CT-1	CT-2	In-fert-1	In-fert-2	W
Plant height (cm)					
FSC	25.6	22.4	31.6	30.7	25.5
FSC-RHB	24.7	20.4	32.9	33.3	22.2
RHB	9.4	8.9	23.2	21.4	8.0
LSD ($p < 0.05$) ($n = 3$)	2.20				
Stem diameter (mm)					
FSC	0.4	0.4	0.4	0.4	0.4
FSC-RHB	0.4	0.4	0.4	0.5	0.4
RHB	0.2	0.2	0.4	0.4	0.2
LSD ($p < 0.05$) ($n = 3$)	0.01				
Chlorophyll content (CCI)					
FSC	7.2	7.1	14.4	12.4	7.0
FSC-RHB	8.2	7.3	16.9	16.9	8.5
RHB	3.5	3.6	11.2	10.3	2.7
LSD ($p < 0.05$) ($n = 3$)	1.51				
Shoot dry matter (mg/plant)					
FSC	455.0	368.3	503.3	515.0	365.0
FSC-RHB	358.3	253.3	728.3	711.7	281.7
RHB	25.0	18.3	238.3	291.7	16.7
LSD ($p < 0.05$) ($n = 3$)	63.69				
Root dry matter (mg/plant)					
FSC	71.7	65.0	66.7	71.7	96.7
FSC-RHB	75.0	61.7	121.7	153.3	97.7
RHB	8.3	6.7	70.0	83.3	5.0
LSD ($p < 0.05$) ($n = 3$)	18.260				

FSC = 100% FS-based compost, FSC-RHB = 50% FS-based compost: 50% rice husk biochar mix, RHB = 100% rice husks biochar, CT-1 = 1:5 (compost tea stock:water), CT-2 = 1:10 (compost tea stock:water), In-fert-1 = 400 mg N/L:176 mg P/L:332 mg K/L, In-fert-2 = 200 mg N/L:176 mg P/L:332 mg K/L) and W = control (water)

Interactions between medium type and inorganic fertilizer solutions show that total (shoot and root) DM were significantly reduced at the highest compost concentration (FSC) and the highest nitrogen (N) concentration in the inorganic fertilizer solutions (In-fert-1). Leaf chlorophyll content was significantly higher in FSC-RHB medium than in the others, but no significant difference was observed between In-fert-1 and In-fert-2.

DISCUSSIONS

The temperature of the co-compost treatment piles was generally above 45 °C for the first few days but did not exceed 55 °C. According to Singh et al. (2011), the primary process criteria used for ensuring the microbiological safety of composts have been narrowly defined as time—temperature conditions. For complete sanitation of composts, the USEPA standard demands temperature of >55 °C for 2 weeks with five turns (USEPA 1993). Though the temperatures in this study did not meet any of the international requirements, they were higher than the maximum 45 °C temperature observed by Kala et al. (2012). The lower temperatures observed in this study could be due to the composting methods employed (open windrows) or to the dissipation of heat due to the small volume of the piles. This confirms earlier reports by Kala et al. (2012) that heap sizes affect the temperature generation during composting. The possible implication of the pile temperatures on pathogen reduction saw the level of *E. coli* in all piles declining below the USEPA class A compost standard by the third week of composting. The complete deactivation of *E. coli* in all piles at the end of composting (week 12) indicates that complete deactivation might have occurred during the cooler curing phases of composting and not during the active thermophilic phases. This confirms earlier reports by Droffner and Brinton (1995) that *Salmonell*a and *E. coli* were found to survive for 59 days at about 60 °C, although pathogens were destroyed during the cooler curing process in industrial compost. Results from this study suggest that high temperatures might not be the only factor responsible for *E. coli* deactivation. This phenomenon is in agreement with Dumontet et al. (1999) that other factors, such as pH, presence of metabolic, and antagonistic compounds produced by indigenous micro flora, accumulation of toxic NH_3, and microbiological competition for nutrients, are also involved in the deactivation. However, it is not clear from this study which of the measured factor or set of factors might have led to the inactivation of *E. coli* as none of the factors tested showed any significant correlation with the *E. coli* counts. Deactivation might, however, be due to competition among the microbes in the system. According to Kone et al. (2007), exposure to temperature over 45 °C for at least 5 days is known to inactivate *Ascaris* eggs which are very predominant in faecal sludge in developing countries. Similarly, the same reason could account for the complete elimination of Ascaris eggs from all compost piles at the end of the composting process. The decrease in carbon dioxide (CO_2) evolution rates over the 12 weeks period showed the microbial activities subsiding as the co-compost piles neared stability. This was as a result of a decrease in the amount of biodegradable substrate and N available (Hsu and Lo 1999). However, the differences in CO_2 production rates observed for the different piles show that the rates may have been affected by the differences in the easily biodegradable materials and the feedstock ratios. This confirms an earlier report by Ayuso et al. (1996) which stated that microbial activities to a large extent depend on the nature of the organic matter added to soil or the compost mixtures. A C:N ratio less than 20 could be considered as a satisfactory maturation level of compost (Heerden et al. 2002). This assertion was confirmed by this study, where the final C:N ratios of the piles ranged between 12 and 20. Ammonium (NH_4-N) concentration decreased as compost piles approached maturity. In all cases, the Ammonium (NH_4-N) levels showed negative correlations (correlation coefficients ranging between 55% and 71% (R^2) with time. This is in agreement with Laos et al. (2002) that NH_4-N concentration decreases towards stable values at the end of the thermophilic stage of composting. However, a reverse situation in the case of nitrate (NO_3- N) concentration with positive correlation coefficients ranging between 65 and 90% (R^2) was observed. These observations also agree with observations by other authors (Sun 2006; Huang et al. 2004; Banegas et al. 2007). The production of NO_3-N is very important, because it is the most available N for plant uptake. Hence, N-mineralization is a crucial process, because it converts organic N into ammonium (NH_4^+) and nitrates (NO_3^-). In this study, N-mineralisation was very low (<11%), making organic N the major form of nitrogen in the piles. This makes the compost a slow release nitrogen source (as also expected in other compost), contrary to the mineral fertilizer, where all the nitrogen is in the plant available form. However, the co-compost may have a longer residual effect on the soil as the microbes continue to breakdown organic N to the available form. In general, the piles with higher percentages of the DFS had significantly higher total N than in the other treatment piles at the end of co-composting. This indicates that the increase in DFS component during composting with EFB and CPH could result in higher N content of the final co-composts. This confirms earlier reports by Kala et al. (2009). In the present study, total N varying from 1.05 to 1.76% falls within an earlier report by Kulhman et al. (1989). Nitrogen losses during composting depend on the materials (feedstock) used and on the pH values of the mixtures (Sanchez-Monedero et al. 2001). It may also vary depending on several environmental factors, such as aeration, moisture content, and temperature (Bishop and Godfrey 1983), and also the frequency of turning. High total nitrogen losses during composting—as seen in this study—are a common trend, and such as the results of this research are largely attributable to decreases in NH_4-N (Rao Bhamidimarri and Pandey 1996). de Bertoldi et al. (1985) reported that the initial C/N ratio of the material also affects loss of N during composting. This could probably explain why pile 2DFS:1EFB:1CPH recorded the least losses in N at the end of composting. P-losses in this study were surprisingly higher (82–88%) than the N-losses. The feedstock ratio did not seem to have any effect on P loss.

The losses were probably due to the higher solubility and subsequent leaching of phosphoric compounds in the compost materials as the process met with the minor raining season in the region. Thus, it may not be advisable to do co-composting of such or similar feedstock in the open without sheltering from rains as this might lead to a significant loss in nutrients. Total K on the other hand, which ranged between 0.7 and 1.0%, was higher than the recommended level (0.3%) suggested by Council of the European Communities (1986) in compost. The total K level recorded indicated that the major sources of K in the co-compost may have been CPH followed by EFB. 2 DFS:2 EFB:1CPH co-compost was selected as the best combination ratio, because it had relatively higher NPK content, had most pathogen reduction, though it had the least N-mineralised percentage. Even though this assertion may be subjective, the primary aim of composting DFS is to make it safe for handling and land application followed by the other agronomic benefits, such as organic matter and nutrient content.

Tomato seeds germinated and emerged faster in the RHB medium than in the FS-based compost amended media (FSC and FSC-RHB), indicating that the RHB medium may be preferred for seed germination. In addition, RHB had the lowest bulk density and the highest water holding capacity (WHC) amongst the three media types considered in this study. These may have provided the most favourable air and water infiltration for seed emergence. This agrees with an earlier report by Ball (1985) who reported that, a loose, porous medium is preferred for seed germination. Faecal sludge-based compost (FSC) recorded the lowest seed germination and emergence in this study and this may have been due to presence of certain inhibitory properties in the compost such as the high electrical conductivity (EC). This also confirms an earlier report by Wong and Chu (1985) that chemical constituents and properties of compost extracts, such NH_3, salts, ethylene oxide, heavy metals, and pH, affected seed germination and root elongation. Early vegetative growth of the tomato transplants after emergence was observed to be higher in the FS-based compost amended media (FSC and FSC-RHB) than in the RHB medium. This can be attributed to the relatively higher plant nutrients in the compost than the rice husks. According to Roe et al. (1997), plants that have adequate nutrition and water generally tend to have a higher shoot/root ratio than plants that are deficient in either. In this study, the lowest shoot/root ratio of transplants was recorded for RHB, while relatively higher ratios were recorded for the FS-based compost amended media, indicating that conditions in terms of water and nutrient availability for growth of transplants were more favourable in the compost amended media. However, between FSC and FSC-RHB, an interesting interaction may have occurred between the physical characteristics of the medium and its nutrient supplying capabilities. Transplants in the FSC were taller and slender compared to those in the FSC-RHB which had thicker stems. However, the other selected growth parameters measured were not significantly different. Stem diameter and root systems, according to Arenas (1999), are variables deemed most important by Florida Transplant Growers when it comes to transplant quality. After 21 days of treatment (sub-irrigation), transplants in the compost amended medium still showed superior growth qualities over transplants in RHB. This could be attributed to the relatively higher nutrient contents in the compost compared to RHB as discussed above. Similar observations were reported by Diaz-Perez et al. (2006). However, between transplants in FSC-RHB and FSC, there were no significant differences between the measured growth parameters except in the case of leaf chlorophyll content and root DM which were significantly ($p < 0.05$) higher in the FSC-RHB. This could mean that the relatively low bulk density of the FSC-RHB compared to the FSC medium may have provided lower resistance to the spread of the transplant root system to take up much water and nutrients. In addition, the FSC-RHB substrate may have provided better plant anchorage and nutrient uptake hence the more chlorophyll content found in the leaves. The direct cause of improved plant growth following compost extract (compost tea) application is often not clear (Shrestha et al. 2012). According to Merrill et al. (1997), the mineral nutrients extracted from compost can improve soil fertility directly. Other claims include a role for the extracted microbiota in improved mineralisation of soil organic matter and solubilisation of soil minerals, chelation of ions (Janzen et al. 1995), suppression/biocontrol of certain plant root and foliar diseases (Bernal-Vicente et al. 2008), and microbial production of plant growth promoting hormones, such as auxins (Garcia et al. 2002), or cytokinin-like substances (Arthur et al. 2001). In this study, transplants fertilized with CT-1 performed significantly ($p < 0.05$) better in terms of vegetative growth than CT-2. This can be due to higher nutrient content (nutritional effect) of CT-1. Similar results were also observed by Shrestha et al. (2012), where it was noted that growth benefit of compost tea was not directly biological in nature. They further observed that the high dose of compost extract may have had a positive impact on plant growth. Furthermore, responses to CT-2 were in most cases not significantly different ($p < 0.05$) from that of the control (water) indicating that CT-2 may have been too diluted resulting in lower nutrient content. Both inorganic fertilizer solutions (In-fert-1 and In-fert-2) had a higher positive effect on tomato transplant growth and leaf chlorophyll content than was observed in the compost tea treatments (CT-1 and CT-2) and control (water), indicating that a relatively higher nitrogen (N) concentrations may be preferred by the tomato transplants for growth and chlorophyll formation. However, there were no significant differences ($p < 0.05$) in transplants growth responses to In-fert-1 and In-fert-2 except for root DM, which was significantly higher in In-fert-2. This could probably mean that increasing the N concentration beyond 200 mg N/L may not necessarily result in any significant transplant growth and hence may not be cost effective. Similar results were also obtained by Kang and van Iersel (2004), where increasing concentration of N beyond 210 mg N/L did not further increase growth of *Salvia splendens.* In general, the root systems of the transplants in the compost tea treatments (CT-1 and CT-2) were poorly developed compared to transplants in the inorganic fertilizer solutions and control (water). This might be attributable to some sort of toxicity as a result of microbial activities around the root region. Comparing interactions between growing medium and inorganic fertilizer solution showed superior transplant qualities over that of the compost tea solutions and growth media. Interactions between compost tea solutions (CT-1 and CT-2) and growing medium showed that the highest total DM (shoot DM + root DM) was observed in the FSC medium fertilized with CT-1. This could be attributed to the higher nutrient contents in both FSC medium and CT-1 solution. A different observation was, however, made for the interactions between the inorganic fertilizer solutions and the medium types. Total DM was significantly ($p < 0.05$) reduced beyond FSC-RHB and In-fert-2 solution. This reduction in total DM could be as a result of the relatively higher EC levels arising from both the FSC medium and In-fert-2 solution, resulting in some form of toxicity or inhibition. Electrical conductivity (EC) levels approaching 3.0 dS/cm are considered higher than desirable (Davidson et al. 2000). This confirms earlier reports by Lumis et al. (2000) that reduced growth at the high fertilizer rate is the result of substrate EC and nitrogen levels being too high.

CONCLUSION

In conclusion, the study showed that co-composting DFS with EFB and CPH can be used as an alternative option for managing these wastes and to produce a suitable soil amendment. However, the quality of the compost could be further improved by reducing nutrient losses. Available options which were not tested in this study include carrying out the composting under sheds to protect the compost from rains and reduce leaching of nutrients, adding more bulking agents, such as biochar to trap nitrogen which could be lost through volatilization. Increasing the heap sizes could also increase the heap temperatures and further improve the sanitization of the compost. The fact that 2DFS:2EFB:1CPH was the best combination according to the criteria used indicates that this optimal ration also supports the idea of using this as a waste management solution as more human waste is used which subsequently can improve sanitation. Tomato seeds in this study germinated faster in RHB medium than in the other two FS-based compost amended growth medium, i.e., FSC and FSC-RHB.

However, early vegetative growth of the transplants was higher in quality in the FSC and FSC-RHB growth media than in the RHB medium. FS-based compost is, therefore, a suitable growing medium for tomato. Though the compost tea solutions (CT-1 and CT-2) affected the growth of tomato transplants positively, its effects could not be compared with the inorganic fertilizer solutions. Indeed, transplants fertilized with In-fert-2 showed the highest growth response. Doubling the nitrogen concentration in the inorganic fertilizer did not result in any significant growth of tomato transplants. The compost tea solution CT-1 is, however, recommended for organic production of tomatoes. However, further studies into the optimal rate and frequency of application of compost teas on tomato are recommended.

ACKNOWLEDGEMENTS

This study was made possible by financial support from African Water Facility (AWF)—the African Developments Bank (AfDB), the University of Berkeley, the International Water Management Institute (IWMI), and the CGIAR Research Program on Water, Land, and Ecosystems (WLE) through the project dubbed "Design for reuse—harvesting the value of effluents for sustaining the nutrients for sustaining the operation of sanitation facilities".

REFERENCES

1. Abu Qdais H, Al-Widyan M (2016) Evaluating composting and co-composting kinetics of various agro-industrial wastes. Int J Recycle Org Waste Agricult 5:273–280. doi:10.1007/s40093-016-0137-3
2. Adamtey N (2005) Evaluation of agricultural and agro-industrial residues for composting for agricultural use in Ghana: a case study in the Kwaebibirim District. Dissertation, University of Ghana
3. APHA-AWWA-WEF (2001) Standard methods for the examination of water and wastewater, 22nd edn. American Public Health Association, Washington D.C., p 1360. ISBN: 978-087553-013-0
4. Arenas M (1999) Coconut (*Cocus nucifera*) pith or coir as an alternative media for tomato transplant production. Dissertation, University of Florida
5. Arthur GD, Jäger AK, Van Staden J (2001) The release of cytokinin-like compounds from *Gingko biloba* leaf material during composting. Environ Exp Bot 45:55–61
6. Atiyeh RM, Edwards CA, Subler S, Metzger JD (2001) Pig manure vermicompost as a component of a horticultural bedding plant medium: effects on physicochemical properties and plant growth. Bioresour Technol 78:11–20
7. Ayuso M, Pascual JA, García C, Hernandez T (1996) Evaluation of urban waste for agricultural use. Soil Sci Plant Nutr 42(1):105–111
8. Ball V (ed) (1985) Ball redbook: greenhouse growing, 14th edn. Reston Publishing Co., Reston

9. Banegas V, Moreno JL, Moreno JI, Garcı'a C, Leon G, Herna'ndez T (2007) Composting anaerobic and aerobic sewage sludge using two proportions of sawdust. Waste Manag 27:1317–1327. doi:10.1016/j.wasman.2006.09.008
10. Bernal-Vicente A, Ros M, Tittarelli F, Intrigliolo F, Pascual JA (2008) Citrus compost and its water extract for cultivation of melon plants in greenhouse nurseries. Evaluation of nutriactive and biocontrol effects. Bioresour Technol 99:8722–8728
11. Bishop PL, Godfrey C (1983) Nitrogen variations during sludge composting. BioCycle 24:34–39
12. Black CA (ed) (1965) Methods of soils analysis, Part 2 - chemical and microbiological properties, Agronomy monograph No.9. American society of Agronomy, Madison, Wisconsin USA
13. Boot NLD, Scott RE (2008) Faecal sludge management in Accra, Ghana; Strengthening links in the chain. In: 33rd WEDC International Conference, Accra, Ghana
14. Bremmer JM (1965) Inorganic forms of nitrogen. In: Black CAA (ed) Methods of soil analysis - chemical and microbiological properties, Part 2. American society of Agronomy, Madison, Winsconsin USA, pp 1179–1206
15. Bustamante MA, Paredes C, Moral R, Agulló E, Pérez-Murcia MD, Abad M (2008) Composts from distillery wastes as peat substitutes for transplant production. Resource Conserv Recycle 52:792–799
16. Chang NB, Davila E (2007) Minimax regret optimization analysis for a regional solid waste management system. Waste Manag 27(6):820–832
17. Cofie OO, Kranjac-Berisavljevic G, Drechsel P (2004) The use of human waste for peri -urban agriculture in Northern Ghana. Renew Agric Food Syst 20(2):73–80
18. Cofie O, Kone D, Rothenberger S, Moser S, Zubruegg C (2009) Co-composting of faecal sludge and organic solid waste for agriculture: process dynamics. Water Res (Oxf) 43(18):4665–4675
19. Council of the European Communities (CEC) (1986) Council directive on the protection of the environment, and in particular of the soil, when sewage sludge is used in agriculture (86/278/EEC). Official Journal of the European Communities No. L 181/16-27
20. Davidson H, Mecklenburg R, Peterson C (2000) Nursery Management: administration and culture. Prentice Hall, Upper Saddle River
21. Day M, Shaw K (2000) Biological, chemical and physical processes of composting. In: Stofella PJ, Kahn BA (eds) Compost utilization in horticultural cropping systems, Lewis Publishers, Boca Raton, USA, PP 17–50
22. de Bertoldi M, Vallini G, Pera A (1985) Technological aspects of composting including modeling and microbiology. In: Gasser JKR (ed) Composting of agricultural and other wastes. Elsevier Applied Science Publishers, New York, pp 27–41
23. Diaz-Perez JC, Granberry DM, Germishuizen P (2006) Transplant growth and stand establishment of Bell pepper (*Capsicum annuum* L.) plants as affected by compost-amended substrate. International symposium on seed, transplants and stand establishment of horticultural crops; Translating seed and seedling physiology into technology. ISHS Acta Horticulturae 782: IV
24. Dick WA, McCoy EL (1993) Enhancing soil fertility by addition of compost. In: Hoitink HAJ, Keener HM (eds) Science and engineering of composting: design, environmental, microbial and utilization aspects. Wooster, Ohio, pp 622–633
25. Diver S (2001) Notes on compost teas: a supplement to the ATTRA publication: compost teas for plant disease control. Appropriate Technology Transfer for Rural Areas (ATTRA). Available: http://www.attra.ncat.org. Accessed 4 May 2013
26. Droffner ML, Brinton WF (1995) Survival of *E. coli* and Salmonella populations in aerobic thermophilic composts as measured with DNA gene probes. Int J Hygiene. Environ Med 197(5):387–397
27. Dumontet S, Dinel H, Baloda SB (1999) Pathogen reduction in sewage sludge by composting and other biological treatments: a review. Biol Agric Hortic 16:409–430

28. Eklind Y, Ramert B, Wivstad M (2001) Evaluation of growing media containing farmyard manure compost, household waste compost or chicken manure for the propagation of lettuce (*Lactuca sativa* L.) transplants. Biol Agric Hortic 19:157–181
29. Estevez-Schwarz I, Seoane-Labandeira S, Nunez-Delgado A, Lopez-Mosquera ME (2012) Production and characterization of compost made from garden and other waste. Pol J Environ Stud 21(4):855–864
30. Garcia MI, Cruz Sosa F, Saavedra AL, Hernandez MS (2002) Extraction of auxin-like substances from compost. Crop Res 24:323–327
31. Gerson R, Honma S (1978) Emergence response of pepper at low soil temperature. Euphytica 27:151–156
32. Grigatti M, Giorgioni ME, Ciavatta C (2007) Compost-based growing media: influence on growth and nutrient use of bedding plants. Bioresour Technol 98:3526–3534
33. Heerden IV, Cronje C, Swart SH, Kotze JM (2002) Microbial, chemical and physical aspects of citrus waste composting. Bioresour Technol 81:71–76
34. Hsu JH, Lo SL (1999) Chemical and Spectroscopic analysis of organic matter transformations during composting of pig manure. J Environ Pollut 104:189–196
35. Huang GF, Wong JWC, Wu QT, Nagar BB (2004) Effect of C/N on composting of pig manure with sawdust. Waste Manag 24:805–813. doi:10.1016/j.wasman.2004.03.011
36. Janzen RA, Cook FD, McGill WB (1995) Compost extract added to microcosms may simulate community-level controls on soil microorganisms involved in element cycling. Soil Biol Biochem 27:181–188
37. Jeong Y, Kim J (2001) A new method for conservation of nitrogen in aerobic composting process. Bioresour Technol 79:129–133
38. Kala DR, Rosenani AB, Fauziah CI, Thohirah LA (2009) Composting oil palm wastes and sewage sludge for use in potting media of ornamental plants. Malays J Soil Sci 13:77–91
39. Kala DR, Rosenani AB, Thohirah LA, Fauziah I, Ahmad SH (2012) Oil palm waste – sewage sludge compost as a peat substitute in a soilless potting medium for Chrysanthemum. Glob J Sci Front Res Agric Biol 12(2): Version 1
40. Kang J, van Iersel MW (2004) Nutrient solution concentration affects Shoot: root ratio, Leaf Area ratio, and growth of sub irrigated Salvia (*Salvia splendens*). HortScience 39(1):49–54
41. Kone D, Cofie O, Zurbrugg C, Gallizzi K, Moser D, Drescher S, Strauss M (2007) Helminth eggs inactivation efficiency by faecal sludge dewatering and co-composting in tropical climates. Water Res 41(19):4397–4402
42. Kulhman LR, Dale B, Groenhof AC, Young TFE (1989) Windrow composting of sewage sludge and other wastes. Presented at the international recycling congress, Geneva, Switzerland, 19–22 Jan 1993
43. Laos F, Mazzarino MJ, Walter I, Roselli L, Satti P, Moyano S (2002) Composting of fish offal and biosolids in northwestern Patagonia. Bioresour Technol 81:179–186. doi:10.1016/S0960-8524(01)00150-X
44. Lumis G, Purvis P, Taurins L (2000) Flood irrigation of container-grown Euonymus and Thuja as affected by fertilizer rate and substrate. J Environ Hort 18(1):13–17
45. Mann HT (1976) Sanitation without sewers-the aqua privy. Building research establishment, Overseas building notes No. 168, water research centre, Stevenage, UK, p 6
46. Merrill R, Hoberechi K, McKeon J (1997) Organic teas from composts and manures. In: Organic farming research foundation project report. Cabrillo Community College
47. Okalebo JR, Gathua KW, Woomer PJ (2002) Laboratory methods of soil and plant analysis: a working manual, 2nd edn. TSBF-CIAT and SACRED Africa, Nairobi
48. Ostos JC, Lopez-Garrido R, Murillo JM, Lopez R (2008) Substitution of peat for municipal solid waste and sewage sludge-based composts in nursery growing media: effects on growth and nutrition of the native shrub *Pistacia lentiscus* L. Bioresour Technol 99:1793–1800

49. Pant AP (2009) Vermicompost extracts influence growth, mineral nutrients, phytonutrients and antioxidant activity in pak choi (*Brassica rapa* cv. Bonsai, Chinensis group) grown under vermicompost and chemical fertilizer. J Sci Food Agric 89:2383–2392
50. Papafotiou M, Phsyhalou M, Kargas G, Chatzipavlidis I, Chronopoulos J (2004) Olive-mill wastes compost as growing medium component for the production of poinsettia. Sci Hortic 102:167–175
51. Perez-Murcia MD, Moral R, Moreno-Caselles J, Perez-Espinosa A (2006) Use of composted sewage sludge in growth media for broccoli. Bioresour Technol 97:123–130
52. Rao Bhamidimarri SM, Pandey SP (1996) Aerobic thermophilic composting of piggery solid wastes. Water Sci Technol 33(8):89–94. doi:10.1016/0273-1223(96)00265-X
53. Ribeiro HM, Romero AM, Pereira H, Borges P, Cabral F, Vasconcelos E (2007) Evaluation of a compost obtained from forestry wastes and solid phase of pig slurry as a substrate for seedlings production. Bioresour Technol 98:3294–3297. doi:10.1016/j.biortech.2006.07.002
54. Roe NE, Stoffella PJ, Graetz D (1997) Composts from various Municipal solid waste feedstock affect vegetable crops. I. Emerge and seedling growth. J Am Soc Hort Sci 122(3):427–432
55. Sanchez-Monedero MA, Roig A, Paredes C, Bernal MP (2001) Nitrogen transformation during organic waste composting by the Rutgers system and its effects on pH, EC and maturity of the composting mixtures. Bioresour Technol 78:301–308
56. Scheuerell JS (2004) Compost tea production practices, microbial properties, and plant disease suppression. 1 International conference on soil and compost eco-biology, Leon–Spain, 15–17 Sept 2004
57. Scheuerell S, Mahaffee W (2002) Compost tea: principles and prospects for plant disease control. Compost Sci Util 10:313–338
58. Schwartzbrod J, Gaspard P (1998) Quantification and viability determination for helminth eggs in sludge (Modified EPA France, 1993 Method). Faculty of Pharmacy, University "Henri Poincare" Department of Microbiology B.P.403 F-54001 Nancy Cedex/France, pp 3–6
59. Shrestha K, Walsh KB, Midmore DJ (2012) Microbially enhanced compost extract: does it increase solubilisation of minerals and mineralisation of organic matter and thus improve plant nutrition? J Bioremed Biodegrad 3:149. doi:10.4172/2155-6199.1000149
60. Singh R, Kim J, Sheperd MW Jr, Luo F, Jiang X (2011) Determining thermal inactivation of *Escherichia coli* 0157:h7 in fresh compost by simulating early phases of the composting process. Appl Environ Microbiol 77(12):4126–4135
61. Strauss M, Montangero A (2002) Faecal sludge treatment. A report of Swiss Federal institute for environmental science and technology. IHE Delft 2002:9–10
62. Sun X (2006) Nitrogen transformation in food-waste composting. Dissertation, University of Regina
63. Toledano J, Kumar S, Danielou M (2004) Government of Ghana-ministry of food and agriculture. The World Bank-Africa Region-Rural Development, Tree Crops Development Initiative, Reconnaissance Mission Report, Pretoria
64. USDA, USCC (2001) Test methods for the examination of composting and compost (TMECC). The United States Department of Agriculture and the United States Composting Council. Accessed 31 May 2013. http://tmecc.org/tmecc//
65. USEPA (1993) Standards for the use or disposal of sewage sludge 40 code of federal regulations part 503. US Environmental Protection Agency, Washington, DC
66. Vengadaramana A, Jashothan PTJ (2012) Effect of organic fertilizers on the water holding capacity of soil indifferent terrains of Jaffna peninsula in Sri Lanka. J Nat Prod Plant Resour 2(4):500–503

67. Walkley A, Black AI (1934) An examination of the degtjareff method for determining soil organic matter and proposed modification of the chromic acid titration method. Soil Sci 37:29–38

68. Wong MH, Chu LM (1985) The responses of edible crops treated with extracts of refuse compost of different ages. Agric Waste 14:63–74

69. Zhang L, Sun XY, Tian Y, Gong XQ (2013) Effects of brown sugar and calcium superphosphate on the secondary fermentation of green waste. Bioresour Technol 131:68–75. doi:10.1016/j.biortech.2012.10.059

CHAPTER 18

Comparison of Traditional and Modern Deficit Irrigation Techniques in Corn Cultivation using Treated Municipal Wastewater

Sayyed-Hassan Tabatabaei[1] • Rohollah Fatahi Nafchi[1] • Payam Najafi[2] • Mohammad Mehdi Karizan[1] • Zohreh Nazem[1]

[1] *Department of Water Engineering, Faculty of Agriculture, Shahrekord University, Shahrekord, Iran*

[2] *Department of Water Engineering, Faculty of Agriculture, Islamic Azad University, Isfahan Branch (Khorasgan), Isfahan, Iran*

ABSTRACT

Purpose

Declining water resources, increasing drought period and increasing irrigated area cause a shift to deficit irrigated production which is not based on full water requirement. This study was conducted to investigate the impacts of two different deficit irrigation methods, including traditional deficit irrigation (TDI) and partial root-zone drying (PRD) on water use efficiency (WUE) in corn cultivations located at the lands close to the Shahrekord wastewater treatment plant.

Methods

A factorial design was employed, consisted of fifteen treatments and three replications. The first factor was three water applications, including 60, 80 and 100 (control) percent of plant water requirement. The second factor was three water sources, including 100% fresh water (FW), 50% water and 50% wastewater (5050), 100% wastewater (WW). T-Tape irrigation type system was used for irrigation of corn (*Zea mays* L.). The third factor was water management in three levels: full irrigation (FI), PRD and TDI.

Results

The result showed that dry and fresh weight, leaf area index (LAI), dry biomass percentage and WUE were affected by water requirement at 5% probability level expect of height plant that is affected at 1% probability level. Water quality was affected on all the study indicators of corn plant expect of WUE. WUE among PRD by deficit irrigation treatments were higher than TDI and it was the minimum in FI. The dry and wet weight and LAI were maximum at FI and then PDR_{80} than TDI_{80}. The height of corn plants were high at FI and TDI_{80} than PRD_{80}.

Conclusion

Finally, wastewater application in treatment named 5050-PRD_{80} and 5050-TDI_{80}compensated water deficit in WUE, LAI and dry biomass percentage. PRD method recognized more suitable than TDI for corn plants.

Keywords: Deficit irrigation Partial root-zone drying Full irrigation Wastewater

INTRODUCTION

Water shortage is assumed as major agricultural production constraint. Therefore, the relationship between crop yield and water use efficiency (WUE) has constantly been the focus of many agricultural researchers in the arid and semi-arid regions (Zhang and Oweis 1999).

In all agricultural systems, low WUE can occur when soil evaporation is high (considering crop evapotranspiration), early growth rate is slow, and water application does not correspond to crop demand and when shallow roots are unable to utilize deep water in the soil profile (Gallardo et al. 1996). Also, in cases of drought, nutrients are not easily available from inorganic sources of soil nutrients for plant (Chemura 2014). New innovations are particularly important in water-scarce regions with the aim of saving irrigation water, and thereby increasing crop WUE (Gencoglan et al. 2006). To do so, one way is to develop new irrigation scheduling techniques such as traditional deficit irrigation (TDI), which are not necessarily based on full water requirement.

Traditional deficit irrigation method is a practice of applying less water than crop evapotranspiration demand with the intent of imposing a managed level of water stress to the crop (Grant et al. 2008). The strategy is to maintain plants under certain water deficit for a prescribed duration of the growth season. The aim is to control reproductive and vegetative growth and to improve WUE and crop quality. In addition, correct application of TDI requires a thorough assessment of yield reduction's economic impact caused by drought stress (Geerts and Raes 2009). Maximum yield could be obtained with the fulfillment of complete crop water requirements. However, practicing TDI would increase in the irrigated area or frequency of cultivation. TDI provides a possible solution to the dilemma of sustained and/or increased productions (Geerts et al. 2008).

The effects of deficit and supplemental irrigation are diverse at different crop growth stages and in different soil moisture conditions. So, it is indispensable to supply enough water for crops at the stages; that they are most sensitive to water stress, and thereafter, irrigation water volume could be reduced in other stages. It is then important to monitor soil moisture and supply water based on the soil condition. Partial root zone drying (PRD) is now considered as a general technique which is applied various methods at the field. The PRD is a more recent irrigation technique developed from TDI (Yazar et al. 2009). These include the controlled alternative drip irrigation of a part of root zone, applied in a vertical soil profile, controlled alternative border irrigation, controlled alternative furrow irrigation (FI), controlled alternative surface irrigation, and subsurface irrigation.

Zegbe-Domínguez et al. (2003) illustrated that leaf water potential, total plant fresh weight and total dry weight of tomato crop were lower in the TDI than the FI and PRD. In the PRD irrigation, WUE improved by 83%. In the TDI, dry mass that classified into stems and leaves was higher than dry mass of FI and PRD; nevertheless, dry mass that categorized into fruits was lower in PRD and TDI than the dry mass of FI. Various methods of water conservation have been investigated in agriculture, such as alternative PRD, TDI, and drip irrigation. In that respect, researchers have shown that WUE can be enhanced (Li et al. 2010). In general, these techniques are a trade-off: a lower yield for a higher WUE. This suggests that the PRD is a better water saving option than TDI. Saved water through PRD can be used for irrigation of additional cultivated areas (Patel and Rajput 2013). Optimum WUE in TDI is obtained by an economic analysis using yield and cost functions as described by English and James (1990).

Mohammad and Mazahreh (2003) studied the effect of wastewater irrigation on soil chemical properties and found that secondary treated wastewater could improve soil fertility parameters. However, more efficient treatment is recommended to reduce the salt content. In addition, proper irrigation management and periodic soil quality parameters' monitoring are required to minimize negative effects on the soil.

Sahin et al. (2014) showed root and sugar yields reduced as 22 and 24% in PRD treatment than FI, respectively. But irrigation water saving was about 42% by PRD treatment. They suggested that PRD technique could be useful strategy by drip-irrigated sugar beet in areas with water shortage regions.

Cantore et al. (2014) obtained enough yield and high quality tuber from 50% less water-applied treatment by TDI as like the FI treatment.

In Isfahan region, Iran, municipal and industrial treated wastewater used as a water resource or irrigation of different production (Tabatabaei et al. 2012). So, it is more necessary to select the best and new irrigation techniques with the management to improve efficiency of irrigation system (Najafi et al. 2016).

Therefore, the objective of this paper was to evaluate the type of irrigation system including PRD and to compare it with FI system as TDI which they both are water-saving irrigation practices. Furthermore, their effects on plant's indicators, such as dry and fresh weight, plant height, dry matter percentage and WUE were investigated.

MATERIALS AND METHODS

Site description

The Shahrekord wastewater treatment station is located 7 km south-west of the Shahrekord city at 32°20′N latitude and 50°51′E longitude. Shahrekord has a semi-arid climate using combination climatologic methods (Kham-chin Moghadam and Rezaei Pajand 2009). The average height of this region above the sea level and rainfall are about 2060 m and 321.5 mm, respectively (without any rainfall during experimental period). The average temperature of this city is 12.5 °C and the maximum and minimum relative humidity are 71 and 16%, respectively.

Soil and wastewater characteristic

Field experiments were carried out during the July–October 2010 (growth period). In general, the soil texture is clay in which its characteristics and also chemical analysis of treated wastewater by sequence batch reactor (SBR) method and water sources are presented in Tables 1 and 2 (Karizan 2011).

Table 1 Chemical and physical properties of soil

Depth (cm)	**EC (dS/m)**	**pH**	**Soil texture**	ρ_b **(g/cm³)**	**FC (%)**	**PWP (%)**
0–45	1.8	7.9	Clay	1.19	37	18

Table 2 Chemical analysis of pretreated municipal wastewater and water sources using as irrigation water

Sources	**EC (dS/m)**	**pH**	**TDS (mg/l)**	**TSS (mg/l)**	**SAR (mol/l)$^{1/2}$**	**BOD (mg/l)**	**N-NO_3^-(mg/l)**	***K* (mg/l)**	**PO_4^{-3}(mg/l)**
Wastewater	0.75	7.8	450	30	2.2	18.75	16	14	30.5
Water	0.3	7.45	38	0	0.13	3.96	2.61	0.21	15

Experimental design

The field selected for the research was plowed and fertilized using macro- and micro-fertilizers in July 2010 based on the soil tests. Each treatment was designed in the format of furrow irrigation style. Experimental plot size was 3 × 4 m with 1 m distance between each plot. The distance between the planting rows was 75 cm and the distance between corn plants was 10 cm (Fig. 1). Research was implemented based on the factorial (split plot) design in three replicates and fifteen treatments. Quality and water management were the two factors, so that three water quality levels made up the main plots (Table 3). Water used for irrigation contained three quality levels: fresh water (FW), a mixture of 50% fresh water and 50% wastewater (5050), and finally 100% wastewater (WW).

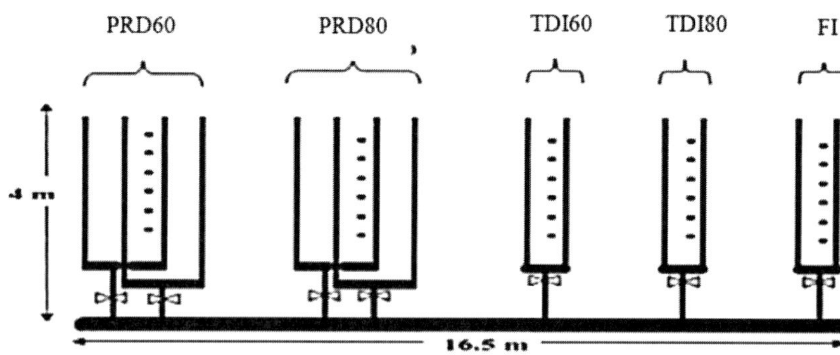

Figure. 1 Layout of the system. *PRD* partial root-zone drying, *TDI* traditional deficit irrigation, *FI* full irrigation. ⋈ Valve; *asterisks* corns; ▬ lateral

Table 3 Fifteen treatments applied in this study

Treatments		
WW-FI_{100}	5050-FI_{100}	FW-FI_{100}
WW-PRD_{80}	5050-PRD_{80}	FW-PRD_{80}
WW-TDI_{80}	5050-TDI_{80}	FW-TDI_{80}
WW-PRD_{60}	5050-PRD_{60}	FW-PRD_{60}
WW-TDI_{60}	5050-TDI_{60}	FW-TDI_{60}

FW 100% fresh water, *5050* 50% fresh water plus 50% wastewater, *WW* 100% wastewater, *FI* full irrigation, *PRD* partial root-zone drying, *TDI* traditional deficit irrigation, *60 AND 80* 80 and 60 percent of the estimated crop water requirement

Water management was the third factor determined in three levels: full irrigation (FI), partial root-zone drying (PRD) and the traditional deficit irrigation (TDI). Three levels of water application factor were implemented in 100, 80 and 60 percent of estimated crop water requirement (creating water stress of zero, 20 and 40 percentages, respectively). The reference evapotranspiration (ET_0) (mm/day) was estimated using the FAO Penman–Monteith equation and employing daily data from a nearby automatic weather station. Crop water requirements were calculated by multiplying the ET_0 values with the corn crop coefficients (K_c) given by Kheirabi et al. (2009) and E_{pan} as 5–8.5 mm/day. Concerning irrigation system, drip irrigation (T-tape) was used and was conducted 8 days after planting in late July. Corn crops were irrigated every three days. After the growth period, the plant physiological parameters such as shoot's fresh and dry weight and leaf area index (LAI) were measured. Plant height was measured in each treatment using a meter scale and the biomass (the above part of the ground). Thereafter, it was cut into small pieces and dried at 70 °C to reach a constant weight for the estimation of dry matter. Dry biomass percentage (DBP) was calculated using the following equation.

DBP = (Sample total dry weight / Sample total fresh weight) × 100

Soil samples were collected from 0 to 45 cm soil depth. Air-dried soil samples were sieved through a 2-mm sieve to determinate physicochemical soil properties. Electrical conductivity (EC) and acidity of soil (pH), wastewater and water were identified by Jenway EC and pH meters (Page et al. 1991), respectively. Cation like as Ca and Mg was determined by titration method with 0.01 N EDTA (Klute 1986) and sodium (Na) and potassium (K) were measured by flame photometry. Sodium adsorption ratio (SAR) was defined by the relation.

Soil's phosphor was measured using the method described by Olsen et al. (1954), soil texture was determined by hydrometer (Lee and Bauder 1986), total nitration by Kjeldahl digestion (Page et al. 1991), bulk density by clod and particle density by pycnometer (Klute 1986), total suspended solids (TSS) by gravimetric analysis (APHA 1975), total dissolved solids' (TDS) by Canadian Council of Ministers of the Environment (2011).

Afterwards, analysis of variance was performed and treatments' averages were compared with the least significant difference test (LSD) with 5%

probability level. Statistical analysis was performed using "Statistical software" and also "Excel software" was applied for data management.

RESULTS AND DISCUSSIONS

Dry and fresh weight of corn shoots

The results showed that fresh weight of corn shoots in different treatments had significant difference at 5% level of probability. Fresh weight' mean of crop that irrigated by FI (5940 g/plant) was significantly higher than the fresh weight' mean of crop planted under TDI_{60} irrigation treatment (492.33 g) (Fig. 2). Water stress in TDI was the main reason to reduce shoot' fresh weight at 5% probability level. Although, wet and dry weight of corn plans of TDI_{60} plus PRD_{60} and TDI_{80} plus PRD_{80} together in each group were in the same probability level. Although, both PRD treatments (PRD_{60} and PRD_{80}) were higher than TDI treatments (TDI_{60} and TDI_{80}) (Fig. 2).

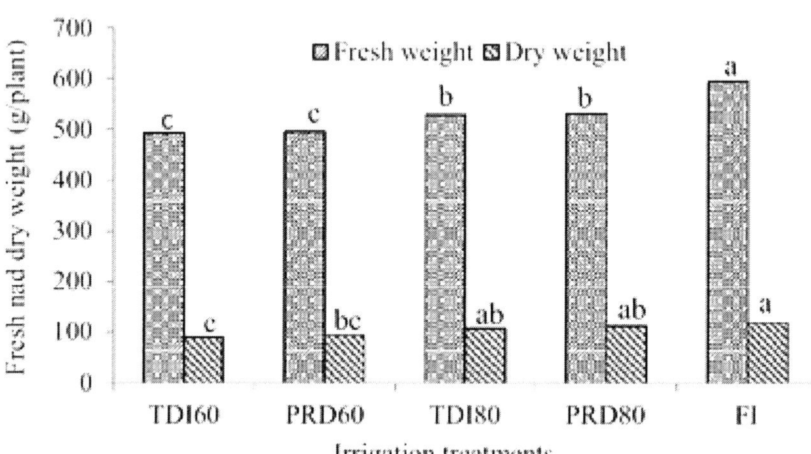

Figure. 2 Effect of irrigation treatments on dry and fresh weight of corn shoots ($p < 0.05$)

Nagaz et al. (2013) indicated that shoot lost its fresh weight rapidly in water stress situation. On the other hand, the corn plants irrigated with PRD had greater fresh weight in comparison with traditional irrigation system (Li et al. 2010); nevertheless, it was not significant at 5% probability level.

More dry weight of the corn shoots (119.7 g/plant) were observed in FI comparing TDI_{60}(91.44 g/plant). Also, dry weight of TDI_{80} with 108. 65 g/plant decreased to 91.44 (g/plant) in TDI_{60} (Fig. 2). Water deficiency primarily affects crops by reduction of dry matter accumulation (Karam et al. 2003) in which happens through TDI diminishing dry matter development (Lopez et al. 1996) because of water tension, stomata conductance and photosynthesis rate (Nazemosadat and Kazemini 2008). In this experiment, irrigation with 5050-treatment reached a peak of 556.3 g fresh weight/plant. It was then followed by WW-treatment used as irrigation water. FW irrigation treatment with 505.33 g fresh weight/plant was applied at the end of grouping (Fig. 3). The rate of fresh weight was raised from 4 to 10% using 5050-treatment as the first treatment and using WW-treatment as the second treatment, respectively ($p < 0.01$). Also, it was seen that 5050-treatment and WW-treatments resulted in 30.3 and 17.2% increase in dry weight ($p < 0.01$).

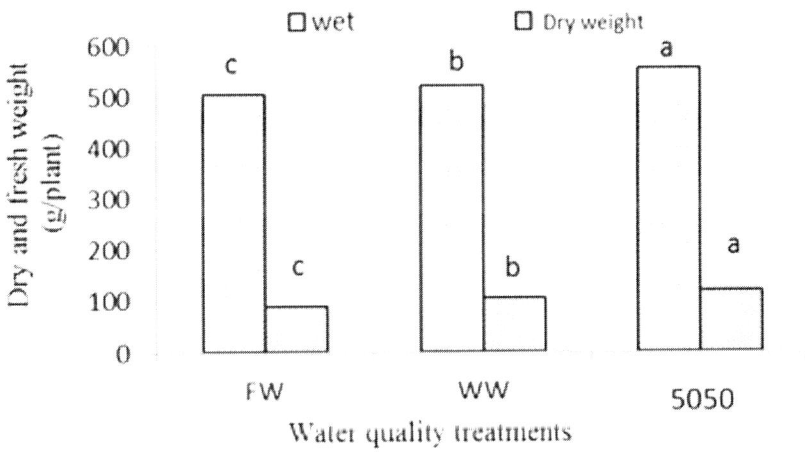

Figure. 3 Effect of water quality on dry and fresh weight of corn shots ($p < 0.01$)

Also, Rousan et al. (2008) reported that dry weight was increased more under FI with fresh water and wastewater than deficit irrigation. This trend was perceived to be similar in the yield response of Dorjia et al. (2004). In another study implemented by Fulai et al. (2006), the results showed that deficit irrigation method reduced leaves' dry matter and in contrast increased roots' dry matter of potato' stems. Dorjia et al. (2004) in the course of their research on hot pepper find that total fruit's fresh mass per plant in PRD was 586.5 g higher than its amount in TDI; whereas it was 724.1 g lower than its amount in control irrigation treatment. However, there was no considerable difference in total fruit' dry mass among the treatments. These results were consistent with our findings as well. The ratio of $N-NO_3$ (mg/l) in wastewater was six times bigger than their volumes in FW-treatment; therefore, using WW-treatment as irrigation water increased more $N-NO_3$ concentration of corn' shoots. On the other hand, analysis of variance shoots' fresh weight demonstrated that interaction between water quality and TDI method was significant at 1% probability level. The results uncovered that the least and the most of yield were achieved using TDI_{60} (585.3 g) and 5050-FI (468.7 g) treatments, respectively. Also, WW-FI reduced two percent of yield in comparison with FW-FI due to nutrient toxicity. However, there was a slight rise in yield of WW-PRD_{80} and WW-PRD_{60} treatments.

Overall, PRD irrigation method using wastewater as irrigation water brought about greater yield than using FW-TDI method. Unlike fresh weight, irrigation with wastewater did not have any significant difference in dry weight instead; it indicated that there was only an increase in water percentage. Findings showed that the biggest dry weight was occurred in 5050-FI. FW-TDI_{60} was located at the end of grouping. Available nutrients of wastewater transmitted a signal from roots to shoots for strengthening of plant stomata (Li et al. 2010).

Dry biomass percentage (DBP)

In this study, the ratio of samples' total dry weight to samples' total fresh weight of samples is DBP. The amounts of DBP in all treatments were significantly different at 1% probability level. Concerning this parameter, PRD_{80} had the biggest DBP (21.29%) and TDI_{80} (with 18.51%) was placed at the end of this category (Fig. 4) because of its soil moisture stress and photosynthesis rate. Also, the DBP of TDI_{80} and FI treatments were 20.56 and 20.23 and there was no significant difference with the maximum DBP in PRD_{60} (Fig. 4). As it was already said, water deficiency primarily affects crops through prevention of dry matter accumulation (Karam et al. 2003) due to reduced dry matter development (Lopez et al. 1996). Dorjia et al. (2004) represented that the DBP to roots was equal among all different treatments; nevertheless, a higher allocation was revealed in stems and leaves in TDI and PRD plants comparing the control plants. In addition, wastewater irrigation might have an impact on the dissolution of the nutrient may be due to reduce the microbial activity in the soil and resulting in slower release of plant available nutrients (Shammi et al. 2016). In this research, applying lower irrigation water amount by TDI than FI techniques cause to increase more DBP.

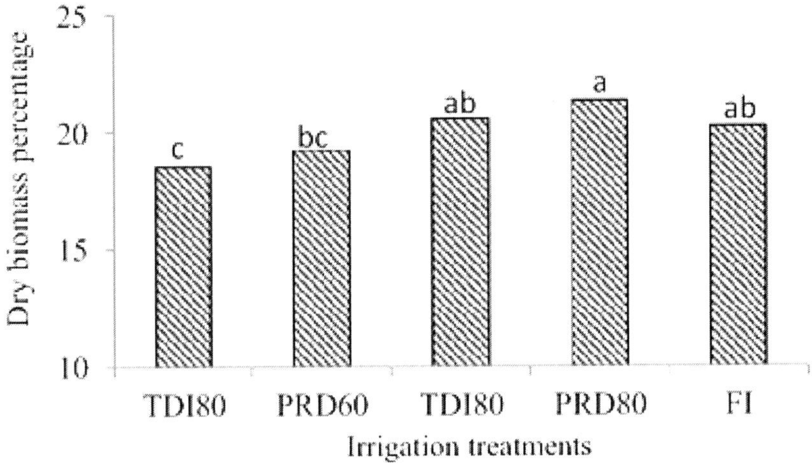

Figure. 4 Effect of irrigation treatments on dry biomass percentage ($p < 0.05$)

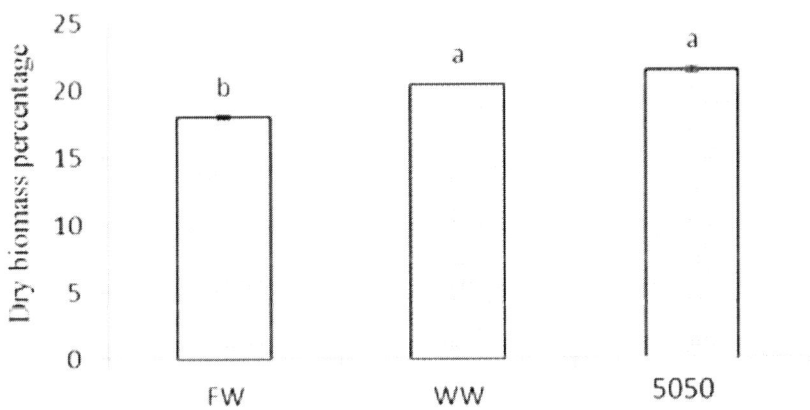

Figure. 5 Effect of water quality on dry biomass percentage ($p < 0.05$)

In addition, analysis of variance showed that DBP had a significant difference at 5% probability level based on different irrigation water methods (Fig. 5).

The DBP means the ratio of total dry weight to total fresh weight of samples. The highest and lowest rates were reported applying 5050-WW with 21.47% and FW-treatment with 18.01%, respectively.

Rousan et al. (2008) illustrated that regardless of the water quality, TDI results in lower DBP. In their view, the reason was that amount of applied irrigation water was significantly less than the tomato crop's water requirement. Other researchers attributed improved plant growth with wastewater to improved soil condition, and in particular to enhanced soil organic matter content (Mohammad and Mazahreh 2003; Yonts 2010).

Leaf area index (LAI)

Leaf area is greatly changed with soil water availability. The effect of soil dryness on physiological indices was small over a certain range of soil water deficit started from 0 to approximately 40% (Patanè 2011). Reducing leaves' number and size are a result of cell behavior sensitive to water shortage (Sawia et al. 2002; Atti et al. 2004).

The findings indicated that water irrigation volume had a significant impact on LAI at 5% probability level (Fig. 6). According to the graph, a reduction of 20 and 40% in irrigation water happened as a result of TDI consequently diminished 9.3 and 25% of LAI. In addition, statistics demonstrate that PRD_{60} method decrease more effectively the LAI than the TDI_{60} method (Fig. 6). This could be due to transmitted signal from the roots to the shoots ordering them to decrease leaf growth and increase production of abscisic acid (ABA), and finally to close leaf stomata. Similar results were reported by Liu et al. (2006) and Shahnazari et al. (2008). The findings also represented that in FI treatment the maximum and minimum LAI across the period were 7.17 and 5.2, respectively.

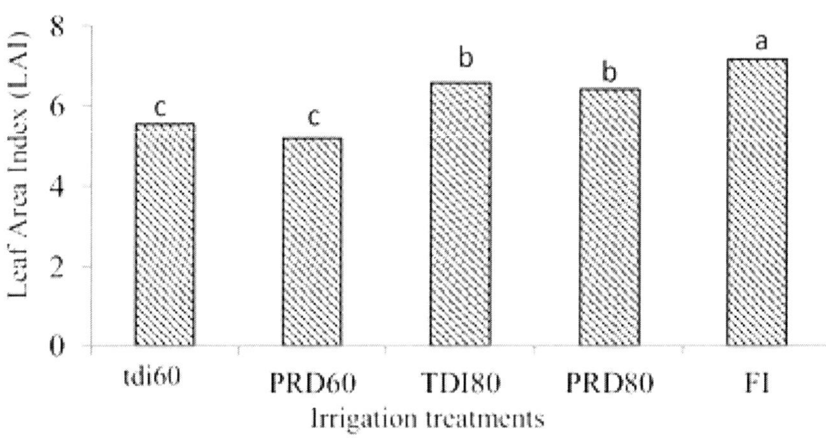

Figure. 6 Effect of irrigation treatments on leaf area index (LAI) ($p < 0.05$)

According to the research and based on the statistical comparison, water quality did not have any special effect on LAI. Nonetheless, the largest and least LAI belonged to 5050-treatment and FW-treatment with 6.24 and 6.13, respectively (Fig. 7).

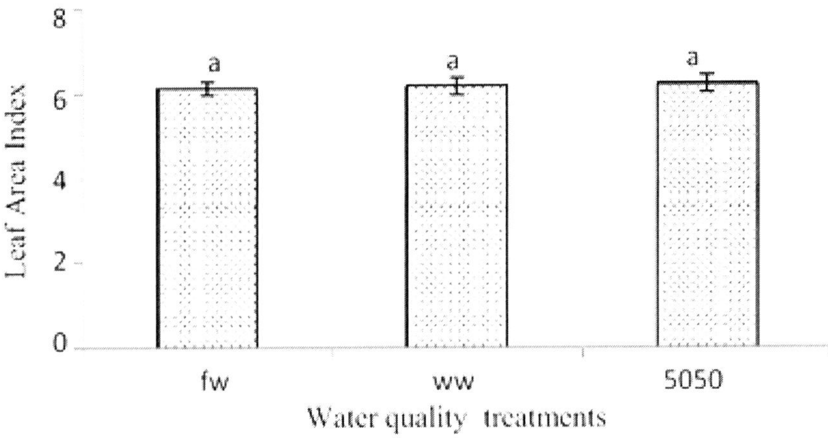

Figure. 7 Effect of water quality on leaf area index (LAI)

Also, this finding is similar to the results of Hirich et al. (2011) in which a local variety of chickpea was planted under TDI (50% full irrigation) with treated domestic wastewater. The TDI applied for corn plants during vegetative growth lead to a greater proportion of its root systems developed deeper in the soil profile, which could lead to better use of stored soil water. Corn responded to drier soil conditions by increasing the proportion of roots deeper in the soil (Benjamin and Nielsen 2006). On the other hand, terminal water stress decreased the total plant dry matter including LAI and leaf weight (Sharma et al. 2007; Labidi et al. 2009; Behboudian et al. 2001).

Based on the results, a decrease in LAI was seen in all treatments expect FW-FI. This result similar to Yonts (2010) results that indicated the leaf area index of sugar beet planted under TDI was approximately 14% less than FI.

Also, the most and least LAI were distinguished in 5050-FI (7.5) and 5050-PRD_{60} treatments (4.8).

Plant height

One of the growth indexes that are affected strongly by water shortage is plant height (Atti et al. 2004; Sawia et al. 2002). In this research, plant height was measured three times (stages) but the results of two stages of height of corn plants were showed in Fig. 8.

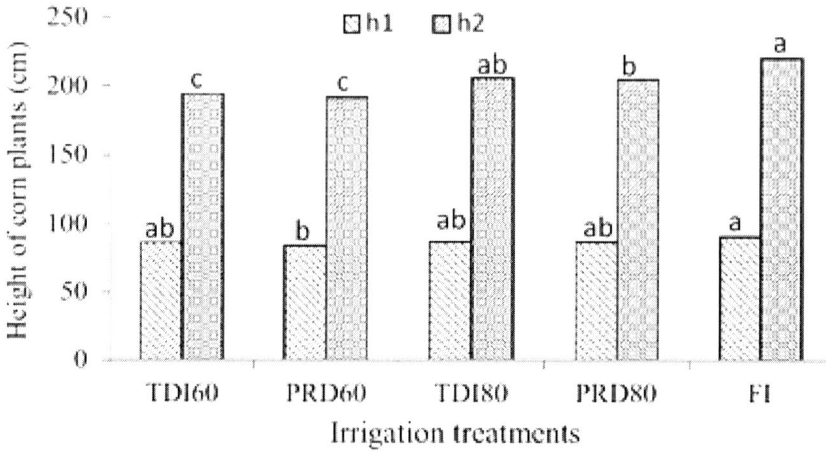

Figure. 8 Effect of irrigation treatments on height of corn plants ($p < 0.01$). h1: 15 days before applied irrigation treatments. h2: 50 days in the end of experiment

Analysis of variance showed that height of corn plants had a significant difference at 1% probability level based on the irrigation methods. The TDI_{60} and PRD_{60} were at the same statistical level in the end of experiment. Although there is significant difference between these treatments on height of corn plants on 15 days before applied irrigation treatments (Fig. 8). On the other hand, three treatments included TDI_{80}, PRD_{80} and FI did not have significant difference in the first stage of measuring height of corn plants but using FI method caused to increase height of corn plants than TDI and PRD irrigation techniques.

Full irrigation had the maximum height of corn plants among irrigation treatments (Fig. 8). Also, PRD_{80}, TDI_{80}, PRD_{60} and TDI_{60} caused 7.1, 6.4, 12.8 and 12% of height reduction, respectively, in comparison with FI. The reduced plant height in all irrigation treatments was significant expect TDI_{80} at 1% probability level. However, there was not seen any significant difference between PRD and TDI irrigation methods. This outcome is similar to the results of Du et al. (2006) on cotton plants. It seems there is a genetic correlation between yield and plant height, cultivated plants with water stress resistance is able to maintain against reduced plant height.

Yonts (2010) reported that sugar beet plant height was 127–305 mm less in deficit irrigation method than full irrigation treatments. Also, variance analysis of interactive effects between water quality and water quantity of Yonts' study showed that there was no significant difference at 5% probability level.

The analysis of variance showed that height of corn plants had a significant difference at the 5% probability level based on water quality. The results indicate that using 5050-FI was the best composition of irrigation water to increasing height of corn plant (Fig. 9).

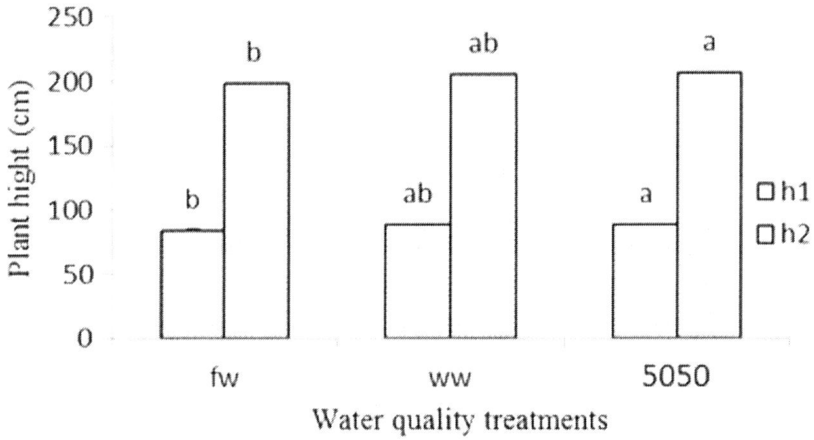

Figure. 9 Effect of water quality on height of corn plant ($p < 0.05$). h1: height of corn plants on 15 days before applied irrigation treatments. h2: height of corn plants on 50 days in the end of experiment

Chemura (2014) showed in the research on coffee, irrigation water amounts in terms of height have significant effect ($p < 0.05$). Coffee plants were tallest and thickest plant stems when supplied the more irrigation water levels (1000 ml) and shortest and thinnest under the lowest irrigation amounts (500 ml) because of inorganic fertilizers are the most effective at high irrigation levels while organic manure performs better than inorganic fertilizers under low irrigation water amount.

Water use efficiency (WUE)

Water use efficiency, if defined as the biomass accumulation over water consumed may be fairly constant for a given species in given climate. WUE can be enhanced by less irrigation. However, such enhancement is largely a trade-off against lower biomass production (Zhang and Yang 2004). The maximum WUE was belonged to PRD_{60} with 3.09 (kg dry matter/m^3 water using) compared to the minimum WUE in FI treatment with 2.47 (kg dry matter/m^3 water using). Sahin et al. (2014) indicated that PRD technique increased by 34.9% WUE compared to FI of sugar beet (*Beta vulgaris* L.) because of increase root yield during 2-year study.

Overall, PRD had the WUE much more than TDI and it was significant at 5% probability level. It was similar to the results of Shahnazari et al. (2008) and Liu et al. (2006) on potato; Gencoglan et al. (2006) on green bean; De la Hera et al. (2006) on fruit quality in field-grown wine grapes. On the other hand, Hu et al. (2009) and Wang et al. (2008) reported that there was a significant rise in WUE for potato plant, when used the PRD to TDI method in pot.

Water use efficiency like other indicators dry and fresh weight shoots; plant height was being more affected by the amount of irrigation water than TDI methods. It is clear from the data that the amount of WUE in TDI_{60} to TDI_{80} and then TDI_{80} to FI were significant at 5% probability level (Table 4).

Table 4 Effects of irrigation treatments and water quality on water use efficiency (kg/m^3)

	TDI_{60}	**PRD_{60}**	**TDI_{80}**	**PRD80**	**FI**
Irrigation treatments	2.97ab	3.1a	2.74c	2.86bc	2.47d

	FW	**WW**	**5050**
Water quality	2.44c	2.88b	3.17a

FW 100% fresh water, *5050* 50% fresh water plus 50% wastewater, *WW* 100% wastewater, *FI* full irrigation, *PRD* partial root-zone drying, *TDI* deficit irrigation; 80 and 60 percent of the estimated crop water requirement

In other research, WUE was higher in TDI and PRD plants than in control plants (Zegbe-Domínguez et al. 2003). Acar et al. (2014) illustrated by use of DI by drip system average 25% volume of water saving can be made.

The analysis of variance showed that WUE had a significant difference at the 5% probability level based on irrigation water quality (Table 4). The 5050-treatment with 3.17 kg dry matter/m^3 using water was leader in this group. Irrigation with FW-treatment with 2.435 kg dry matter/m^3 water was in the end. The results indicated that irrigation with treatment included WW-treatment than 5050-treatment as another quality irrigation water has a significant decline on fresh water because of two main reasons: first of all, nutrient toxicity using WW-treatment as irrigation water and second, water stress by increasing osmotic potential due to high level of salts in this irrigation water source. The soil conditions such as arid climate and heavy soil texture caused to increase these effects if using wastewater as irrigation water (Table 1). On the other hand, wastewater has high amount of essential nutrients like N and P, while they are useful for plant growing.

It is clear that the minimum WUE belonged to use FW-FI and then irrigation with WW-FI. According to analysis of variance of interactive effects between water quality and water quantity, 5050-PRD_{80} and 5050-PRD_{60} with 3.36 and 3.3 kg/m^3 has the biggest WUE. Whereas FW-FI and FW-TDI_{80} with 2.31 and 2.26 kg/m^3 in contrast to 5050-PRD_{80} and 5050-PRD_{60} had the smaller WUE (Table 4). The results showed that using 5050-treatment caused to compensation of 40% water deficit for WUE because of higher concentration of N–NO_3(16 mg/l) in wastewater in comparison with of N–NO_3 lower concentration (2.61 mg/l) in FW-treatment, increased dry matter development.

CONCLUSIONS

This experiment was done to assess TDI and PRD as water-saving irrigation techniques. In both TDI and PRD treatments than FI, dry and, fresh weight, height of plant and LAI of plants were reduced because of water stress but PRD and TDI method were caused to increase WUE and DBP. Using PRD_{80} method is more suitable than TDI method. After FI technique, corn plant that irrigated by PRD_{80} method has the bigger dry and fresh weight, DBP and WUE than TDI. So, it outcome in PRD_{80}, we can store irrigation water during study with less effects on corn plants yield because of better water distribution in hemispheric wetted volume than TDI as other water-saving irrigation technique. It seems transmitted signal from roots to shoots to decline the rate of leaf growth and increasing to produce ABA and to close stomata. The height of corn plants was more affected by declining the volume of irrigation water, therefore FI > PRD_{80} > TDI_{80} > PRD_{60} > TDI_{60}. WUE as the main aim was the maximum by PRD_{60} to FI as the minimum WUE. Based on the results, it was parented that applying PRD method has higher performance than TDI. Also, in comparison of three water qualities, using 5050-treatment caused to compensation of water deficit for WUE, LAI, and yield indices but irrigation with WW-treatment alone had unsuitable effects because of nutrient toxicity and undesirable effects of increasing osmotic potential due to high level of salts. Changing of water quality has significant difference on dry and fresh weight, DBP, height of plant and WUE expect LAI. Although PRD and TDI methods were found to be feasible for corn cultivation but using PRD system was considered as a better strategy for water-saving than TDI.

ACKNOWLEDGEMENTS

The authors would like to thank the faculty of Agriculture at Shahrekord University for its financial supports of this research project.

REFERENCES

1. Acar B, Topak R, Yavu D, Kalender MA (2014) Is Drip irrigation technique sustainable solution in agriculture for semi-arid regions? A case study of Middle Anatolian region. Turk Int J Agric Econ Dev 2(2):1–8
2. APHA (1975) Method 208 D. Total non-filterable residue dried at 103–105 °C (total suspended matter) in standard methods for the examination of water and wastewater, 14th edn. American Public Health Association, Washington, DC, p 460
3. Atti S, Bonnell R, Smith DL, Prasher S (2004) Response of an indeterminate soybean to chronic water deficit during reproductive development under greenhouse conditions. Can Water Resour J 29(4):209–224
4. Behboudian H, Ma MQ, Palta CNT, Jairo A (2001) Reactions of chickpea to water stress: yield and seed composition. Sci Food Agric J 81(13):1288–1291
5. Benjamin JG, Nielsen DC (2006) Water deficit effects on root distribution of soybean, field pea and chickpea. Field Crops Res J 97:248–253
6. Canadian Council of Ministers of the Environment (2011) Protocol manual for water quality sampling in Canada. Canadian Council of Ministers of the Environment, P 180
7. Cantore V, Wassar F, Yamaç SS, Sellami MH, Albrizio R, Stellacci AM, Todorovic M (2014) Yield and water use efficiency of early potato grown under different irrigation regimes. Int J Plant Prod 8(3):409–428
8. Chemura A (2014) The growth response of coffee (*Coffea arabica* L.) plants to organic manure, inorganic fertilizers and integrated soil fertility management under different irrigation water supply levels. Int J Recycl Org Waste Agric 3:59
9. De la Hera ML, Romero P, Gomez-Plaza E, Martinez A (2006) Is partial root-zone drying an effective irrigation technique to improve water use efficiency and fruit quality in field-grown wine grapes under semi-arid conditions? Agric Water Manag J 87:261–274
10. Dorjia K, Behboudiana MH, Zegbe-Dominguez JA (2004) Water relations, growth, yield, and fruit quality of hot pepper under deficit irrigation and partial root-zone drying. Sci Hortic J 104:137–149
11. Du T, Kang SH, Zhang J, Li F, Hu X (2006) Yield and physiological responses of cotton to partial root-zone irrigation in the oasis field of northwest China. Agric Water Manag J 84:41–52
12. English M, James L (1990) Deficit irrigation. II: Observation in Colombia basin. J Irrig Drain Eng 116(3):413–426

13. Fulai L, Shahnazari A, Andersen MN, Jacobsen SE, Jensen CR (2006) Physiological responses of potato (*Solanum tuberosum* L.) to partial root-zone drying: ABA signaling leaf gas exchange and water use efficiency. Exp Bot J 57:3727–3735
14. Gallardo M, Jackson LE, Schulbach K, Snyder RL, Thompson RB, Wyland LJ (1996) Production and water use in lettuces under variable water supply. Irrig Sci J 16:125–137
15. Geerts S, Raes D (2009) Review: deficit irrigation as an on-farm strategy to maximize crop water productivity in dry areas. Agric Water Manag J 96:1275–1284
16. Geerts S, Raes D, Garcia M, Condori O, Mamani J, Miranda R, Cusicanqui J, Taboada C, Yucra E, Vacher J (2008) Could deficit irrigation be a sustainable practice for quinoa (*Chenopodium quinoa* Willd.) in the Southern Bolivian Altiplano? Agric Water Manag J 95:909–917
17. Gencoglan C, Altunbey H, Gencoglan S (2006) Response of green bean (*P. vulgaris* L.) to subsurface drip irrigation and partial root-zone-drying irrigation. Agric Water Manag J 84:274–280
18. Grant OM, Stoll M, Jones HG (2004) Partial rootzone drying does not affect fruit yield of raspberries. J Hortic Sci Biotechnol 79(1):125–130
19. Hirich A, Choukr-Allah R, Jacobsen Sven-Erik, Hamdy A, Youssfi L, El-Omari H (2011) Improving water productivity of chickpea by the use of deficit irrigation with treated domestic wastewater. World Acad Sci Eng Technol 59:1352–1357
20. Hu T, Kang S, Li F, Zhang F (2009) Effects of partial root-zone drying irrigation on the nitrogen absorption and utilization of maize. Agric Water Manag J 96:208–214
21. Karam F, Breidy J, Stephan C, Rouphael J (2003) Evapotranspiration, yield and water use efficiency of drip irrigated corn in the Bekaa Valley of Lebanon. Agric Water Manag J 63:125–137
22. Karizan MM (2011) Use of deficit irrigation techniques (DI and PRD) in irrigated with municipal wastewater branch under drip irrigation (T-tape) on corn silage. M.Sc. thesis, Department of Water Engineering, Faculty of Agriculture, Shahrekord University, Shahrekord, Iran
23. Kham-chin Moghadam F, Rezaei Pajand H (2009) Realize maximum daily rainfall by using De Martonne method and compare by linear combination method. Tech Eng J Islam Azad Univ Mashhad Branch 2:93–103
24. Kheirabi J, Darbandi S, Entesari M, Salamat A, Moghadesi F (2009) Evaluation and comparison Penman–Monteith equation with 24-FAO methods in Iran. Iranian National Committee on Irrigation and Drainage Publication. Tehran, Iran, p 208
25. Klute A (1986) Physical and mineralogical methods In: Methods of soil analysis. Part 1. American Society of Agronomy-Soil Science Society of America publication, 2nd edn. p 1189
26. Labidi N, Henda M, Messedi D, Slama I, Chedly A (2009) Assessment of intervarietal differences in drought tolerance in chickpea using both nodule and plant traits as indicators. Plant Breed Crop Sci J 1:80–86
27. Lee GW, Bauder JW (1986) Particle size analysis. In: Klute A (ed) Methods of soil analysis, Part 1. American Society of Agronomy-Soil Science Society of America publication, pp 383–411
28. Li F, Yu J, Nong M, Kang S, Zhang J (2010) Partial root-zone irrigation enhanced soil enzyme activities and water use of maize under different ratios of inorganic to organic nitrogen fertilizers. Agric Water Manag J 97:231–239
29. Liu F, Shahnazari A, Andersen MN, Jacobsen SE, Jensen CR (2006) Effects of deficit irrigation (DI) and partial root drying (PRD) on gas exchange, biomass partitioning, and water use efficiency in potato. Sci Hortic J 109:113–117
30. Lopez FB, Johansen C, Chauhan YS (1996) Effects of timing of drought stress on phenology, yield and yield components of short-duration pigeonpea. Agron Crop Sci J 177:311–320
31. Mohammad MJ, Mazahreh N (2003) Changes in soil fertility parameters in response to irrigation of forage crops with secondary treated wastewater. Soil Sci Plant Anal J 34:1281–1294

32. Nagaz K, Mokh FE, Masmoudi MM, Mechlia NB (2013) Soil salinity, yield and water productivity of lettuce under irrigation regimes with saline water in arid conditions of Tunisia. Int J Agron Plant Prod 4:892–900
33. Najafi P, Asgari K, Samadi N (2016) Heavy metal elimination from industrial wastewater using natural substrate on pitcher irrigation. Int J Recycl Org Waste Agric 5:333–337
34. Nazemosadat SMJ, Kazemini SA (2008) The effects of water stress and photosynthesis active radiation on photosynthesis rate of wheat. Agric Sci Nat Resour J 14(6):34–44
35. Olsen SR, Cole CV, Watunable FS, Dean LA (1954) Estimation of available P in soil by extraction with Na-bicarbonate. Circular, No.939 USDA. US. Government, Washington DC
36. Page AL, Miller RH, Keeney DR (1991) Chemical and biological properties. In: Methods of soil analysis, part 2. American Society of Agronomy-Soil Science Society of America publication, 2 edn, p 1159
37. Patanè C (2011) Leaf area index, leaf transpiration and stomata conductance as affected by soil water deficit and VPD in processing tomato in semi-arid mediterranean. Clim Agron Crop Sci J 197:165–176
38. Patel N, Rajput TBS (2013) Effect of deficit irrigation on crop growth, yield and quality of onion in subsurface drip irrigation. Plant Prod J 7:417–436
39. Rousan LM, Rusan MJ, Amayreh J (2008) Irrigation with treated wastewater under full and deficit soil moisture conditions. In: Eds: Al-Mattareneh H, Sidek LM, Youssef MZ (eds) Proceeding of the international conference on construction and building technology (ICCBT): innovations in water resources and environmental engineering, Kuala Lumpour, Malaysia, vol 38, pp 405–416
40. Sahin U, Ors S, Kiziloglu FM, Kuslu Y (2014) Evaluation of water use and yield responses of drip-irrigated sugar beet with different irrigation techniques. Chil J Agric Res 74:302–310
41. Sawia M, Omaima U, EL-Beharya A, Abdolhamid AF (2002) Effect of irrigation level and N source on green bean productivity. Egypt Appl Sci J 47:326–347
42. Shahnazari A, Ahmadi SH, Lærke PE, Liu F, Plauborg F, Jacobsen SE, Jensen CR, Andersen MN (2008) Nitrogen dynamics in the soil-plant system under deficit and partial root-zone drying irrigation strategies in potatoes. Eur J Agric 28:65–73
43. Shammi M, Abul Kashem M, Mostafizu Rahman Md, Delwar Hossain Md, Rahman R, Khabir Uddin M (2016) Health risk assessment of textile effluent reuses as irrigation water in leafy vegetable Basella alba. Int J Recycl Org Waste Agric 5:113–123
44. Sharma KD, Pannu RK, Tyagi PK, Chaudhary BD, Singh DP (2007) Response of chickpea genotypes to plant water relations and yield under soil moisture stress. Agrometeo J 9:42–48
45. Tabatabaei SH, Najafi P, Mirzaei SMJ, Nazem Z, Heidarpour M, Hajrasoliha S, Afyuni M, Beigi Harchegani H, Landi E, Akasheh L, Zamanian M, Barani M, Amini H (2012) Compost' leachate recycling through land treatment and application of natural Zeolite. Int J Recycl Org Waste Agric 1:2. doi:10.1186/2251-7715-1-2. http://www.ijrowa.com/content/1/1/2
46. Wang J, Kang S, Li F, Zhang F, Li Z, Zhang J (2008) Effects of controlled alternate partial root-zone drying irrigation on soil microorganism and growth of maize. Plant Soil J 302:45–52
47. Yazar A, Gokgel F, Sezen MS (2009) Corn yield response to partial root zone drying and deficit irrigation strategies applied with drip system. Plant Soil Environ J 55:494–503
48. Yonts CD (2010) Development of season long deficit irrigation strategies for Suger-beets. The University of Nebraska Panhandle Research and Extension Center, Scottsbluff, p 69361
49. Zegbe-Domínguez J, Behboudian A, Lang MHA, Clothier BB (2003) Deficit irrigation and partial root zone drying maintain fruit dry mass and enhance fruit quality in

'Petopride' processing tomato (*Lycopersicon esculentum*, Mill.). Sci Hortic J 98:505–510
50. Zhang H, Oweis T (1999) Water-yield relations and optimal irrigation scheduling of wheat in the Mediterranean region. Agric Water Manag J 38:195–211
51. Zhang J, Yang J (2004) Crop yield and water use efficiency. In: Bacon MA (ed) Water use efficiency in plant biology. Blackwell Publishing, Oxford, pp 189–218

CHAPTER 19

Process Optimization Based on Carrot Powder Color Characteristics

Yuanjuan Gong [a] , Guomiao Deng [a] , Chungsu Han [b] , Xiaofeng Ning [b, *]

[a]Department of Agricultural Mechanization Engineering, Shenyang Agricultural University, Shenyang, Liaoning 110866, China

[b]KSAM Member Department of Biosystems Engineering, Chungbuk National University, Cheong-ju 361763, Korea

ABSTRACT

The carotenoid content indicates the processing quality of carrot powder. The carotenoid content of carrot powder was analyzed using digital image processing technology and color analysis technology, and an estimation model between the carotenoid content and color characteristics was established. The carrot was dried using different drying methods. The verification test result shows that model fitting is prominent and the test error is small and promptly improves the reference of testing for the carotenoid content of carrot powder. The processing parameters of the carrot powder with the highest carotenoid content are as follows: the carrot was frozen and thawed using the freeze drying method; the powder particle size is 80–120 mesh.

Keywords: Carrot powder Carotenoid content Color characteristics Estimation model Optimization of process

1. INTRODUCTION

Carrots have high nutrition, health care and medical values. When a carrot is turned into a powder with microstructure, its surface area and porosity significantly increase and the product dispersivity, solubility and functionality noticeably strengthen, which makes it easily digested and absorbed in human bodies (Zheng and Xia, 2006; Ma et al., 2008; Gong et al., 2006). Carrots are rich in cellulose, sugar, protein and other biologic activity materials, which lead to greater stickiness, higher toughness and worse heat-resistance during the pulverizing process, so pulverizing and classification become much more difficult (Gai, 1999). In particular, when the material granularity gradually decreases during the pulverizing process, its surface area sharply increases; particles with high superficial energy are easily agglomerated; when the particles are much smaller, the agglomeration trend is more prominent; and when the particles are pulverized to some granularity, the dynamic equilibrium between pulverization and agglomeration appears. Furthermore, the pulverizing process deteriorates because of the increased particle agglomeration (Gong and Zeng, 2009). Therefore, ultrafine grinding technology has practical significance regarding carrot powder.

Carotenoid contains a hydrocarbon with an unsaturated double bond or its oxygen derivatives, which are unstable and sensitive to oxygen, light, heat and acid. The decomposition speed of carotenoid increases with increasing temperature. Carotenoid easily breaks down when it is exposed to light, and carrot micro-powder has much higher chances of contacting light to oxidize and break down when the particle size decreases and the surface area increases. It is necessary to study the effects on the carotenoid content in carrot powder under different process conditions to improve the quality of carrot ultra-micro powder. However, the routine testing process of extracting carotenoid in carrot powder using an extracting agent and subsequently measuring the carotenoid content by testing the absorbency is notably complicated and increases the testing error (Fan et al., 2001). Therefore, a rapid test of the carotenoid content in carrot powder has practical significance.

Suzuki. T's study (Suzuki, 1995) confirmed that there is a significant correlation between the color characteristics $G/(R + G + B)$ and the chlorophyll content in broccoli seedling leaves. Adamsen F.J has proven that the chlorophyll content in wheat leaves is well correlated to G/R (Adamsen et al., 1999). Li Y used a digital camera to obtain color images of tomatoes and to extract its color feature values using the image processing software. The experiment indicates that color features are highly correlated to the nitrogen index (Li et al., 2007). Cai HC has verified that G/R and $R/(R + G + B)$ are the main color characteristic parameters of the carotenoid content (Cai et al., 2006). The carotenoid content of carrot powder was analyzed using digital image processing technology and color analysis technology, and an estimation model between the carotenoid content and the color characteristics was established. Based on these results, in an experimental study with different pre-processing methods (fresh carrot, cured carrot and froze thawing carrot were included), carrot was dried using different techniques: vacuum freeze drying, vacuum drying, microwave drying and hot-air drying; then, carrot powder with different particle sizes were processed. In the orthogonal test analysis, the processing parameters of the carrot powder with the highest carotenoid content were determined. The study conclusion on carrot powder production processing has guiding significance.

2. MATERIALS AND METHODS

2.1. Material

The carrot used in the experiments is red core No. 4, and the average moisture content of the raw carrot is 91%.

2.2. Methods

2.2.1. Equipments

FD-5 vacuum freezing dryer (Beijing Boyikang Technology Company); WD850B microwave oven (Guangdong Shunde Gelanshi Group); DZF-6090 vacuum dryer (Shanghai Jinghong Laboratory Equipment Company); JY2002 electronic balance (Shanghai Exact Science Instrument Company, 0.001 g); 101-0A electrothermal constant-temperature dry box (Shanghai Sunshine Experiment Instrument Company); DWF-1088 multi-function food processor (Yangzhou Wenfeng electric appliance company); centrifuge (Japan Rili Company); electric heating constant-temperature water-bath pot (Changzhou Guohua Electric Appliance company).

2.2.2. Preparation of carrot powder

The raw carrot were sliced with the size of 7.1 × 5.5 mm (w × t) and then did the pre-treatments of cooking and freezing-thawing. Subsequently, the carrot were dried using vacuum drying, freeze drying, heated-air drying and the microwave drying method, respectively. Finally the carrot powder can be obtained through grinding and screening.

Heated-air drying, microwave drying, vacuum drying and freeze drying were compared in this study (Che and Zhu, 2006; Tao, 2004). The carrot's average moisture content after drying is 5%. The freeze drying conditions were: pressure of the drying chamber of 150 Pa; temperature of the heater plate of 40 °C. The vacuum drying conditions were: pressure of the drying chamber of 0.07 MPa; temperature of the heater plate of 65 °C. The microwave drying condition was: 20 min with moderate heat (the power is 540 W) at first and low heat (the power is 180 W) subsequently. For heated-air drying, the temperature was 65 °C. After drying, the carrot was mashed with the multi-function pulverizer and screened through 80–120 meshes, 160–200 meshes and more than 240 meshes; then, the carrot powder sample was obtained.

2.2.2.1. Pretreatments of the carrot were as follows

Freezing-thawing treatment: a type of cell disruption technology with low-temperature freezing and high-temperature melting was applied to the carrot. By freezing-thawing, the carrot slices and the carotenoid extraction could be improved (Chen et al., 2004). The freezing-thawing treatment in the experiment was as follows. First, raw carrot slices were placed in the cold trap (−62 °C) of the freeze drier for 5 h. Then, they were brought from the cold trap to the natural environment and maintained for 1 h. Then, they were placed in the cold trap again for 5 h. Finally, they were thawed in room-temperature condition. The melted carrot slices are experimental materials for the freezing-thawing treatment.

Cooked treatment: To release the nutrition ingredient, the carrot cell wall was destroyed using microwaves at high temperature (Chinese National Standards, 1990a,b). For the cooked treatment in this study, a 750 W microwave was used to process raw carrot slices for 3 min. After the processing, the carrot was cooked.

2.2.3. Equipment of image extraction and selection of the color characteristic parameter

Equipment of image extraction: The digital camera model used was CANON-DIGITALIXUS960. The image processing software used was Adobe Photoshopcs8.0.

Determination of the color characteristic parameter: There are commonly RGB and HSI color coordinate systems for digital image processing. The RGB color coordinate system is based on the three colors: red (R), green (G) and blue (B); other colors are blending weighted using the three basic colors.

In the RGB color coordinate system, if we only consider the chroma instead of the light intensity, we only know the relative value of R, G and B. The relative values of r, g and b are called chromaticity coordinates, the calculation formulas of which are as follows:

$$r = \frac{R}{R+G+B}, g = \frac{G}{R+G+B}, b = \frac{B}{R+G+B} \quad (1)$$

Extract method of color characteristic parameter: The digital model was used, where digital images of the carrot powder of various processes were taken in indoor light. Approximately 50 g carrot powder was used for taking a photograph of digital images.

The exposure time and color images were controlled using the digital camera automatic exposure programmer. An image resolution of 2048 by 1536 pixels was used. RGB color images were obtained and placed in a computer in bmp format, each of which had a size of 9.0 MB. Adobe Photoshopcs8.0 was used for image analysis and color feature extraction. The image of carrot powder was selected by the "magic wand" tool in the "tools" panel. Then, the histogram programmer in the drop-down menu of "window" was selected to read the image mean of red (R), green (G) and blue (B) in each channel. Because the background besides the powder digital image was artificially removed in the aforementioned process of image treatment, which reduces the effect of "image noise" in the image-processing results, different color characteristics of the digital image can be taken instead of the actual color information of carrot powder (Li et al., 2007).

2.2.4. Measurement of carotenoid content in carrot powder

Base on Chinese standard GB12291-90, the carotenoid of carrot powder was extracted using an extraction solvent (chloroform:methanol (v:v) = 2:1). The absorbance was measured at λ max of 458 nm, when the extraction solvent was treated as a blank space (Chinese National Standards, 1990a,b; Cao et al., 2003; Gong, 2008).

According to the test results, we calculated the carotenoid content as follows.

$$M = \frac{A \times 400}{m} \quad (2)$$

where M is the carotenoid content (mg/kg); A is the maximum absorbance of the measurement; m is the measurement of the sample; and 400 is the conversion factor.

2.2.5. Statistical analysis

The single variable linear regression and the multivariate linear regression analysis methods were used to analyze the correlation between the carotenoid content and the color characteristics in carrot powder. Differences among the mean values of the experimental data were determined using Duncan's multiple range test and one-way analysis of variance (one-way ANOVA) with a significance level of 0.05. Six samples that were randomly selected from the entire samples were set as the calibration to build the prediction model (Gong et al., 2012, 2007). The other 6 samples were set as survey to examine the prediction model.

3. RESULTS AND DISCUSSION

3.1. Correlation analysis of the carotenoid content and color feature

Extraction of the color features of carrot powder: Table 1 shows the extracted color characteristic parameters of 6 pieces of carrot powder images using various processes.

Table 1. Variation of color characteristic parameters of carrot powder with the processes.

Drying methods	R/(R + G + B)	G/(R + G + B)	B/(R + G + B)	G/R	B/R	R/G	B/G	G/B	R/B
Freeze drying	0.61	0.31	0.08	0.50	0.13	2.01	0.27	3.75	7.52
Heated-air drying	0.52	0.34	0.14	0.65	0.27	1.53	0.41	2.44	3.74
Vacuum drying	0.67	0.29	0.04	0.43	0.05	2.32	0.12	8.03	18.65
Microwave drying	0.61	0.31	0.08	0.51	0.13	1.96	0.25	4.06	7.94
Freezing-thawing microwave drying	0.68	0.28	0.04	0.41	0.06	2.41	0.13	7.41	17.89
Cooked vacuum drying	0.62	0.32	0.06	0.52	0.09	1.91	0.17	5.84	11.15

The correlation between the carotenoid content and the color characteristics in carrot powder: The regression analysis, which is a data analysis method, was used to analyze the relationship between the color characteristics and the carotenoid content of 6 types of carrot powder of various processing techniques (Liao et al., 2004). Table 2 shows the correlation between the color characteristic parameters of carrot powder and the carotenoid content. Table 2 indicates that the carotenoid content in carrot powder is highly correlated to G/(R + G + B), which has a significance test level of $\alpha = 0.01$. In addition, the correlation between the carotenoid content and color characteristic parameters of R/(R + G + B), G/R, R/G has a prominent test level of $\alpha = 0.05$. Therefore, the prediction model of carotenoid content can be established with the aforementioned color characteristic parameters.

Table 2. Correlation analysis of the color features and carotenoid content of carrot powder.

Color features	R/(R + G + B)	G/(R+G + B)	B/(R + G + B)	G/R	B/R	R/G	B/G	G/B	R/B
Correlation coefficient of Carotenoid r	0.76	0.90	0.63	0.84	0.67	0.82	0.57	0.49	0.57
Variance analysis $F_{0.05}$	5.51*	17.32**	2.68	9.90*	3.19	8.49*	1.97	1.27	1.92

3.2. Establishment of the color feature prediction model of carotenoid content

Based on the statistical analysis results in Table 2, the carotenoid content prediction model was built using the linear regression method, which selects G/(R + G + B), R/(R + G + B), G/R and R/G as the characteristic parameters of the key color in the digital image.

$$Y_1 = \frac{1.16 - 3.03 \times G}{R + G + B} \quad (r = 0.90, F_{0.05} = 17.32) \tag{3}$$

$$Y_2 = \frac{-0.37 + 0.96 \times R}{R + G + B} \quad (r = 0.76, F_{0.05} = 5.51) \tag{4}$$

$$Y_3 = 0.58 - \frac{0.72 \times G}{R} \quad (r = 0.84, F_{0.05} = 9.90) \tag{5}$$

$$Y_4 = -0.16 + \frac{0.19 \times R}{G} \quad (r = 0.82, F_{0.05} = 8.49 \tag{6}$$

The single variable linear regression analysis between the carotenoid content and the color characteristic parameters in carrot powder indicates that the correlation coefficient α of the model has a significance test level of 0.05 and the correlation coefficient α of G/(R + G + B) has a significance test level of 0.01, which can better estimate the carotenoid content.

The multivariate linear regression analysis, which is one of the data analyses, was used to further analyze the relationship between the carotenoid content and the color features. The multivariate regression model is set as follows:

$$Y_5 = 6.1 + \frac{3.86 \times R}{R + G + B} - \frac{16.41 \times G}{R + G + B} + \frac{0.015 \times G}{R} - \frac{1.59 \times R}{G} \quad (r = 0.99, F_{0.01} = 50.4) \tag{7}$$

where Y_5 is the predicted total carotenoid content (g/kg).

The correlation analysis shows that the model is significant and that the mathematical model fits well, which can be applied to the prediction.

3.3. Verification of color feature model of carotenoid content

A freeze-thawed and cooked carrot was made into carrot powder after drying using freezing, microwave and heated-air drying methods. A CANON-DIGITALIXUS960 digital camera was used to take digital images of the carrot powder of various pre-treatments and drying methods in indoor light. Adobe Photoshopcs8.0 was used for image analysis and color feature extraction. The extracted characteristic values of red (R), green (G) and blue (B) were put into the prediction model. The predictive value was calculated and compared with the real value. The result was shown in Table 3. Finally, the conclusions of the error analysis indicate that the most error values of the 5 prediction models are below ±10%, especially the error of the multivariate regression model is below ±5%, and the estimated accuracy is high, which can be applied to estimate the carotenoid content of carrot powder.

Table 3. Comparison of experimental and predicted value of carotenoid content.

Drying method	Experimental value (mg/kg)	Predicted value (mg/kg)					Error				
		Y1	Y2	Y3	Y4	Y5	Y1	Y2	Y3	Y4	Y5
F-T-F-D	302	298	291	282	281	296	1.431%	3.911%	6.649%	7.071%	2.193%
F-T-H-D	201	213	212	200	214	201	-6.072%	-4.937%	0.351%	-6.393%	0.088%
F-T-V-D	283	297	291	293	296	283	-4.748%	-2.746%	-3.481%	-4.418%	0.169%
C–F-D	158	148	139	143	145	197	6.560%	13.892%	9.533%	8.441%	-4.618%
C-M-D	132	146	132	139	147	133	-10.333%	0.643%	-4.991%	-11.130%	-0.516%
C–H-D	100	136	113	114	112	98	-14.302%	-11.526%	-13.376%	-11.335%	2.273%

F-T-F-D: Freezing thawing- Freeze drying; F-T-H-D:Freezing thawing- Heated air drying; F-T-V-D:Freezing thawing- Vacuum drying;C–F-D:Cooked- Freeze drying; C-M-D:Cooked- Microwave drying;C–H-D:Cooked –Heated air drying.

3.4. Analysis of influencing factors of the carotenoid content

Table 4 shows the changes in the carotenoid content for different particle sizes of carrot powder, which were obtained using different drying methods. The variance analysis of the data in Table 4 and the results in Table 5 show that the P value for particle sizes and drying methods were 0.000 and 0.019, respectively; thus, different particle sizes and drying methods significantly affect the carotenoid content of carrot powder.

Table 4. Effect of drying methods and grain size on the carotenoid content (mg/kg).

	Heated-air drying	Microwave drying	Vacuum drying	Freeze drying
80˜120 meshes	114.4	231.2	264	344.8
160˜200 meshes	85.6	110	193.6	200.8
>240 meshes	58.4	73.2	175.6	144.1

Table 5. Analysis of variance on the carotenoid content of different drying methods and grain sizes.

Differences source	SS	df	MS	F	P
Grain size	33,729.15	2	16,864.57	5.14	0.000
Drying methods	39,845.39	3	13,281.8	4.76	0.019
error	7232.45	6	1205.41		
total	80,806.99	11			

Fig. 1 shows the changes in the carotenoid content of raw carrot powder with different particle sizes. Fig. 1 shows that the carotenoid content of smashed carrots, which were dried using four drying methods, evidently decreases with decreasing particle size because carotenoid easily breaks down when it is exposed to light and carrot micro-powder has significantly increased chances of contacting light to oxidize and break down when the particle size decreases and surface area increases.

Fig. 2 shows the changes in carotenoid content of raw carrot powders that were prepared using different drying methods. Fig. 2 shows that the freeze-dried carrot powder has the highest carotenoid content, followed by the vacuum-dried powder, and the heated-air dried powder has the lowest carotenoid content because carotenoid contains a hydrocarbon with an unsaturated double bond or its oxygen derivatives, which are unstable and sensitive to oxygen, light, heat and acid. The decomposition speed of carotenoid increases with increasing temperature, and the heated-air drying process has s higher temperature, so carotenoid was lost.

Fig. 3 shows the changes in the carotenoid content of a carrot powder that was crushed using 160˜200 meshes with different pretreatment methods. The freezing-thawed carrot powder has a higher carotenoid content than fresh and cured powder. Carotene is released because the carrot cells are destroyed in the freezing-thawing process; a type of carotenoid is released because of the ripening method, which destroys the cell walls of the carrot cells with the microwave high temperature. The decomposition speed of carotenoid increases with increasing temperature in high-temperature processes, so the freezing-thawed carrot has the highest carotenoid content.

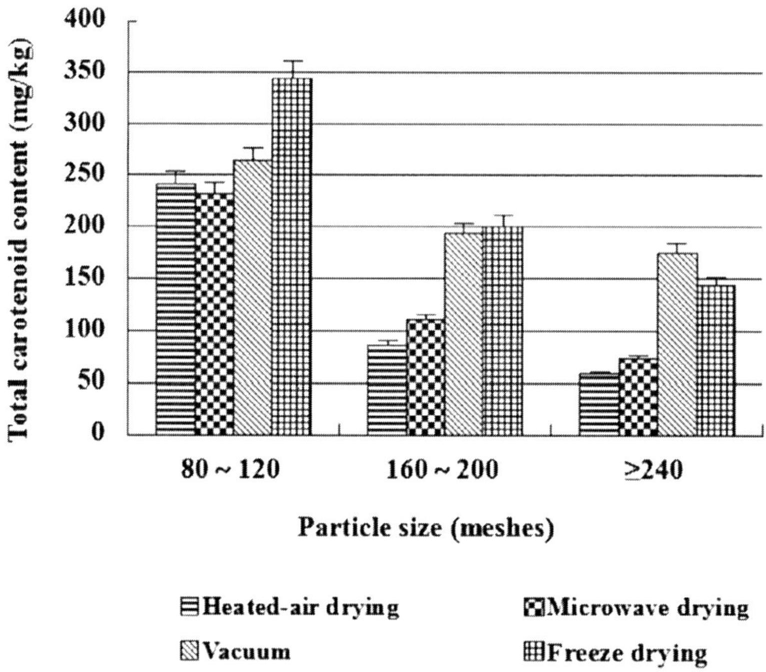

Figure. 1. Effect of the grain size on the carotenoid content.

3.5. Orthogonal test of the influence factors of the carotenoid content

The orthogonal test analyzes the effects of three factors (drying method, pretreatment method and carrot powder particle size) on the carotenoid content of carrot powder. An orthogonal test with three factors and four levels L16 (34) was designed. Table 6 shows the factors and levels of the test.

Table 6. Factors and levels.

Level	A Drying method	B Pretreatment method	C Particle size/meshes
1	vacuum drying	fresh carrot	1(80~120)
2	Freeze drying	cured carrot	2(160~200)
3	Heated-air drying	froze thawing carrot	3(>240)
4	Microwave drying	fresh carrot	4(>240)

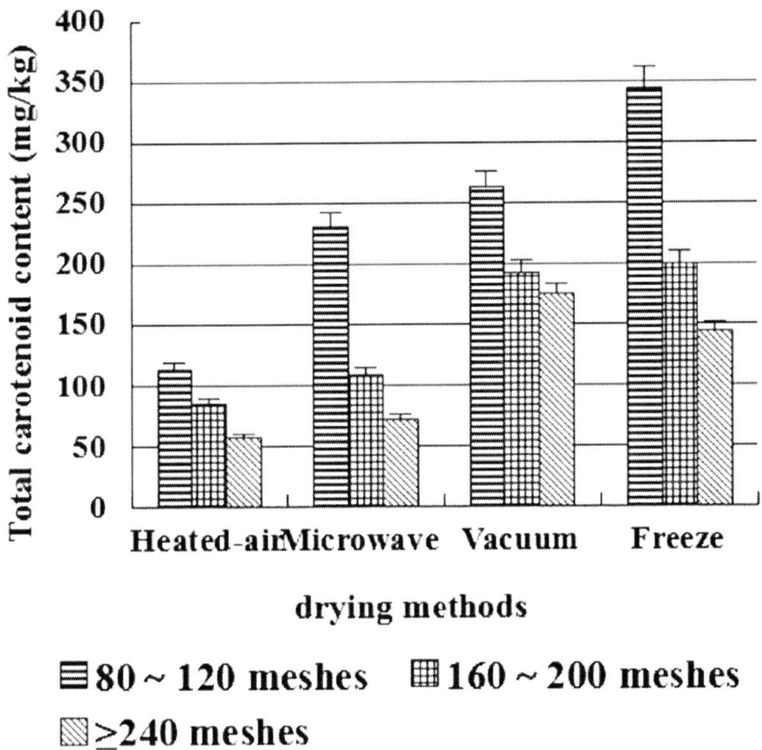

Figure. 2. Effect of the drying methods on the carotenoid content.

The test was performed according to the design of the test project and a certain test index. Table 7 shows the scheme and results of the experiment, and Table 8 shows the result analysis.

Table 7. Orthogonal design and results.

Number	Drying method A	Pretreatment B	Particle size (meshes) C	A × B	A × C	Carotenoid content (mg/kg)
1	1(Vacuum drying)	1(fresh)	1(80~120)	1	1	281^{ab}
2	1(Vacuum drying)	2(cured)	2(160~200)	2	2	223^{cd}
3	1(Vacuum drying)	3(froze thawing)	3(>240)	3	3	200^{d}
4	1(Vacuum drying)	4(fresh)	4(>240)	4	4	211^{d}
5	2(Freeze drying)	1(fresh)	2(160~200)	3	4	277^{ab}
6	2(Freeze drying)	2(cured)	1(80~120)	4	3	238^{c}
7	2(Freeze drying)	3(froze thawing)	4(>240)	1	2	289^{a}
8	2(Freeze drying)	4(fresh)	3(>240)	2	1	258^{b}
9	3(Heated-air drying)	1(fresh)	3(>240)	4	2	202^{d}
10	3(Heated-air drying)	2(cured)	4(>240)	3	1	154
11	3(Heated-air drying)	3(froze thawing)	1(80~120)	2	4	253^{b}
12	3(Heated-air drying)	4(fresh)	2(160~200)	1	3	198^{de}
13	4(Microwave drying)	1(fresh)	4(>240)	2	3	172^{e}
14	4(Microwave drying)	2(cured)	3(>240)	1	4	205^{d}
15	4(Microwave drying)	3(froze thawing)	2(160~200)	4	1	258^{b}
16	4(Microwave drying)	4(fresh)	1(80~120)	3	2	222^{cd}

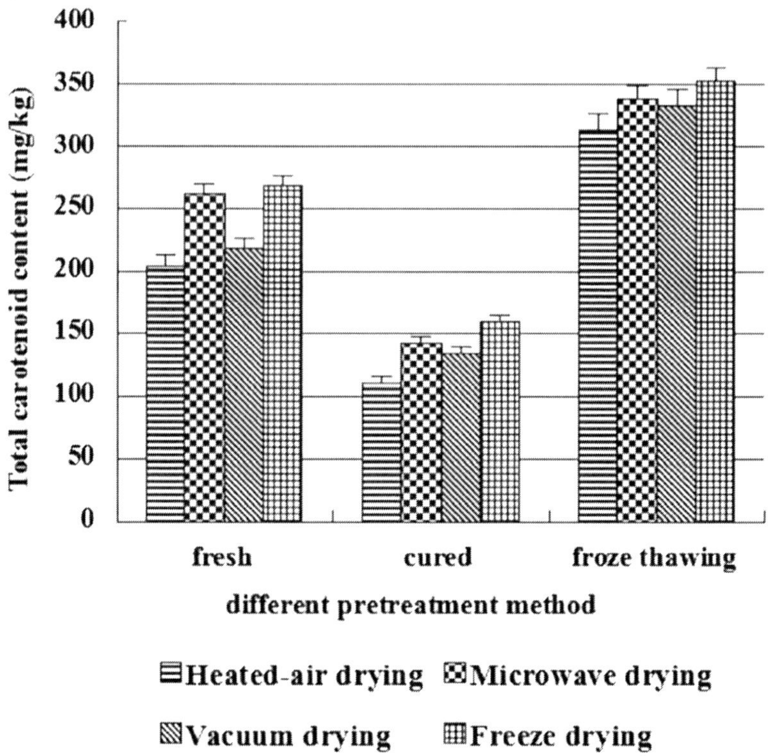

Figure. 3. Effect of the pretreatment methods on the carotenoid content.

Table 8. Results of the nutrient components in carrot power and the analysis.

Index	Carotenoid content (mg/kg)				
Factors	A	B	C	A × B	A × C
K_1	915.6	1820.4	993.6	982.8	950.4
K_2	1071.6	819.6	956.4	906.0	945.6
K_3	806.4	1010.4	1700.4	853.2	807.6
K_4	856.8	—	—	908.4	946.8
k_1	228.9	227.6	248.4	245.7	237.2
k_2	267.9	204.9	239.1	226.5	236.4
k_3	201.6	252.6	212.2	213.3	202.1
k_4	214.2	—	—	227.1	236.7
R	66.3	47.7	36.2	32.4	35.1
Factors optimal level	A_2	B_3	C_1	$(A \times B)_1$	$(A \times C)_1$
Primary and secondary factors	A 、B 、C 、 A × C 、A × B				
Optimal scheme	$A_2B_3C_1$				

It could be known from analysis of the Table 8, the drying method showed the most pronounced effect on the carotenoid content of carrot powder and followed by pretreatment way and carrot powder particle size. The carotenoid content after freeze drying is high and that after heated-air drying is the lowest. Moreover, after drying and crushing, the freezing-thawed carrot powder has a higher carotenoid content than the microwave-cured powder. When the particle size decreases, the carotenoid content of carrot powder decreases. The carrot powder with 80 ˜ 120 meshes has the highest carotenoid content because when the carrot powder particle size decreases, its surface area increases and the degree of oxidation increases. The interaction functions of A × B and A × C do not affect the test results; the optimum scheme for A2B3C1 is vacuum freeze drying, freeze-thawed carrot powder, and the particle size is 80 ˜ 120 meshes. These conditions yield the highest carotenoid content.

4. CONCLUSION

The analysis of the correlation between the color characteristic parameters and the carotenoid content shows that the carotenoid content in carrot powder is highly interrelated to $G/(R+G+B)$ and has an extremely significant test level ($\alpha = 0.01$). Moreover, the correlations between the carotenoid content in carrot powder and the three color characteristic parameters $R/(R+G+B)$, G/R and R/G have a significant test level ($\alpha = 0.05$). Therefore, these color characteristic parameters can be used to establish a prediction model of carotenoid content in carrot powder.

Based on the experimental analysis of the carotenoid content in carrot powder using digital image and color feature analysis technology, an estimation model was established for the quantitative relation between the carotenoid content and its color feature in carrot powder. The performed test indicates that the model fits well and the experimental error is tiny, which can provide a foundation for a rapid test of the carotenoid content in carrot powder.

It could be known from analysis of the Table 7, the drying method showed the most pronounced effect on the carotenoid content of carrot powder and followed by pretreatment way and carrot powder particle size. The carotenoid content after freeze drying is high, and that after heated-air drying was the lowest. Moreover, after drying and crushing, the freezing-thawed carrot powder has a higher carotenoid content than the microwave-cured powder. When the particle size decreases, the carotenoid content of carrot powder decreases. The carrot powder with 80 ˜ 120 meshes has the highest carotenoid content. Thus, with vacuum freeze drying, freeze-thawing and a particle size of 80 ~ 120 meshes, carrot powder has the highest carotenoid content.

ACKNOWLEDGMENTS

This work was supported by the Natural Science Foundation of LiaoNing Province (2013020062).

REFERENCES

1. Adamsen, F.J., Paul, J. Pinter, Barncs, Jr E.M., 1999. Measuring wheat senescence witha digital camera. Crop Sci. 39, 719e724.
2. Cai, H.C., Cui, H.X., Song, W.T., 2006. Initial research on the photosynthetic pigmentcontent and color characteristics of cucumber early flowering. Trans. Chin. Soc.Agric. Eng. 22, 34e37.
3. Cao, L.K., Huang, W., Wang, J.H., Zhang, X.J., 2003. Superfine grinding for dilupidatingwalls of maize pollen. Trans. Chin. Soc. Agric. Eng. 19, 209e211.
4. Che, Z.M., Zhu, X.L., 2006. Study on the conditions to increase the content of bcarotenein carrot juice. J. Xihua Univ. Nat. Sci. Ed. 25, 83e85.
5. Chen, J., Xie, X.X., Pang, Y.M., 2004. The rapid test of the carotenoid content in thecarrot powder. J. Xinjiang Agric. Sci. 41, 103e105.
6. Chinese National Standards, 1990a. The Evaluation of the Whole Carotenoid Contentin Fruits or Vegetables. China Standard Press. GB12291e1990.
7. Chinese National Standards, 1990b. The Measuring Method of Food Carotene. China Standard Press. GB12389e1990.
8. Fan, L.Y., Ren, C.M., Liu, Y.Q., 2001. Study on the process of extracting beta carotenefrom carrot. Food Sci. 22, 53e55.
9. Gai, G.S., 1999. Recent development in superfine crushing and classification. China Powder Sci. Technol. 5, 22e26.
10. Gong, Y.J., 2008. Experimental Study on Carrot Fine Pulverizing Technology andRelate Parameters Shenyang Agricultural University, China.
11. Gong, Y.J., Zeng, C., 2009. Influence factors on physical properties and nutrientcomponent for carrot micro-powder. Trans. Chin. Soc. Agric. Mach. 40, 124e128.
12. Gong, Y.J., Zeng, R.X., Tian, S.B., 2006. Review of refining processing and comprehensiveutilization of carrot. Trans. Chin. Soc. Agric. Eng. 4, 199e203.
13. Gong, Y.J., Zhang, B.H., Li, C.H., 2007. Experimental study on technological parametersfor burdock fine pulverizing. J. Agric. Eng. 22, 256e258.
14. Gong, Y.J., Meng, F.Z., Qin, J.W., Lin, J., 2012. Experimental study on dry characteristics and grinding technology for North Cordyceps. Trans. Chin. Soc. Agric. Mach. 5, 142e146.
15. Li, Y., Wang, J., Ma, T.F., 2007. Preliminary study of evaluation about processing tomato leaf nitrogen on basic of color feature. J. Northwest Agric. 16, 175e179.
16. Liao, X.J., Hu, X.S., Liu, W., 2004. Study on physical and chemical properties of 50 mesh and 100 mesh carrot powder and their application. Food Sci. 2, 61e64.
17. Ma, R.S., Li, Y.L., Gong, Y.J., 2008. Effects of different drying methods on physical and chemical properties of carrot power. Sci. Technol. Food Ind. 29, 107e109.
18. Suzuki, T., 1995. Measurement of growth of plug seedings by image processing in broccoli. Acta Hortic. 399, 333e343.

19. Tao, J.X., 2004. Technology for processing carrot juice using acidifying and cookingsoftening method. Trans. Chin. Soc. Agric. Eng. 20, 228e230.
20. Zheng, Y.Y., Xia, Y.B., 2006. Nutrition and health care production with carrot and the foreground of exploitation. Packag. Food Mach. 24, 35e37.

Index